Ship Hydrostatics and Stability

A.B. Biran

Technion – Faculty of Mechanical Engineering

BUTTERWORTH
HEINEMANN

AMSTERDAM BOSTON HEIDELBERG LONDON NEW YORK OXFORD
PARIS SAN DIEGO SAN FRANCISCO SINGAPORE SYDNEY TOKYO

Butterworth-Heinemann is an imprint of Elsevier
Linacre House, Jordan Hill, Oxford OX2 8DP, UK
30 Corporate Drive, Suite 400, Burlington, MA 01803, USA

First edition 2003
Reprinted 2005, 2006, 2007

British Library Cataloguing in Publication Data
A catalogue record for this book is available from the British Library

Library of Congress Cataloging-in-Publication Data
A catalog record for this book is available from the Library of Congress

ISBN: 978-0-7506-4988-9

For information on all Butterworth-Heinemann publications
visit our website at books.elsevier.com

Transferred to Digital Printing in 2009

To my wife Suzi

Contents _____

Preface

This book is based on a course of Ship Hydrostatics delivered during a quarter of a century at the Faculty of Mechanical Engineering of the Technion–Israel Institute of Technology. The book reflects the author's own experience in design and R&D and incorporates improvements based on feedback received from students.

The book is addressed in the first place to undergraduate students for whom it is a first course in Naval Architecture or Ocean Engineering. Many sections can be also read by technicians and ship officers. Selected sections can be used as reference text by practising Naval Architects.

Naval Architecture is an age-old field of human activity and as such it is much affected by tradition. This background is part of the beauty of the profession. The book is based on this tradition but, at the same time, the author tried to write a modern text that considers more recent developments, among them the theory of *parametric resonance*, also known as *Mathieu effect*, the use of personal computers, and new regulations for intact and damage stability.

The Mathieu effect is believed to be the cause of many marine disasters. German researchers were the first to study this hypothesis. Unfortunately, in the first years of their research they published their results in German only. The German Federal Navy – *Bundesmarine* – elaborated stability regulations that allow for the Mathieu effect. These regulations were subsequently adopted by a few additional navies. Proposals have been made to consider the effect of waves for merchant vessels too.

Very powerful personal computers are available today; their utility is enhanced by many versatile, user-friendly software packages. PC programmes for hydro-static calculations are commercially available and their prices vary from several hundred dollars, for the simplest, to many thousands for the more powerful. Programmes for particular tasks can be written by a user familiar with a good software package. To show how to do it, this book is illustrated with a few examples calculated in Excel and with many examples written in MATLAB. MATLAB is an increasingly popular, comprehensive computing environment characterized by an interactive mode of work, many built-in functions, imme-diate graphing facilities and easy programming paradigms. Readers who have access to MATLAB, even to the Students' Edition, can readily use those exam-ples. Readers who do not work in MATLAB can convert the examples to other programming languages.

Several new stability regulations are briefly reviewed in this book. Students and practising Naval Architects will certainly welcome the description of such rules and examples of how to apply them.

This book is accompanied by a selection of freely downloadable MATLAB files for hydrostatic and stability calculations. In order to access this material please visit www.bh.com/companions/ and follow the instructions on the screen.

About this book

Theoretical developments require an understanding of basic calculus and analytic geometry. A few sections employ basic vector calculus, differential geometry or ordinary differential equations. Students able to read them will gain more insight into matters explained in the book. Other readers can skip those sections without impairing their understanding of practical calculations and regulations described in the text.

Chapter 1 introduces the reader to basic terminology and to the subject of hull definition. The definitions follow new ISO and ISO-based standards. Translations into French, German and Italian are provided for the most important terms.

The basic concepts of hydrostatics of floating bodies are described in Chapter 2; they include the conditions of equilibrium and initial stability. By the end of this chapter, the reader knows that hydrostatic calculations require many integrations. Methods for performing such integrations in Naval Architecture are developed in Chapter 3.

Chapter 4 shows how to apply the procedures of numerical integration to the calculation of actual hydrostatic properties. Other matters covered in the same chapter are a few simple checks of the resulting plots, and an analysis of how the properties change when a given hull is subjected to a particular class of transformations, namely the properties of affine hulls.

Chapter 5 discusses the statical stability at large angles of heel and the curve of statical stability.

Simple models for assessing the ship stability in the presence of various heeling moments are developed in Chapter 6. Both static and dynamic effects are considered, as well as the influence of factors and situations that negatively affect stability. Examples of the latter are displaced loads, hanging loads, free liquid surfaces, shifting loads, and grounding and docking. Three subjects closely related to practical stability calculations are described in Chapter 7: Weight and trim calculations and the inclining experiment.

Ships and other floating structures are approved for use only if they comply with pertinent regulations. Regulations applicable to merchant ships, ships of the US Navy and UK Navy, and small sail or motor craft are summarily described in Chapter 8.

The phenomenon of parametric resonance, or Mathieu effect, is briefly described in Chapter 9. The chapter includes a simple criterion of distinguishing between stable and unstable solutions and examples of simple simulations in MATLAB.

Ships of the German Federal Navy are designed according to criteria that take into account the Mathieu effect: they are introduced in Chapter 10.

Chapters 8 and 10 deal with intact ships. Ships and some other floating structures are also required to survive after a limited amount of flooding. Chapter 11 shows how to achieve this goal by subdividing the hull by means of watertight bulkheads. There are two methods of calculating the ship condition after damage, namely the method of lost buoyancy and the method of added weight. The difference between the two methods is explained by means of a simple example. The chapter also contains short descriptions of several regulations for merchant and for naval ships.

Chapters 8, 10 and 11 inform the reader about the existence of requirements issued by bodies that approve the design and the use of ships and other floating bodies, and show how simple models developed in previous chapters are applied in engineering calculations. Not all the details of those regulations are included in this book, neither all regulations issued all over the world. If the reader has to perform calculations that must be submitted for approval, it is highly recommended to find out which are the relevant regulations and to consult the complete, most recent edition of them.

Chapter 12 goes beyond the traditional scope of Ship Hydrostatics and provides a bridge towards more advanced and realistic models. The theory of linear waves is briefly introduced and it is shown how real seas can be described by the superposition of linear waves and by the concept of spectrum. Floating bodies move in six degrees of freedom and the spectrum of those motions is related to the sea spectrum. Another subject introduced in this chapter is that of tank stabilizers, a case in which surfaces of free liquids can help in reducing the roll amplitude.

Chapter 13 is about the use of modern computers in hull definition, hydrostatic calculations and simulations of motions. The chapter introduces the basic concepts of computer graphics and illustrates their application to hull definition by means of the MultiSurf and SurfaceWorks packages. A roll simulation in SIMULINK, a toolbox of MATLAB, exemplifies the possibilities of modern simulation software.

Using this book

Boldface words indicate a key term used for the first time in the text, for instance **length between perpendiculars**. Italics are used to emphasize, for example *equilibrium of moments*. Vectors are written with a line over their name: $\overline{KB}$, $\overline{GM}$. Listings of MATLAB programmes, functions and file names are written in typewriter characters, for instance `mathisim.m`.

Basic ideas are exemplified on simple geometric forms for which analytic solutions can be readily found. After mastering these ideas, the students should practise on real ship data provided in examples and exercises, at the end of each chapter. The data of an existing vessel, called *Lido 9*, are used throughout the book to illustrate the main concepts. Data of a few other real-world vessels are given in additional examples and exercises.

I am closing this preface by paying a tribute to the memory of those who taught me the profession, Dinu Ilie and Nicolae Pârâianu, and of my colleague in teaching, Pinkhas Milkh.

Acknowledgements _____

The first acknowledgements should certainly go to the many students who took the course from which emerged this book. Their reactions helped in identifying the topics that need more explanations. Naming a few of those students would imply the risk of being unfair to others.

Many numerical examples were calculated with the aid of the programme system ARCHIMEDES. The TECHNION obtained this software by the courtesy of Heinrich Söding, then at the Technical University of Hannover, now at the Technical University of Hamburg. Included with the programme source there was a set of test data that describe a vessel identified as *Ship No. 83074*. Some examples in this book are based on that data.

Sol Bodner, coordinator of the Ship Engineering Program of the Technion, provided essential support for the course of Ship Hydrostatics. Itzhak Shaham and Jack Yanai contributed to the success of the programme.

Paul Münch provided data of actual vessels and *Lido Kineret, Ltd* and the *Özdeniz Group, Inc.* allowed us to use them in numerical examples. Eliezer Kantorowitz read initial drafts of the book proposal. Yeshayahu Hershkowitz, of Lloyd's Register, and Arnon Nitzan, then student in the last graduate year, read the final draft and returned helpful comments. Reinhard Siegel, of AeroHydro, provided the drawing on which the cover of the book is based, and helped in the application of MultiSurf and SurfaceWorks. Antonio Tiano, of the University of Pavia, gave advice on a few specialized items. Dan Livneh, of the Israeli Administration of Shipping and Ports, provided updating on international codes of practice. C.B. Barrass reviewed the first eleven chapters and provided helpful comments.

Richard Barker drew the attention of the author to the first uses of the term Naval Architecture. The common love for the history of the profession enabled a pleasant and interesting dialogue.

Naomi Fernandes of MathWorks, Baruch Pekelman, their agent in Israel, and his assistants enabled the author to use the latest MATLAB developments.

The author thanks Addison-Wesley Longman, especially Karen Mosman and Pauline Gillet, for permission to use material from the book *MATLAB for Engineers* written by him and Moshe Breiner.

The author thanks the editors of Elsevier, Rebecca Hamersley, Rebecca Rue, Sallyann Deans and Nishma Shah for their cooperation and continuous help. It was the task of Nishma Shah to bring the project into production. Finally, the author appreciates the way Padma Narayanan, of Integra Software Services, managed the production process of this book.

1
Definitions, principal dimensions

1.1 Introduction

The subjects treated in this book are the basis of the profession called **Naval Architecture**. The term *Naval Architecture* comes from the titles of books published in the seventeenth century. For a long time, the oldest such book we were aware of was Joseph Furttenbach's *Architectura Navalis* published in Frankfurt in 1629. The bibliographical data of a beautiful reproduction are included in the references listed at the end of this book. Close to 1965 an older Portuguese manuscript was rediscovered in Madrid, in the Library of the Royal Academy of History. The work is due to João Baptista Lavanha and is known as *Livro Primeiro da Architectura Naval*, that is 'First book on Naval Architecture'. The traditional dating of the manuscript is 1614. The following is a quotation from a translation due to Richard Barker:

> Architecture consists in building, which is the permanent construction of any thing. This is done either for defence or for religion, and utility, or for navigation. And from this partition is born the division of Architecture into three parts, which are Military, Civil and Naval Architecture.

> And Naval Architecture is that which with certain rules teaches the building of ships, in which one can navigate well and conveniently.

The term may be still older. Thomas Digges (English, 1546–1595) published in 1579 an *Arithmeticall Militarie Treatise*, named *Stratioticos* in which he promised to write a book on 'Architecture Nautical'. He did not do so. Both the British Royal Institution of Naval Architects – RINA – and the American Society of Naval Architects and Marine Engineers – SNAME – opened their websites for public debates on a modern definition of Naval Architecture. Out of the many proposals appearing there, that provided by A. Blyth, FRINA, looked to us both concise and comprehensive:

> Naval Architecture is that branch of engineering which embraces all aspects of design, research, developments, construction, trials

and effectiveness of all forms of man-made vehicles which operate either in or below the surface of any body of water.

If Naval Architecture is a branch of Engineering, what is Engineering? In the New Encyclopedia Britannica (1989) we find:

> Engineering is the professional art of applying science to the optimum conversion of the resources of nature to the uses of mankind. Engineering has been defined by the Engineers Council for Professional Development, in the United States, as the creative application of "scientific principles to design or develop structures, machines . . ."

This book deals with the scientific principles of Hydrostatics and Stability. These subjects are treated in other languages in books bearing titles such as *Ship theory* (for example Doyère, 1927) or *Ship statics* (for example Hervieu, 1985). Further scientific principles to be learned by the Naval Architect include Hydrodynamics, Strength, Motions on Waves and more. The 'art of applying' these principles belongs to courses in Ship Design.

1.2 Marine terminology

Like any other field of engineering, Naval Architecture has its own vocabulary composed of technical terms. While a word may have several meanings in common language, when used as a technical term, in a given field of technology, it has one meaning only. This enables unambigous communication within the profession, hence the importance of clear definitions.

The technical vocabulary of people with long maritime tradition has peculiarities of origins and usage. As a first important example in English let us consider the word **ship**; it is of Germanic origin. Indeed, to this day the equivalent Danish word is *skib*, the Dutch, *schep*, the German, *Schiff* (pronounce 'shif'), the Norwegian *skip* (pronounce 'ship'), and the Swedish, *skepp*. For mariners and Naval Architects a ship has a soul; when speaking about a ship they use the pronoun 'she'.

Another interesting term is **starboard**; it means the right-hand side of a ship when looking forward. This term has nothing to do with stars. Pictures of Viking vessels (see especially the Bayeux Tapestry) show that they had a steering board (paddle) on their right-hand side. In Norwegian a 'steering board' is called 'styri bord'. In old English the Nordic term became 'steorbord' to be later distorted to the present-day 'starboard'. The correct term should have been 'steeringboard'. German uses the exact translation of this word, 'Steuerbord'.

The left-hand side of a vessel was called *larboard*. Hendrickson (1997) traces this term to 'lureboard', from the Anglo-Saxon word 'laere' that meant empty, because the steersman stood on the other side. The term became 'lade-board' and 'larboard' because the ship could be loaded from this side only. Larboard sounded too much like starboard and could be confounded with this. Therefore, more than 200 years ago the term was changed to **port**. In fact, a ship with a steering board on the right-hand side can approach to port only with her left-hand side.

1.3 The principal dimensions of a ship

In this chapter we introduce the principal dimensions of a ship, as defined in the international standard ISO 7462 (1985). The terminology in this document was adopted by some national standards, for example the German standard DIN 81209-1. We extract from the latter publication the symbols to be used in drawings and equations, and the symbols recommended for use in computer programs. Basically, the notation agrees with that used by **SNAME** and with the *ITTC Dictionary of Ship Hydrodynamics* (RINA, 1978). Much of this notation has been used for a long time in English-speaking countries.

Beyond this chapter, many definitions and symbols appearing in this book are derived from the above-mentioned sources. Different symbols have been in use in continental Europe, in countries with a long maritime tradition. Hervieu (1985), for example, opposes the introduction of Anglo-Saxon notation and justifies his attitude in the Introduction of his book. If we stick in this book to a certain notation, it is not only because the book is published in the UK, but also because English is presently recognized as the world's *lingua franca* and the notation is adopted in more and more national standards. As to spelling, we use the British one. For example, in this book we write 'centre', rather than 'center' as in the American spelling, 'draught' and not 'draft', and 'moulded' instead of 'molded'.

To enable the reader to consult technical literature using other symbols, we shall mention the most important of them. For ship dimensions we do this in Table 1.1, where we shall give also translations into French and German of the most important terms, following mainly ISO 7462 and DIN 81209-1. In addition, Italian terms will be inserted and they conform to Italian technical literature, for example Costaguta (1981). The translations will be marked by 'Fr' for French, 'G' for German and 'I' for Italian. Almost all ship hulls are symmetric with respect with a longitudinal plane (plane xz in Figure 1.6). In other words, ships present a 'port-to-starboard' symmetry. The definitions take this fact into account. Those definitions are explained in Figures 1.1 to 1.4.

The outer surface of a steel or aluminium ship is usually not smooth because not all plates have the same thickness. Therefore, it is convenient to define the hull surface of such a ship on the inner surface of the plating. This is the **Moulded surface** of the hull. Dimensions measured to this surface are qualified as **Moulded**. By contrast, dimensions measured to the outer surface of the hull or of an appendage are qualified as **extreme**. The moulded surface is used in the first stages of ship design, before designing the plating, and also in test-basin studies.

Table 1.1 Principal ship dimensions and related terminology

English term	Symbol	Computer notation	Translations
After (aft) perpendicular	AP		Fr perpendiculaire *arrière*, G hinteres Lot, I perpendicolare addietro
Baseline	BL		Fr ligne de base, G Basis, I linea base
Bow			Fr proue, l'avant, G Bug, I prora, prua
Breadth	B	B	Fr largeur, G Breite, I larghezza
Camber			Fr bouge, G Balkenbucht, I bolzone
Centreline plane		CL	Fr plan longitudinal de symétrie, G Mittschiffsebene, I Piano di simmetria, piano diametrale
Depth	D	DEP	Fr creux, G Seitenhöhe, I altezza
Depth, moulded			Fr creux sur quille, G Seitenhöhe, I altezza di costruzione (puntale)
Design waterline	DWL	DWL	Fr flottaison normale, G Konstruktionswasserlinie (KWL), I linea d'acqua del piano di costruzione
Draught	T	T	Fr tirant d'eau, G Tiefgang, I immersione
Draught, aft	T_A	TA	Fr tirant d'eau arrière, G Hinterer Tiefgang, I immersiona a poppa
Draught, amidships	T_M		Fr tirant d'eau milieu, G mittleres Tiefgang, I immersione media
Draught, extreme			Fr profondeur de carène hors tout, G größter Tiefgang, I pescaggio
Draught, forward	T_F	TF	Fr tirant d'eau avant, G Vorderer Tiefgang, I immersione a prora
Draught, moulded			Fr profondeur de carène hors membres,
Forward perpendicular	FP		Fr perpendiculaire avant, G vorderes Lot, I perpendicolare avanti

Table 1.1 *Cont.*

English term	Symbol	Computer notation	Translations
Freeboard	f	FREP	Fr franc-bord, G Freibord, I franco bordo
Heel angle	ϕ_s	HEELANG	Fr bande, gîte, Krängungswinkel I angolo d'inclinazione trasversale
Length between perpendiculars	L_{pp}	LPP	Fr longueur entre perpendiculaires, G Länge zwischen den Loten, I lunghezza tra le perpendicolari
Length of waterline	L_{WL}	LWL	Fr longueur à la flottaison, G Wasserlinielänge, I lunghezza al galleggiamento
Length overall	L_{OA}		Fr longueur hors tout, G Länge über allen, I lunghezza fuori tutto
Length overall submerged	L_{OS}		Fr longueur hors tout immergé, G Länge über allen unter Wasser, I lunghezza massima opera viva
Lines plan			Fr plan des formes, G Linienriß, I piano di costruzione, piano delle linee
Load waterline	DWL	DWL	Fr ligne de flottaison en charge, G Konstruktionswasserlinie, I linea d'acqua a pieno carico
Midships			Fr couple milieu, G Hauptspant, I sezione maestra
Moulded			Fr hors membres, G auf Spanten, I fuori ossatura
Port		P	Fr bâbord, G Backbord, I sinistra
Sheer			Fr tonture, G Decksprung, I insellatura
Starboard		S	Fr tribord, G Steuerbord, I dritta
Station			Fr couple, G Spante, I ordinata
Stern, poop			Fr arrière, poupe, G Hinterschiff, I poppa
Trim			Fr assiette, G Trimm, I differenza d'immersione
Waterline	WL	WL	Fr ligne d'eau, G Wasserlinie, I linea d'acqua

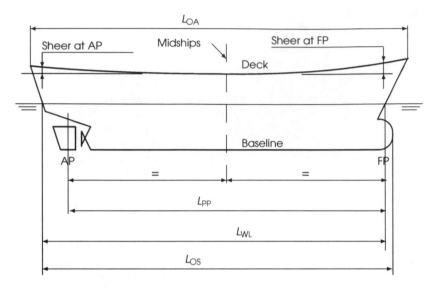

Figure 1.1 Length dimensions

Figure 1.2 How to measure the length between perpendiculars

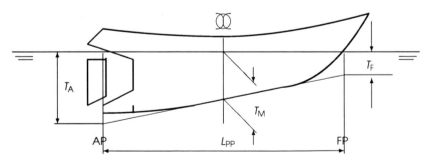

Figure 1.3 The case of a keel not parallel to the load line

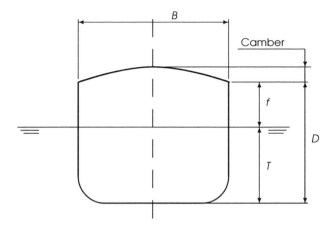

Figure 1.4 Breadth, depth, draught and camber

The **baseline**, shortly BL, is a line lying in the longitudinal plane of symmetry and parallel to the designed summer load waterline (see next paragraph for a definition). It appears as a horizontal in the lateral and transverse views of the hull surface. The baseline is used as the longitudinal axis, that is the x-axis of the system of coordinates in which hull points are defined. Therefore, it is recommended to place this line so that it passes through the lowest point of the hull surface. Then, all z-coordinates will be positive.

Before defining the dimensions of a ship we must choose a reference waterline. ISO 7462 recommends that this **load waterline** be the **designed summer load line**, that is the waterline up to which the ship can be loaded, in sea water, during summer when waves are lower than in winter. The qualifier 'designed' means that this line was established in some design stage. In later design stages, or during operation, the load line may change. It would be very inconvenient to update this reference and change dimensions and coordinates; therefore, the 'designed' datum line is kept even if no more exact. A notation older than ISO 7462 is DWL, an abbreviation for 'Design Waterline'.

The **after perpendicular,** or **aft perpendicular**, noted AP, is a line drawn perpendicularly to the load line through the after side of the rudder post or through the axis of the rudder stock. The latter case is shown in Figures 1.1 and 1.3. For naval vessels, and today for some merchant vessels ships, it is usual to place the AP at the intersection of the aftermost part of the moulded surface and the load line, as shown in Figure 1.2. The **forward perpendicular**, FP, is drawn perpendicularly to the load line through the intersection of the fore side of the stem with the load waterline. Mind the slight lack of consistency: while all moulded dimensions are measured to the moulded surface, the FP is drawn on the outer side of the stem. The distance between the after and the forward perpendicular, measured parallel to the load line, is called **length between perpendiculars** and its notation is L_{pp}. An older notation was LBP. We call **length overall**, L_{OA}, the length between the ship extremities. The **length overall submerged**, L_{OS}, is the maximum length of the submerged hull measured parallel to the designed load line.

We call **station** a point on the baseline, and the transverse section of the hull surface passing through that point. The station placed at half L_{pp} is called **midships**. It is usual to note the midship section by means of the symbol shown in Figure 1.5 (a). In German literature we usually find the simplified form shown in Figure 1.5 (b).

The **moulded depth**, D, is the height above baseline of the intersection of the underside of the deck plate with the ship side (see Figure 1.4). When there are several decks, it is necessary to specify to which one refers the depth.

The **moulded draught**, T, is the vertical distance between the top of the keel to the designed summer load line, usually measured in the midships plane (see Figure 1.4). Even when the keel is parallel to the load waterline, there may be appendages protruding below the keel, for example the sonar dome of a warship. Then, it is necessary to define an **extreme draught** that is the distance between the lowest point of the hull or of an appendage and the designed load line.

Certain ships are designed with a keel that is not parallel to the load line. Some tugs and fishing vessels display this feature. To define the draughts associated with such a situation let us refer to Figure 1.3. We draw an auxiliary line that extends the keel afterwards and forwards. The distance between the intersection of this auxiliary line with the aft perpendicular and the load line is called **aft draught** and is noted with T_A. Similarly, the distance between the load line and the intersection of the auxiliary line with the forward perpendicular is called **forward draught** and is noted with T_F. Then, the draught measured in the midship section is known as **midships draught** and its symbol is T_M. The difference between depth and draft is called **freeboard**; in DIN 81209-1 it is noted by f.

The **moulded volume of displacement** is the volume enclosed between the submerged, moulded hull and the horizontal waterplane defined by a given draught. This volume is noted by ∇, a symbol known in English-language literature as *del*, and in European literature as *nabla*. In English we must use two words, 'submerged hull', to identify the part of the hull below the waterline. Romance languages use for the same notion only one word derived from the Latin 'carina'. Thus, in French it is 'carène', while in Catalan, Italian, Portuguese, Romanian, and Spanish it is called 'carena'.

In many ships the deck has a transverse curvature that facilitates the drainage of water. The vertical distance between the lowest and the highest points of the

Figure 1.5 (a) Midships symbol in English literature, (b) Midships symbol in German literature

deck, in a given transverse section, is called **camber** (see Figure 1.4). According to ISO 7460 the camber is measured in mm, while all other ship dimensions are given in m. A common practice is to fix the camber amidships as 1/50 of the breadth in that section and to fair the deck towards its extremities (for the term 'fair' see Subsection 1.4.3). In most ships, the intersection of the deck surface and the plane of symmetry is a curved line with the concavity upwards. Usually, that line is tangent to a horizontal passing at a height equal to the ship depth, D, in the midship section, and runs upwards towards the ship extremities. It is higher at the bow. This longitudinal curvature is called **sheer** and is illustrated in Figure 1.1. The deck sheer helps in preventing the entrance of waves and is taken into account when establishing the load line in accordance with international conventions.

1.4 The definition of the hull surface

1.4.1 Coordinate systems

The DIN 81209-1 standard recommends the system of coordinates shown in Figure 1.6. The x-axis runs along the ship and is positive forwards, the y-axis is transversal and positive to port, and the z-axis is vertical and positive upwards. The origin of coordinates lies at the intersection of the centreline plane with the transversal plane that contains the aft perpendicular. The international standards ISO 7460 and 7463 recommend the same positive senses as DIN 81209-1 but do not specify a definite origin. Other systems of coordinates are possible. For example, a system defined as above, but having its origin in the midship section, has some advantages in the display of certain hydrostatic data. Computer programmes written in the USA use a system of coordinates with the origin of coordinates in the plane of the forward perpendicular, FP, the x-axis positive

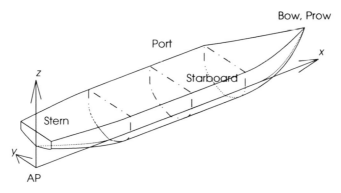

Figure 1.6 System of coordinates recommended by DIN 81209-1

afterwards, the y-axis positive to starboard, and the z-axis positive upwards. For dynamic applications, taking the origin in the centre of gravity simplifies the equations. However, it should be clear that to each loading condition corresponds one centre of gravity, while a point like the intersection of the aft perpendicular with the base line is independent of the ship loading. The system of coordinates used for the hull surface can be also employed for the location of weights. By its very nature, the system in which the hull is defined is fixed in the ship and moves with her. To define the various **floating conditions**, that is the positions that the vessel can assume, we use another system, fixed in space, that is defined in ISO 7463 as x_0, y_0, z_0. Let this system initially coincide with the system x, y, z. A vertical translation of the system x, y, z with respect to the space-fixed system x_0, y_0, z_0 produces a draught change.

If the ship-fixed z-axis is vertical, we say that the ship floats in an upright condition. A rotation of the ship-fixed system around an axis parallel to the x-axis is called **heel** (Figure 1.7) if it is temporary, and **list** if it is permanent. The heel can be produced by lateral wind, by the centrifugal force developed in turning, or by the temporary, transverse displacement of weights. The list can result from incorrect loading or from flooding. If the transverse inclination is the result of ship motions, it is time-varying and we call it **roll**.

When the ship-fixed x-axis is parallel to the space-fixed x_0-axis, we say that the ship floats on **even keel**. A static inclination of the ship-fixed system around an axis parallel to the ship-fixed y-axis is called **trim**. If the inclination is dynamic, that is a function of time resulting from ship motions, it is called **pitch**. A graphic explanation of the term trim is given in Figure 1.7. The trim is measured as the difference between the forward and the aft draught. Then, trim is positive if the ship is **trimmed by the head**. As defined here the trim is measured in metres.

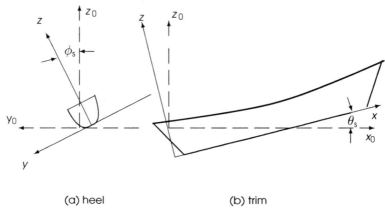

(a) heel   (b) trim

Figure 1.7 Heel and trim

1.4.2 Graphic description

In most cases the hull surface has double curvature and cannot be defined by simple analytical equations. To cope with the problem, Naval Architects have drawn lines obtained by cutting the hull surface with sets of parallel planes. Readers may find an analogy with the definition of the earth surface in topography by *contour lines*. Each contour line connects points of constant height above sea level. Similarly, we represent the hull surface by means of lines of constant x, constant y, and constant z. Thus, cutting the hull surface by planes parallel to the yOz plane we obtain the **transverse** sections noted in Figure 1.8 as St0 to St10, that is **Station 0**, **Station 1**, ... **Station 10**. Cutting the same hull by horizontal planes (planes parallel to the base plane xOy), we obtain the **waterlines** marked in Figure 1.9 as WL0 to WL5. Finally, by cutting the same hull with longitudinal planes parallel to the xOz plane, we draw the buttocks shown in Figure 1.10. The most important buttock is the line $y = 0$ known as **centreline**; for almost all ship hulls it is a plane of symmetry.

Stations, waterlines and buttocks are drawn together in the **lines drawing**. Figure 1.11 shows one of the possible arrangements, probably the most common one. As stations and waterlines are symmetric for almost all ships, it is sufficient to draw only a half of each one. Let us take a look to the right of our drawing; we see the set of stations represented together in the **body plan**. The left half of the body plan contains stations 0 to 4, that is the stations of the **afterbody**, while the right half is composed of stations 5 to 10, that is the **forebody**. The set of buttocks, known as **sheer plan**, is placed at the left of the body plan. Beneath is the set of waterlines. Looking with more attention to the lines drawing we find out that each line appears as curved in one projection, and as straight lines in

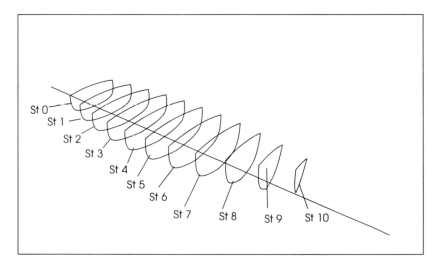

Figure 1.8 Stations

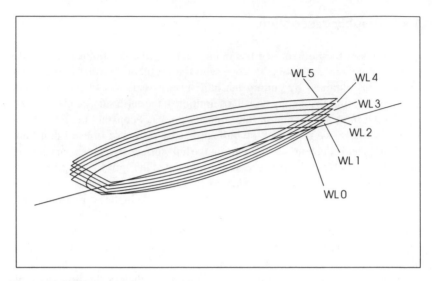

Figure 1.9 Waterlines

the other two. For example, stations appear as curved lines in the body plan, as straight lines in the sheer and in the waterlines plans.

The station segments having the highest curvature are those in the **bilge** region, that is between the bottom and the ship side. Often no buttock or waterlines cuts them. To check what happens there it is usual to draw one or more additional lines by cutting the hull surface with one or more planes parallel to the baseline

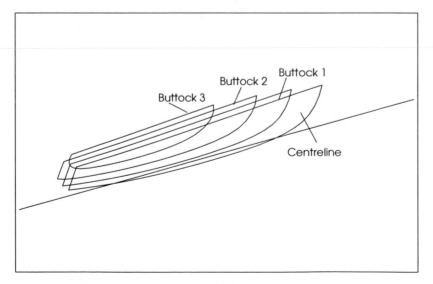

Figure 1.10 Buttocks

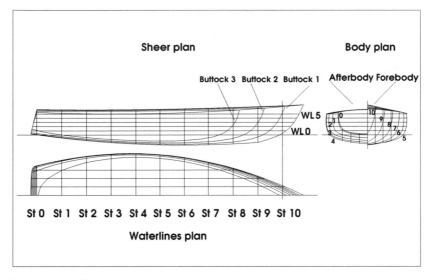

Figure 1.11 The lines drawing

but making an angle with the horizontal. A good practice is to incline the plane so that it will be approximately normal to the station lines in the region of highest curvature. The intersection of such a plane with the hull surface is appropriately called **diagonal**.

Figure 1.11 was produced by modifying under MultiSurf a model provided with that software. The resulting surface model was exported as a DXF file to TurboCad where it was completed with text and exported as an EPS (Encapsulated PostScript) file. Figures 1.8 to 1.10 were obtained from the same model as MultiSurf **contour** curves and similarly post-processed under TurboCad.

1.4.3 Fairing

The curves appearing in the lines drawing must fulfill two kinds of conditions: they must be coordinated and they must be 'smooth', except where functionality requires for abrupt changes. Lines that fulfill these conditions are said to be **fair**. We are going to be more specific. In the preceding section we have used three projections to define the ship hull. From descriptive geometry we may know that two projections are sufficient to define a point in three-dimensional space. It follows that the three projections in the lines drawing must be coordinated, otherwise one of them may be false. Let us explain this idea by means of Figure 1.12. In the body plan, at the intersection of Station 8 with Waterline 4, we measure that **half-breadth** $y(\text{WL4}, \text{St8})$. We must find exactly the same dimension between the centreline and the intersection of Waterline 4 and Station 8 in the waterlines plan. The same intersection appears as a point, marked by a circle,

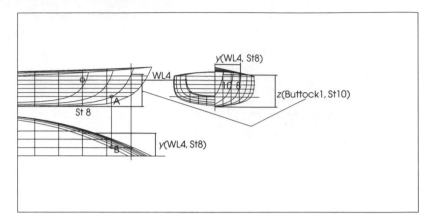

Figure 1.12 Fairing

in the sheer plan. Next, we measure in the body plan the distance z(Buttock1, St10) between the base plane and the intersection of Station 10 with the longitudinal plane that defines Buttock 1. We must find exactly the same distance in the sheer plan. As a third example, the intersection of Buttock 1 and Waterline 1 in the sheer plan and in the waterlines plan must lie on the same vertical, as shown by the segment AB.

The concept of smooth lines is not easy to explain in words, although lines that are not smooth can be easily recognized in the drawing. The manual of the surface modelling program *MultiSurf* rightly relates fairing to the concepts of beauty and simplicity and adds:

> A curve should not be more complex than it needs to be to serve its function. It should be free of unnecessary inflection points (reversals of curvature), rapid turns (local high curvature), flat spots (local low curvature), or abrupt changes of curvature ...

With other words, a 'curve should be pleasing to the eye' as one famous Naval Architect was fond of saying. For a formal definition of the concept of **curvature** see Chapter 13, Computer methods.

The fairing process cannot be satisfactorily completed in the lines drawing. Let us suppose that the lines are drawn at the scale 1:200. A good, young eye can identify errors of 0.1 mm. At the ship scale this becomes an error of 20 mm that cannot be accepted. Therefore, for many years it was usual to redraw the lines at the scale 1:1 in the **moulding loft** and the fairing process was completed there.

Some time after 1950, both in East Germany (the former DDR) and in Sweden, an optical method was introduced. The lines were drawn in the design office at the scale 1:20, under a magnifying glass. The drawing was photographed on glass plates and brought to a projector situated above the workshop. From there

Table 1.2 Table of offsets

	St	0	1	2	3	4	5	6	7	8	9	10
	x	0.000	0.893	1.786	2.678	3.571	4.464	5.357	6.249	7.142	8.035	8.928
WL	z					Half breadths						
0	0.360		0.900	1.189	1.325	1.377	1.335	1.219	1.024	0.749	0.389	
1	0.512	0.894	1.167	1.341	1.440	1.463	1.417	1.300	1.109	0.842	0.496	0.067
2	0.665	1.014	1.240	1.397	1.482	1.501	1.455	1.340	1.156	0.898	0.564	0.149
3	0.817	1.055	1.270	1.414	1.495	1.514	1.470	1.361	1.184	0.936	0.614	0.214
4	0.969	1.070	1.273	1.412	1.491	1.511	1.471	1.369	1.201	0.962	0.648	0.257
5	1.122	1.069	1.260	1.395	1.474	1.496	1.461	1.363	1.201	0.972	0.671	0.295

the drawing was projected on plates so that it appeared at the 1:1 scale to enable cutting by optically guided, automatic burners.

The development of hardware and software in the second half of the twentieth century allowed the introduction of computer-fairing methods. Historical highlights can be found in Kuo (1971) and other references cited in Chapter 13. When the hull surface is defined by algebraic curves, as explained in Chapter 13, the lines are smooth by construction. Recent computer programmes include tools that help in completing the fairing process and checking it, mainly the calculation of curvatures and **rendering**. A rendered view is one in which the hull surface appears in perspective, shaded and lighted so that surface smoothness can be summarily checked. For more details see Chapter 13.

1.4.4 Table of offsets

In shipyard practice it has been usual to derive from the lines plan a digital description of the hull known as **table of offsets**. Today, programs used to design hull surface produce automatically this document. An example is shown in Table 1.2. The numbers correspond to Figure 1.11. The table of offsets contains half-breadths measured at the stations and on the waterlines appearing in the lines plan. The result is a table with two entries in which the offsets (half-breadths) are grouped into columns, each column corresponding to a station, and in rows, each row corresponding to a waterline. Table 1.2 was produced in MultiSurf.

1.5 Coefficients of form

In ship design it is often necessary to classify the hulls and to find relationships between forms and their properties, especially the hydrodynamic properties. The **coefficients of form** are the most important means of achieving this. By their definition, the coefficients of form are non-dimensional numbers.

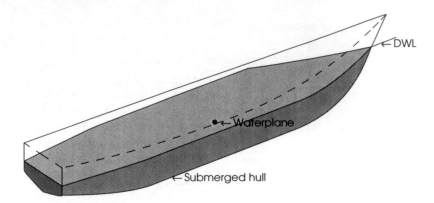

Figure 1.13 The submerged hull

The **block coefficient** is the ratio of the moulded displacement volume, ∇, to the volume of the parallelepiped (rectangular block) with the dimensions L, B and T:

$$C_B = \frac{\nabla}{LBT} \qquad (1.1)$$

In Figure 1.14 we see that C_B indicates how much of the enclosing parallelepiped is filled by the hull.

The **midship coefficient**, C_M, is defined as the ratio of the midship-section area, A_M, to the product of the breadth and the draught, BT,

$$C_M = \frac{A_M}{BT} \qquad (1.2)$$

Figure 1.15 enables a graphical interpretation of C_M.

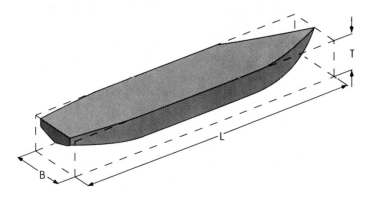

Figure 1.14 The definition of the block coefficient, C_B

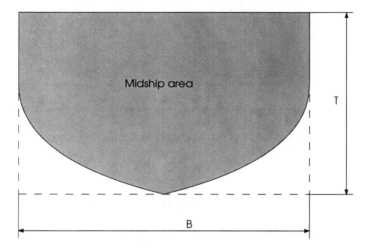

Figure 1.15 The definition of the midship-section coefficient, C_M

The **prismatic coefficient**, C_P, is the ratio of the moulded displacement volume, ∇, to the product of the midship-section area, A_M, and the length, L:

$$C_P = \frac{\nabla}{A_M L} = \frac{C_B LBT}{C_M BTL} = \frac{C_B}{C_M} \tag{1.3}$$

In Figure 1.16 we can see that C_P is an indicator of how much of a cylinder with constant section A_M and length L is filled by the submerged hull. Let us note the **waterplane area** by A_W. Then, we define the **waterplane-area coefficient** by

$$C_{WL} = \frac{A_W}{LB} \tag{1.4}$$

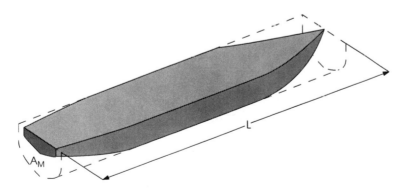

Figure 1.16 The definition of the prismatic coefficient, C_P

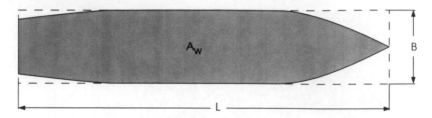

Figure 1.17 The definition of the waterplane coefficient, C_{WL}

A graphic interpretation of the waterplane coefficient can be deduced from Figure 1.17.

The **vertical prismatic coefficient** is calculated as

$$C_{VP} = \frac{\nabla}{A_W T} \tag{1.5}$$

For a geometric interpretation see Figure 1.18.

Other coefficients are defined as ratios of dimensions, for instance L/B, known as **length–breadth ratio**, and B/T known as 'B over T'. The **length coefficient of Froude**, or **length–displacement ratio** is

$$Ⓜ = \frac{L}{\nabla^{1/3}} \tag{1.6}$$

and, similarly, the **volumetric coefficient**, ∇/L^3.

Table 1.3 shows the symbols, the computer notations, the translations of the terms related to the coefficients of form, and the symbols that have been used in continental Europe.

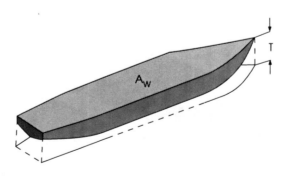

Figure 1.18 The definition of the vertical prismatic coefficient, C_{VP}

Table 1.3 Coefficients of form and related terminology

English term	Symbol	Computer notation	Translations European symbol
Block coefficient	C_B	CB	Fr coefficient de block, δ, G Blockcoeffizient, I coefficiente di finezza (bloc)
Coefficient of form			Fr coefficient de remplissage, G Völligkeitsgrad, I coefficiente di carena
Displacement	Δ		Fr déplacement, G Verdrängung, I dislocamento
Displacement mass	Δ	DISPM	Fr déplacement, masse, G Verdrängungsmasse
Displacement volume	∇	DISPV	Fr Volume de la carène, G Verdrängungs Volumen, I volume di carena
Midship coefficient	C_M	CMS	Fr coefficient de remplissage au maître couple, β, G Völligkeitsgrad der Hauptspantfläche, I coefficiente della sezione maestra
Midship-section area	A_M		Fr aire du couple milieu, G Spantfläche, I area della sezione maestra
Prismatic coefficient	C_P	CPL	Fr coefficient prismatique, ϕ, G Schärfegrad, I coefficiente prismatico o longitudinale
Vertical prismatic coefficient	C_{VP}	CVP	Fr coefficient de remplissage vertical ψ, I coefficiente di finezza prismatico verticale
Waterplane area	A_W	AW	Fr aire de la surface de la flottaison, G Wasserlinienfläche, I area del galleggiamento
Waterplane-area coefficient	C_{WL}		Fr coefficient de remplissage de la flottaison, α, G Völligkeitsgrad der Wasserlinienfläche, I coefficiente del piano di galleggiamento

1.6 Summary

The material treated in this book belongs to the field of Naval Architecture. The terminology is specific to this branch of Engineering and is based on a long maritime tradition. The terms and symbols introduced in the book comply with recent international and corresponding national standards. So do the definitions of the main dimensions of a ship. Familiarity with the terminology and the corresponding symbols enables good communication between specialists all over the world and correct understanding and application of international conventions and regulations.

In general, the hull surface defies a simple mathematical definition. Therefore, the usual way of defining this surface is by cutting it with sets of planes parallel to the planes of coordinates. Let the x-axis run along the ship, the y-axis be transversal, and the z-axis, vertical. The sections of constant x are called **stations**, those of constant z, waterlines, and the contours of constant y, **buttocks**. The three sets must be coordinated and the curves be fair, a concept related to simplicity, curvature and beauty.

Sections, waterlines and buttocks are represented together in the **lines plan**. Line plans are drawn at a reducing scale; therefore, an accurate fairing process cannot be carried out on the drawing board. In the past it was usual to redraw the lines on the moulding loft, at the 1:1 scale. In the second half of the twentieth century the introduction of digital computers and the progress of software, especially computer graphics, made possible new methods that will be briefly discussed in Chapter 13.

In early ship design it is necessary to choose an appropriate hull form and estimate its hydrodynamic properties. These tasks are facilitated by characterizing and classifying the ship forms by means of non-dimensional coefficients of form and ratios of dimensions. The most important coefficient of form is the block coefficient defined as the ratio of the **displacement volume** (volume of the submerged hull) to the product of ship length, breadth and draught. An example of ratio of dimensions is the length–breadth ratio.

1.7 Example

Example 1.1 – Coefficients of a fishing vessel
In INSEAN (1962) we find the test data of a fishing-vessel hull called C.484 and whose principal characteristics are:

L_{WL}	14.251 m
B	4.52 m
T_M	1.908 m
∇	58.536 m^3
A_M	6.855 m^2
A_W	47.595 m^2

We calculate the coefficients of form as follows:

$$C_B = \frac{\nabla}{L_{pp}BT_M} = \frac{58.536}{14.251 \times 4.52 \times 1.908} = 0.476$$

$$C_{WL} = \frac{A_W}{L_{WL}B} = \frac{47.595}{14.251 \times 4.52} = 0.739$$

$$C_M = \frac{A_M}{BT} = \frac{6.855}{4.52 \times 1.908} = 0.795$$

$$C_P = \frac{\nabla}{A_M L_{WL}} = \frac{58.536}{6.855 \times 14.251} = 0.599$$

and we can verify that

$$C_P = \frac{C_B}{C_M} = \frac{0.476}{0.795} = 0.599$$

1.8 Exercises

Exercise 1.1 – Vertical prismatic coefficient
Find the relationship between the vertical prismatic coefficient, C_{VP}, the waterplane-area coefficient, C_{WL}, and the block coefficient, C_B.

Exercise 1.2 – Coefficients of Ship 83074
Table 1.4 contains data belonging to the hull we called *Ship 83074*. The length between perpendiculars, L_{pp}, is 205.74 m, and the breadth, B, 28.955 m. Complete the table and plot the coefficients of form against the draught, T. In Naval Architecture it is usual to measure the draught along the vertical axis, and other data – in our case the coefficients of form – along the horizontal axis (see Chapter 4).

Exercise 1.3 – Coefficients of hull C.786
Table 1.5 contains data taken from INSEAN (1963) and referring to a tanker hull identified as C.786.

Table 1.4 Coefficients of form of Ship 83074

Draught T m	Displacement volume ∇ m^3	Waterplane area A_{WL} m^2	C_B	C_{WL}	C_M	C_P
3	9029	3540.8	0.505	0.594	0.890	0.568
4	12632	3694.2			0.915	
5	16404	3805.2			0.931	
6	20257	3898.7			0.943	
7	24199	3988.6			0.951	
8	28270	4095.8			0.957	
9	32404	4240.4			0.962	

Table 1.5 Data of tanker
hull C.786

L_{WL}	205.468 m
B	27.432 m
T_M	10.750 m
∇	46341 m^3
A_M	0.220
A_{WL}	3.648

Calculate the coefficients of form and check that

$$\frac{C_B}{C_M} = C_P$$

2
Basic ship hydrostatics

2.1 Introduction

This chapter deals with the **conditions of equilibrium** and **initial stability of floating bodies**. We begin with a derivation of **Archimedes' principle** and the definitions of the notions of **centre of buoyancy** and **displacement**. Archimedes' principle provides a particular formulation of the law of *equilibrium of forces* for floating bodies. The law of *equilibrium of moments* is formulated as **Stevin's law** and it expresses the relationship between the *centre of gravity* and the *centre of buoyancy* of the floating body. The study of *initial stability* is the study of the behaviour in the neighbourhood of the position of equilibrium. To derive the condition of initial stability we introduce Bouguer's concept of **metacentre**.

To each position of a floating body correspond one centre of buoyancy and one metacentre. Each position of the floating body is defined by three parameters, for instance the triple {*displacement, angle of heel, angle of trim*}; we call them the **parameters of the floating condition**. If we keep two parameters constant and let one vary, the centre of buoyancy travels along a curve and the metacentre along another. If only one parameter is kept constant and two vary, the centre of buoyancy and the metacentre generate two surfaces. In this chapter we shall briefly show what happens when the displacement is constant. The discussion of the case in which only one angle (that is, either heel or trim) varies leads to the concept of **metacentric evolute**.

The treatment of the above problems is based on the following assumptions:

1. the water is incompressible;
2. viscosity plays no role;
3. surface tension plays no role;
4. the water surface is plane.

The first assumption is practically exact in the range of water depths we are interested in. The second assumption is exact in static conditions (that is without motion) and a good approximation at the very slow rates of motion discussed in ship hydrostatics. In Chapter 12 we shall point out to the few cases in which viscosity should be considered. The third assumption is true for the sizes of floating bodies and the wave heights we are dealing with. The fourth assumption is never true, not even in the sheltered waters of a harbour. However, this hypothesis allows us to derive very useful, general results, and calculate essential properties of ships and other floating bodies. It is only in Chapter 9 that we shall leave the assumption of a plane water surface and see what happens in waves. In fact, the theory of ship hydrostatics was developed during 200 years under the hypothesis of a plane water surface and only in the middle of the twentieth century it was recognized that this assumption cannot explain the capsizing of a few ships that were considered stable by that time.

The results derived in this chapter are general in the sense that they do not assume particular body shapes. Thus, no symmetry must be assumed such as it usually exists in ships (port-to-starboard symmetry) and still less symmetry about two axes, as encountered, for instance, in Viking ships, some ferries, some offshore platforms and most buoys. The results hold the same for single-hull ships as for catamarans and trimarans. The only problem is that the treatment of the problems for general-form floating bodies requires 'more' mathematics than the calculations for certain simple or symmetric solids. To make this chapter accessible to a larger audience, although we derive the results for body shapes without any form restrictions, we also exemplify them on parallelepipedic and other simply defined floating body forms. Reading only those examples is sufficient to understand the ideas involved and the results obtained in this chapter. However, only the general derivations can provide the feeling of generality and a good insight into the problems discussed here.

2.2 Archimedes' principle

2.2.1 A body with simple geometrical form

A body immersed in a fluid is subjected to an upwards force equal to the weight of the fluid displaced.

The above statement is known as *Archimedes' principle*. One legend has it that Archimedes (Greek, lived in Syracuse – Sicily – between 287 and 212 BC) discovered this law while taking a bath and that he was so happy that he ran naked in the streets shouting 'I have found' (in Greek *Heureka*, see entry 'eureka' in Merriam-Webster, 1991). The legend may be nice, but it is most probably not true. What is certain is that Archimedes used his principle to assess the amount of gold in gold–silver alloys.

Archimedes' principle can be derived mathematically if we know another law of general hydrostatics. Most textbooks contain only a brief, unconvincing proof based on intuitive considerations of equilibrium. A more elaborate proof is given here and we prefer it because only thus it is possible to decide whether Archimedes' principle applies or not in a given case. Let us consider a fluid whose specific gravity is γ. Then, at a depth z the **pressure** in the fluid equals γz. This is the weight of the fluid column of height z and unit area cross section. The pressure at a point is the same in all directions and this statement is known as *Pascal's principle*. The proof of this statement can be found in many textbooks on fluid mechanics, such as Douglas, Gasiorek and Swaffiled (1979: 24), or Pnueli and Gutfinger (1992: 30–1).

In this section we calculate the *hydrostatic forces* acting on a body having a simple geometric form. The general derivation is contained in the next section. In this section we consider a simple-form solid as shown in Figure 2.1; it is a parallelepipedic body whose horizontal, rectangular cross-section has the sides B and L. We consider the body immersed to the *draught T*. Let us call the top *face 1*, the bottom *face 2*, and number the vertical faces with 3 to 6. Figure 2.1(b) shows the diagrams of the liquid pressures acting on faces 4 and 6. To obtain the *absolute pressure* we must add the force due to the atmospheric pressure p_0. Those who like mathematics will say that the hydrostatic force on face 4 is the *integral* of the pressures on that face. Assuming that forces are positive in a rightwards direction, and adding the force due to the atmospheric pressure, we obtain

$$F_4 = L \int_0^T \gamma z \mathrm{d}z + p_0 LT = \frac{1}{2}\gamma LT^2 + p_0 LT \tag{2.1}$$

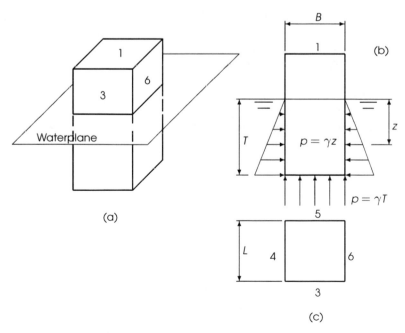

Figure 2.1 Hydrostatic forces on a body with simple geometrical form

Similarly, the force on face 6 is

$$F_6 = -L \int_0^T \gamma z \, \mathrm{d}z - p_0 LT = -\frac{1}{2}\gamma LT^2 - p_0 LT \tag{2.2}$$

As the force on face 6 is equal and opposed to that on face 4 we conclude that the two forces cancel each other.

The reader who does not like integrals can reason in one of the following two ways.

1. The force per unit length of face 4, due to liquid pressure, equals the area of the triangle of pressures. As the pressure at depth T is γT, the area of the triangle equals

$$\frac{1}{2}T \times \gamma T = \frac{1}{2}\gamma T^2$$

Then, the force on the total length L of face 4 is

$$F_4 = L \times \frac{1}{2}\gamma T^2 + p_0 LT \tag{2.3}$$

Similarly, the force on face 6 is

$$F_6 = -L \times \frac{1}{2}\gamma T^2 - p_0 LT \tag{2.4}$$

The sum of the two forces F_4, F_6 is zero.

2. As the pressure varies linearly with depth, we calculate the force on unit length of the face 4 as equal to the depth T times the mean pressure $\gamma T/2$. To get the force on the total length L of face 4 we multiply the above result by L and adding the force due to atmospheric pressure we obtain

$$F_4 = \frac{1}{2}\gamma LT^2 + p_0 LT$$

Proceeding in the same way we find that the force on face 6, F_6, is equal and opposed to the force on face 4. The sum of the two forces is zero. In continuation we find that the forces on faces 3 and 5 cancel one another. The only forces that remain are those on the bottom and on the top face, that is faces 2 and 1. The force on the top face is due only to atmospheric pressure and equals

$$F_1 = -p_0 LB \tag{2.5}$$

and the force on the bottom,

$$F_2 = p_0 LB + \gamma LBT \tag{2.6}$$

The resultant of F_1 and F_2 is an upwards force given by

$$F = F_2 + F_1 = \gamma LBT + p_0 LB - p_0 LB = \gamma LBT \tag{2.7}$$

The product LBT is actually the volume of the immersed body. Then, the force F given by Eq. (2.7) is the weight of the volume of liquid displaced by the immersed body. This verifies Archimedes' principle for the solid considered in this section.

We saw above that the atmospheric pressure does not play a role in the derivation of Archimedes' principle. Neither does it play any role in most other problems we are going to treat in this book; therefore, we shall ignore it in future.

Let us consider in Figure 2.2 a 'zoom' of Figure 2.1. It is natural to consider that the resultant of the forces is applied at the point **P** situated in the centroid of face 2. The meaning of this sentence is that, for any coordinate planes, the moment of the force γLBT applied at the point **P** equals the integral of the moments of pressures. In the same figure, the point **B** is the *centre of volume* of the solid. If our solid would be made of a homogeneous material, the point **B** would be its centre of gravity. We see that **P** is situated exactly under **B**, but at double draught. As a vector can be moved along its line of action, without changing its moments, it is commonly admitted that the force γLBT is applied in the point **B**. A frequent statement is: the force exercised by the liquid is applied in the centre of the displaced volume. The correct statement should be: 'We can consider that the force exercised by the liquid is applied in the centre of the displaced volume'. The force γLBT is called **buoyancy force**.

We have analyzed above the case of a solid that protrudes the surface of the liquid. Two other cases may occur; they are shown in Figure 2.3. We study again the same body as before. In Figure 2.3(a) the body is situated somewhere between the free surface and the bottom. Pressures are now higher; on the vertical faces their distribution follows a trapezoidal pattern. We can still show that the sum of

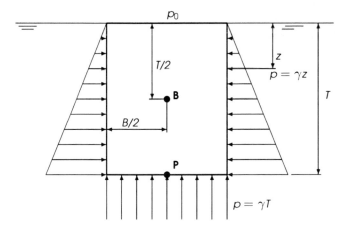

Figure 2.2 Zoom of Figure 2.1

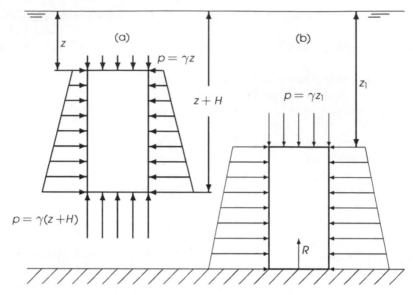

Figure 2.3 Two positions of submergence

the forces on faces 3 to 6 is zero. It remains to sum the forces on faces 1 and 2, that is on the top and the bottom of the solid. The result is

$$\gamma(z + H)LB - \gamma zLB = \gamma LBH \qquad (2.8)$$

As γLBH is the weight of the liquid displaced by the submerged body, this is the same result as that obtained for the situation in Figures 2.1 and 2.2, that is Archimedes' principle holds in this case too.

In Figure 2.3(b) we consider the solid lying on the sea bottom (or lake, river, basin bottom) and assume that no liquid infiltrates under the body. Then no liquid pressure is exercised on face 2. The net hydrostatic force on the body is $\gamma z_1 LB$ and it is directed downwards. Archimedes' principle does not hold in this case. For equilibrium we must introduce a sea-bottom reaction, R, equal to the weight of the body plus the pressure force $\gamma z_1 LB$. The force necessary to lift the body from the bottom is equal to that reaction. However, immediately that the water can exercise its pressure on face 2, a buoyancy force is developed and the body seems lighter. It is as if when on the bottom the body is 'sucked' with a force $\gamma z_1 LB$.

Figure 2.3(b) shows a particular case. Upwards hydrostatic forces can develop in different situations, for example:

- if the submerged body has such a shape that the liquid can enter **under** part of its surface. This is the case of most ships;
- the bottom is not compact and liquid pressures can act through it. This phenomenon is taken into account in the design of dams and breakwaters where it is called uplift.

In the two cases mentioned above, the upwards force can be less than the weight of the displaced liquid. A designer should always assume the worst situation. Thus, to be on the safe side, when calculating the force necessary to bring a weight to the surface one should not count on the existence of the uplift. On the other hand, when calculating a deadweight – such as a concrete block – for an anchoring system, the existence of uplift forces should be taken into account because they can reduce the friction forces (between deadweight and bottom) that oppose horizontal pulls.

2.2.2 The general case

In Figure 2.4 we consider a submerged body and a system of cartesian coordinates, x, y, z, where z is measured vertically and downwards. The only condition we impose at this stage is that no straight line parallel to one of the coordinate axes pierces the body more than twice. We shall give later a hint on how to relax this condition, generalizing thus the conclusions to any body form. Let the surface of the body be S, and let P be the horizontal plane that cuts in S the largest *contour*. The plane P divides the surface S into two surfaces, S_1 situated above P, and S_2 under P. We assume that S_1 is defined by

$$z = f_1(x, y)$$

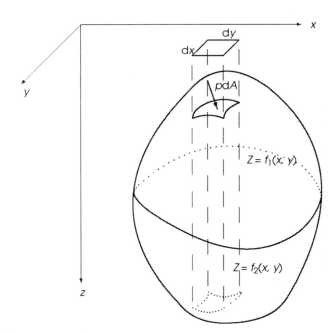

Figure 2.4 Archimedes' principle – vertical force

and S_2 by

$$z = f_2(x, y)$$

The hydrostatic force on an element dA of S_1 is $p\,dA$. This force is directed along the normal, $\mathbf{n}$, to S_1 in the element of area. If the cosine of the angle between $\mathbf{n}$ and the vertical axis is $\cos(\mathbf{n}, z)$, the vertical component of the pressure force on dA equals $\gamma f_1(x, y)\cos(\mathbf{n}, z)dA$. As $\cos(\mathbf{n}, z)dA$ is the projection of dA on a horizontal plane, that is $dxdy$, we conclude that the vertical hydrostatic force on S_1 is

$$\gamma \int\int_{S_1} f_1(x, y)dxdy \tag{2.9}$$

Let us consider now an element of S_2 'opposed' to the one we considered on S_1. We reason as above, taking care to change signs. We conclude that the hydrostatic force on S_2 is

$$-\gamma \int\int_{S_2} f_2(x, y)dxdy \tag{2.10}$$

and the total force on S,

$$F = \gamma \int\int_{S} [f_1(x, y) - f_2(x, y)]dxdy \tag{2.11}$$

The integral in Eq. (2.11) yields the volume of the submerged body. Thus, F equals the weight of the liquid displaced by the submerged body. It remains to show that the horizontal components of the resultant of hydrostatic pressures are equal to zero. We use Figure 2.5 to prove this for the component parallel to the x-axis. The force component parallel to the x-axis acting on the element of area dA is

$$p\cos(\mathbf{n}, x)dA = \gamma z dydz$$

On the other side of the surface, at the same depth z, there is an element of area such that the hydrostatic force on it equals

$$p\cos(\mathbf{n}, x)dA = -\gamma z dydz$$

The sum of both forces is zero. As the whole surface S consists of such 'opposed' pairs dA, the horizontal component in the x direction is zero. By a similar reasoning we conclude that the horizontal component in the y direction is zero too. This is also the result predicted by intuition. In fact, if the resultant of the horizontal components would not be zero we would obtain a 'free' propulsion force.

This completes the proof of Archimedes' principle for a body shape subjected to the only restriction that no straight line parallel to one of the coordinate axes intersects the body more than twice.

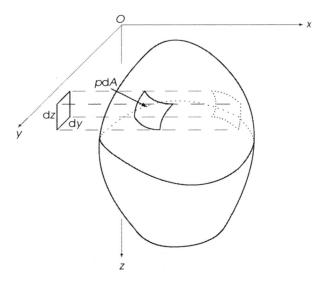

Figure 2.5 Archimedes' principle – force parallel to the *Ox* axis

Could we relieve the above restriction and show that Archimedes' principle holds for a submerged body regardless of its shape? To do this we follow a reasoning similar to that employed sometimes in the derivation of Gauss' theorem in vector analysis (see, for example, Borisenko and Tarapov, 1979). Figure 2.6(a) shows a body that does not fulfill the condition we imposed until now. In fact, in the right-hand part of the body a vertical line can pierce four times the enclosing surface. The dashed line is the trace of the plane that divides the total volume of the body into two volumes, **1, 2**, such that each of them cannot be pierced more than twice by any line parallel to one of the coordinate axis.

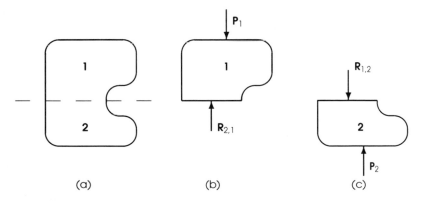

(a) (b) (c)

Figure 2.6 Extending Archimedes' principle

Let us consider now the upper volume, **1**, in Figure 2.6(b). Two forces act on this body:

1. the resultant of hydrostatic pressures, $\mathbf{P}_1$, on the external surface;
2. the force $\mathbf{R}_{2,1}$ exercised by the volume **2** on volume **1**.

Similarly, let us consider the lower volume, **2**, in Figure 2.6(c). Two forces act on this body:

1. the resultant of hydrostatic pressures, $\mathbf{P}_2$, acting on the external surface;
2. the force $\mathbf{R}_{1,2}$ exercised by the volume **1** on volume **2**.

As the forces $\mathbf{R}_{2,1}$ and $\mathbf{R}_{1,2}$ are equal and opposed, putting together the volumes **1** and **2** means that the sum of all forces acting on the total volume is $\mathbf{P}_1 + \mathbf{P}_2$, that is the force predicted by Archimedes' principle. Let us find the x and y-coordinates of the point through which acts the buoyancy force. To do so we calculate the moments of this force about the xOz and yOz planes and divide them by the total force. The results are

$$x_P = \frac{\int\int_S x\gamma z[f_1(x,y) - f_2(x,y)]dxdy}{\int\int_S \gamma z[f_1(x,y) - f_2(x,y)]dxdy} \tag{2.12}$$

$$= \frac{\int\int_S xz[f_1(x,y) - f_2(x,y)]dxdy}{\int\int_S z[f_1(x,y) - f_2(x,y)]dxdy} \tag{2.13}$$

$$y_P = \frac{\int\int_S y\gamma z[f_1(x,y) - f_2(x,y)]dxdy}{\int\int_S \gamma z[f_1(x,y) - f_2(x,y)]dxdy} \tag{2.14}$$

$$= \frac{\int\int_S yz[f_1(x,y) - f_2(x,y)]dxdy}{\int\int_S z[f_1(x,y) - f_2(x,y)]dxdy} \tag{2.15}$$

These are simply the x and y-coordinates of the centre of the submerged volume. We conclude that the buoyancy force passes through the centre of the submerged volume, B (centre of the displaced volume of liquid).

2.3 The conditions of equilibrium of a floating body

A body is said to be in **equilibrium** if it is not subjected to accelerations. Newton's second law shows that this happens if the sum of all forces acting on that body is zero and the sum of the moments of those forces is also zero. Two forces always act on a floating body: the weight of that body and the buoyancy force. In this section we show that the first condition for equilibrium, that is the one regarding the sum of forces, is expressed as Archimedes' principle. The second condition, regarding the sum of moments, is stated as *Stevin's law*.

Further forces can act on a floating body, for example those produced by wind, by centrifugal acceleration in turning or by towing. The influence of those forces is discussed in Chapter 6.

2.3.1 Forces

Let us assume that the bodies appearing in Figures 2.1, 2.3(a) float freely. Then, the weight of each body and the hydrostatic forces acting on it are in equilibrium. Archimedes' principle can be reformulated as:

> The weight of the volume of water displaced by a floating body is equal to the weight of that body.

The weight of the fluid displaced by a floating body is appropriately called displacement. We denote the displacement by the upper-case Greek letter *delta*, that is Δ. If the weight of the floating body is W, then we can express the equilibrium of forces acting on the floating body by

$$\Delta = W \tag{2.16}$$

For the volume of the displaced liquid we use the symbol ∇ defined in Chapter 1. In terms of the above symbols Archimedes' principle yields the equation

$$\gamma\nabla = W \tag{2.17}$$

If the floating body is a ship, we rewrite Eq. (2.17) as

$$\gamma C_{\mathrm{B}}LBT = \Sigma_{i=1}^{n}W_i \tag{2.18}$$

where W_i is the weight of the ith item of ship weight. For example, W_1 can be the weight of the *ship hull*, W_2, of the *outfit*, W_3, of the *machinery*, and so on. The symbol C_{B} and the letters L, B, T have the meanings defined in Chapter 1.

In hydrostatic calculations Eq. (2.18) is often used to find the draught corresponding to a given displacement, or the displacement corresponding to a measured draught. In Ship Design Eq. (2.18) is used either as a **design equation** (see, for example, Manning, 1956), or as an **equality constraint** in design **optimization** problems (see, for example, Kupras, 1976).

Instead of the displacement weight we may work with the **displacement mass**, $\rho\nabla$, where ρ is the **density** of the surrounding water. Then, Eq. (2.18) can be rewritten as

$$\rho C_{\mathrm{B}}LBT = \Sigma_{i=1}^{n}m_i \tag{2.19}$$

where m_i is the mass of the ith ship item. The DIN standards define, indeed, Δ as mass, and use Δ_{F} for displacemnt weight. The subscript 'F' stands for 'force'. In later chapters of this book we shall use the displacement mass rather than the displacement weight.

Table 2.1 Some foreign names for the point **B**

Language	Term	Meaning
French	Centre de carène	Centre of submerged hull
German	Formschwerpunkt	Centre of gravity of solid
Italian	Centro di carena	Centre of submerged hull
Portuguese	Centro do carena	Centre of submerged hull

To remember the meaning of the symbol Δ, let us think that the word 'delta' begins with a 'd', like the word 'displacement' (we ignore the fact that in contemporary-Greek pronunciation 'delta' is actually read as 'thelta'). As to the symbol ∇, it resembles 'V', the initial letter of the word 'volume'.

The point **B** is called in English centre of buoyancy. There are languages in which the name of the point **B** recognizes the fact that **B** is not a centre of pressure. Table 2.1 gives a few examples. This is, of course, a matter of semantics. The line of action of the buoyancy force always passes through the point **B**.

2.3.2 Moments

In this section we discuss the second condition of equilibrium of a floating body: the sum of the moments of all forces acting on it must be zero. This condition is fulfilled in Figure 2.7(a) where the centre of gravity, G, and the centre of buoyancy, B, of the floating body are on the same **vertical** line. The weight of the body and the buoyancy force are equal – that is Δ –, opposed, and act along the same line. The sum of their moments about any reference is zero.

Let us assume that the centre of gravity moves in the same plane, to a new position, G_1, as shown in Figure 2.7(b). The sum of the moments is no longer zero; it causes a clockwise inclination of the body, by an angle ϕ. A volume

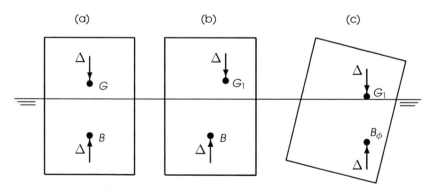

Figure 2.7 Stevin's law

submerges at right, another volume emerges at right. The result is that the centre of buoyancy moves to the right, to a new point that we mark by B_ϕ. The floating body will find a position of equilibrium when the two points G_1 and B_ϕ will be on the same vertical line. This situation is shown in Figure 2.7(c).

There is a possibility of redrawing Figure 2.7 so that all situations are shown in one figure. To do this, instead of showing the body inclined clockwise by an angle ϕ, and keeping the waterline constant, we keep the position of the body constant and draw the waterline inclined counter-clockwise by the angle ϕ. Thus, in Figure 2.8 the waterline corresponding to the initial position is W_0L_0. The weight force, equal to Δ, acts through the initial centre of gravity, G_0; it is vertical, that is perpendicular to the waterline W_0L_0. The buoyancy force, also equal to Δ, acts through the initial centre of buoyancy, B_0: it is vertical, that is perpendicular to the initial waterline.

We assume now that the centre of gravity moves to a new position, G_1. The floating body rotates in the same direction, by an angle ϕ, until it reaches a position of equilibrium in which the new waterline is $W_\phi L_\phi$. The new centre of buoyancy is B_ϕ. The line connecting G_1 and B_ϕ is vertical, that is perpendicular to the waterline $W_\phi L_\phi$. The weight and the buoyancy force act along this line.

Thus, in the case of a floating body, the second condition of equilibrium is satisfied if the centre of gravity and the centre of buoyancy are on the same vertical line. This condition is attributed to Simon Stevin (Simon of Bruges, Flanders, 1548–1620). Stevin is perhaps better known for other studies, among them one on decimal fractions that helped to establish the notation we use today, the discovery in 1586 of the law of composition of forces for perpendicular forces, and a demonstration of the impossibility of perpetual motion.

In Figures 2.7 and 2.8 we assumed that while the body rotates to a new position, no opening, such as a hatch, window, or vent, enters the water. If this

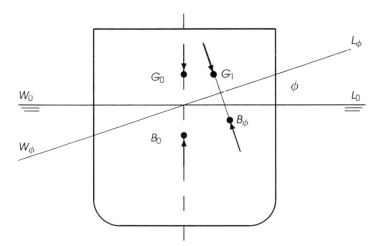

Figure 2.8 Stevin's law

assumption is not correct, the body can either reach equilibrium under more complex conditions (see Chapter 11), or sink.

2.4 A definition of stability

In the preceding section we learnt the conditions of equilibrium of a floating body. The question we ask in the next section is how to determine if a condition of equilibrium is stable or not. Before answering this question we must define the notion of **stability**. This concept is general; we are interested here in its application to floating bodies.

Let us consider a floating body in equilibrium and assume that some force or moment causes a **small change** in its position. Three situations can occur when that force or moment ceases to act.

1. The body returns to its initial position; we say that the condition of equilibrium is **stable**.
2. The position of the body continues to change. We say in this case that the equilibrium is **unstable**. In practical terms this can mean, for example, that the floating body capsizes.
3. The body remains in the displaced position until the smallest perturbation causes it to return to the initial position or to continue to move away from the initial position. We call this situation **neutral equilibrium**.

As an example let us consider the body shown in Figure 2.1. If this body floats freely at the surface we conclude from Eq. (2.17) that the total volume is larger than the weight divided by the specific gravity of the fluid. This body floats in stable equilibrium as to draught. To show this let us imagine that some force causes it to move downwards so that its draught increases by the quantity δT. Archimedes' principle tells us that a new force, $\gamma LB \delta T$, appears and that it is directed upwards. Suppose now that the cause that moved the body downwards decreases slowly. Then, the force $\gamma LB \delta T$ returns the body to its initial position. In fact, as the body moves (slowly) upwards, δT decreases until it becomes zero and then the motion ceases. If the force that drove the body downwards ceases abruptly, the body oscillates around its initial position and, if damping forces are active – they always exist in nature –, the body will eventually come to rest in its initial position.

Next, we assume that some force moved the body upwards so that its draught decreases by δT. A force $-\gamma LB \delta T$ appears now and it is directed downwards. Therefore, if the body is released slowly it will descend until $\delta T = 0$. This completes the proof that the body floating freely **at the surface** is in stable equilibrium with regard to its draught. We mention 'with regard to draught' because, as shown in the next section, the body may be unstable with regard to heel.

When a body floats freely, but is completely submerged, its weight equals exactly its volume multiplied by the specific gravity of the liquid. This body is in neutral equilibrium because it can float **at any** depth. Any small perturbation will move the body from a depth to another one. If the weight of the body is larger than its total volume multiplied by the specific gravity of the liquid, then the body will sink.

Summing-up, we may distinguish three cases.

1. The total volume of a body is larger than its weight divided by the specific gravity of the water:

$$V_{\text{total}} > W/\gamma$$

The body floats at the surface and we can control the draught by adding or reducing weights.
2. The weight of the body exactly equals the total volume multiplied by the specific gravity of the liquid:

$$V_{\text{total}} = W/\gamma$$

The body can float at any depth and we cannot control the position by adding or reducing weights. Any additional weight would cause the body to sink bringing it into case 3. Reducing even slightly its weight will cause the body to come to the surface; its situation changes to case 1.
3. The weight of the body is larger than its volume multiplied by the specific gravity of the water:

$$V_{\text{total}} < W/\gamma$$

The body will sink. To change its position we must either reduce the weight until we reach at least situation 2, or add buoyancy in some way.

In the above analysis we assumed that the specific gravity of the liquid, γ, is constant throughout the liquid volume. This assumption may not be correct if large variations in temperature or salinity are present, or if the liquid volume consists of layers of different liquids. Interesting situations can arise in such cases. Other situations can arise at large depths at which the water density increases while the volume of the floating body shrinks because of the compressibility of its structure. These cases are beyond the scope of this book.

2.5 Initial stability

Figure 2.9(a) is a vertical, transverse section through a ship in upright condition, that is unheeled. If this section passes through the centre of buoyancy, B, we know from Stevin's law that it contains the centre of gravity, G. The water line is

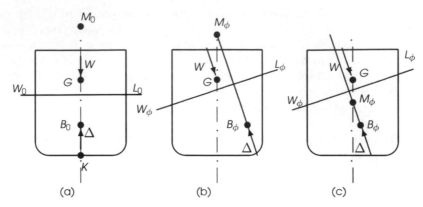

Figure 2.9 The condition of initial stability

W_0L_0. The weight force, W, acts through the centre of gravity, G; the buoyancy force, Δ, through the centre of buoyancy, B_0. The forces W and Δ are equal and collinear and the ship is in an equilibrium condition. Let the ship heel to the starboard with an angle ϕ. For reasons that will become clear in Section 2.8, we assume that the *heel angle is small*. As previously explained, we leave the ship as she is and draw the waterline as inclined to port, with the same angle ϕ. This is done in Figure 2.9(b) where the new waterline is $W_\phi L_\phi$. If the weights are fixed, as they should be, the centre of gravity remains in the same position, G. Because a volume submerges at starboard, and an equal volume emerges at port, the centre of buoyancy moves to starboard, to a new position, B_ϕ. Both forces W and Δ are vertical, that is perpendicular to the waterline $W_\phi L_\phi$. These two forces form a moment that tends to return the ship towards port, that is to her initial condition. We say that the ship is stable.

Figure 2.9(c) also shows the ship heeled towards starboard with an angle ϕ. In the situation shown in this figure the moment of the two forces W and Δ heels the ship further towards starboard. We say that the ship is unstable.

The difference between the situations in Figures 2.9(b) and (c) can be described elegantly by the concept of **metacentre**. This abstract notion was introduced by Pierre Bouguer (French, 1698–1758) in 1746, in his *Traité du Navire*. Let us refer again to Figure 2.9(b). For a ship, the dot-point line is the trace of the port-to-starboard symmetry plane, that is the *centreline*. More generally, for any floating body, the dot-point line is the line of action of the buoyancy force before heeling. The new line of action of the buoyancy force passes through the new centre of buoyancy and is perpendicular to $W_\phi L_\phi$. The two lines intersect in the point M_ϕ. Bouguer called this point *metacentre*.

We can see now the difference between the two heeled situations shown in Figure 2.9:

- in (b) the metacentre is situated *above the centre of gravity, G*;
- in (c) the metacentre is situated *below the centre of gravity, G*.

We conclude that the equilibrium of the floating body is stable if the metacentre is situated above the centre of gravity.

For his contributions of overwhelming importance, Bouguer was sometimes described as 'the father of naval architecture' (quotation in Stoot, 1959). It must be emphasized here that the definition of the metacentre is not connected at all with the form of a ship. Therefore, the fact that in the above figures the metacentre is the intersection of the new line of action of the buoyancy force and the centreline is true only for symmetrical hulls heeled from the upright condition. For a general floating body we can reformulate the definition as follows:

> Let us consider a floating body and its centre of buoyancy B_ϕ. Let the line of action of the buoyancy force be R. If the body changes its inclination by an angle $\delta\phi$, the centre of buoyancy changes its position to $B_{\phi+\delta\phi}$ and the new line of action of the buoyancy force will be, say, S. When $\delta\phi$ tends to zero, the intersection of the lines R and S tends to a point that we call metacentre.

Readers familiar with elementary differential geometry will recognize that, defined as above, the metacentre is the **the centre of curvature of the curve of centres of buoyancy**. The notion of curvature is defined in Chapter 13.

2.6 Metacentric height

In the preceding section we learnt that a surface ship is initially stable if its initial metacentre is above the centre of gravity. For actual calculations we must find a convenient mathematical formulation. We do this with the help of Figure 2.9(a). We choose a reference point, K, at the intersection of the centreline and the baseline and we measure vertical coordinates from it, upwards. Thus defined, K is the origin of z-coordinates. A good recommendation is to choose K as the lowest point of the ship keel; then, there will be no negative z-coordinates. We remember easily the chosen notation because K is the initial letter of the word *keel*.

In the same figure M_0 is the **initial metacentre**, that is the metacentre corresponding to the upright condition. Dropping the subscripts 0 we can write

$$\overline{GM} = \overline{KB} + \overline{BM} - \overline{KG} \tag{2.20}$$

and the condition of initial stability is expressed as

$$\overline{GM} > 0 \tag{2.21}$$

The vector $\overline{GM}$ is called **metacentric height**. The vector $\overline{KB}$ is the z-coordinate of the centre of buoyancy; it is calculated as the z-coordinate of the centroid of the submerged hull as one of the results of *hydrostatic calculations*. The vector $\overline{BM}$ is the metacentric radius whose calculation we are going to discuss in

Subsection 2.8.2. The vector $\overline{KG}$ is the z-coordinate of the centre of gravity of the floating body; it results from **weight calculations**. The quantities $\overline{KB}$ and $\overline{BM}$ depend upon the ship geometry, the quantity $\overline{KG}$ upon the distribution of masses.

2.7 A lemma on moving volumes or masses

Figure 2.10 shows a system of two masses, m_1 and m_2. Let the x-coordinate of the mass m_1 be x_1; that of the mass m_2, x_2. The centre of gravity of the system is G and its x-coordinate is given by

$$x_G = \frac{x_1 m_1 + x_2 m_2}{m_1 + m_2} \tag{2.22}$$

Let us move the mass m_2 a distance d in the x direction. The new centre of gravity is G^* and its x-coordinate,

$$x_G^* = \frac{x_1 m_1 + (x_2 + d)m_2}{m_1 + m_2} = x_G + \frac{dm_2}{m_1 + m_2} \tag{2.23}$$

The product dm_2 is the change of moment caused by the translation of the mass m_2. The centre of gravity of the system moved a distance equal to the change of moment divided by the total mass of the system. A formal statement of this lemma is

> Given a system of masses, if one of its components is moved in a certain direction, the centre of gravity of the system moves in the same direction, a distance equal to the change of moment divided by the total mass.

A similar lemma holds for a system of volumes in which one of them is moved to a new position. The reader is invited to solve Exercise 2.5 and prove the lemma for a three dimensional system of masses.

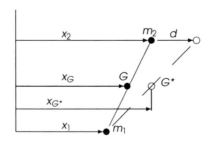

Figure 2.10 Moving a mass in a system of masses

2.8 Small angles of inclination

In this section we prove two very important theorems for bodies that incline at **constant displacement**. This is the case of floating bodies that change their inclination without the addition or loss of weights. Constant displacement means **constant volume of displacement**. In Chapter 1 we mentioned that Romance languages use for the submerged volume terms derived from the Latin word *carina*, for instance *carène* in French, *carena* in Italian. Correspondingly, the theory of bodies inclined at constant volume of displacement is called *Théorie des isocarènes* in French, *Teoria delle isocarene* in Italian. The prefix 'iso' comes from Greek and means 'equal'. Thus, Romance languages use one single term to mean 'bodies inclining at constant volume of displacement'.

A second assumption in this section is that the **angle of inclination is small**. The results developed under this assumption are valid for any floating body. The results are valid for any angle of inclination only for floating bodies belonging to a particular class of forms called **wall-sided**, a concept explained later in this chapter.

2.8.1 A theorem on the axis of inclination

Let us assume that the initial waterplane of the body shown in Figure 2.11 is W_0L_0. Next we consider the same body inclined by a small angle ϕ, such that the new waterplane is $W_\phi L_\phi$. The weight of the body does not change; therefore, also the submerged volume does not change. If so, the volume of the 'wedge'

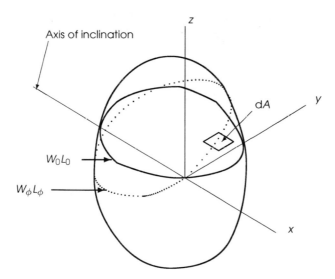

Figure 2.11 Euler's theorem on the axis of inclination

that submerges at right, between the two planes $W_0 L_0$ and $W_\phi L_\phi$, equals the volume of the wedge that emerges at left, between the same two planes. Let us express this mathematically. We take the intersection of the two planes as the x-axis. This is the axis of inclination.

As shown in Figure 2.12, an element of volume situated at a distance y from the axis of inclination has the height $y \tan \phi$. If the base of this element of volume is $dA = dxdy$, the volume is $y \tan \phi dxdy$. Let the area of the waterplane $W_0 L_0$ at the right of the axis of intersection be S_1; that at the left, S_2. Then, the volume that submerges is

$$V_1 = \int \int_{S_1} y \tan \phi dxdy \qquad (2.24)$$

and the volume that emerges,

$$V_2 = - \int \int_{S_2} y \tan \phi dxdy \qquad (2.25)$$

Assuming a small heel angle, ϕ, we can consider the submerging and the emerging volumes as wall sided and write Eqs. (2.24) and (2.25) as we did.

The condition for constant volume is

$$V_1 = V_2$$

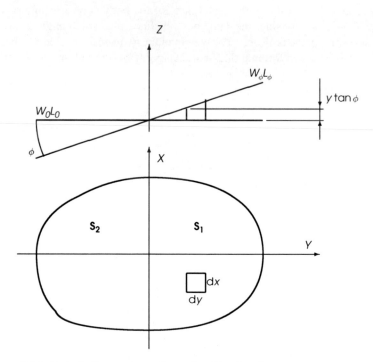

Figure 2.12 Euler's theorem on the axis of inclination

Combining this with Eqs. (2.24) and (2.25) yields

$$\int\int_{S_1} y \tan \phi \, dx dy = -\int\int_{S_2} y \tan \phi \, dx dy \qquad (2.26)$$

and, finally,

$$\int\int_{S} y \, dx dy = 0 \qquad (2.27)$$

where $S = S_1 + S_2$ is the whole waterplane. In words, the first moment of the waterplane area, with respect with the axis of inclination, is zero. This happens only if the axis of inclination passes through the centroid of the waterplane area. We remind the reader that the coordinates of the centroid of a surface S are defined by

$$x_C = \frac{\int\int_S x \, dx dy}{\int\int_S dx dy}, \qquad y_C = \frac{\int\int_S y \, dx dy}{\int\int_S dx dy}$$

Or, as the *Webster's Ninth New Collegiate Dictionary* puts it, 'corresponds to the center of mass of a thin plate of uniform thickness'. The centroid of the waterplane area is known as **centre of flotation** and is noted by F. The corresponding French term is 'centre de gravité de la flottaison', the German term is 'Wasserlinien-Schwerpunkt', and the Italian, 'centro del galleggiamento'.

A statement of the property proven above is

> Let the initial waterplane of a floating body be $W_0 L_0$. After an inclination, at constant volume of displacement, with an angle ϕ, the new waterplane is $W_\phi L_\phi$. The intersection of the two waterplanes is the axis of inclination. If the angle of inclination tends to zero, the axis of inclination tends to a straight line passing through the centroid of the waterplane area.

In practice, this property holds if the angle of inclination is sufficiently small. For heeling of a vessel, this can mean a few degrees, $5°$ for some forms, even $15°$ for others. If the inclination is the trimming of an intact vessel, the angles are usually small enough and this property always holds. The property also holds for larger heel angles if the floating body is **wall sided**. This is the name given to floating bodies whose surface includes a cylinder (in the broader geometrical sense), with generators perpendicular to the initial waterplane. An illustration of such a case is given in Example 2.5. In French and Italian, for example, the term used for wall-sided bodies is *cylindrical floating bodies*.

The term used in some languages, such as French or Italian, for an axis passing through the centroid of an area is **barycentric axis**. This term is economic and we shall use it whenever it will help us to express ideas more concisely.

2.8.2 Metacentric radius

Let us refer again to Figure 2.9. As we shall see, the vector $\overline{B_\phi M_\phi}$ plays an important role in stability. Leaving the subscript ϕ, we generically call $\overline{BM}$ **metacentric radius**; in this section we calculate its magnitude. To do so we must find the shift of the centre of buoyancy, B, for a small angle of inclination ϕ. Here we use the lemma on moving volumes and we calculate

$$\text{change of coordinate} = \frac{\text{change of moment of volume}}{\text{total volume}}$$

As seen from Figures 2.11 and 2.12, the elemental change of volume is $y \tan \phi \mathrm{d}x\mathrm{d}y$. To find the changes of moment respective to the coordinate planes we must multiply the elemental volume by the coordinates of its centroid. To make things easier, we take the origin of coordinates in the initial centre of buoyancy, B_0, measure the x-coordinate parallel to the axis of inclination, positive forwards, the y-coordinate transversely, positive leftwards, and the z-coordinate vertically, positive upwards. The coordinates of the centre of buoyancy B_ϕ are obtained by integrating the changes of moment of the elemental volume, over the waterplane area **S**.

The results are

$$x_B = \frac{\int \int_{\mathbf{S}} xy \tan \phi \mathrm{d}x\mathrm{d}y}{\nabla} = \frac{I_{xy}}{\nabla} \tan \phi \tag{2.28}$$

$$y_B = \frac{\int \int_{\mathbf{S}} y^2 \tan \phi \mathrm{d}x\mathrm{d}y}{\nabla} = \frac{I}{\nabla} \tan \phi \tag{2.29}$$

$$z_B = \frac{\int \int_{\mathbf{S}} \frac{1}{2} y^2 \tan^2 \phi \mathrm{d}x\mathrm{d}y}{\nabla} = \frac{1}{2} \frac{I}{\nabla} \tan^2 \phi \tag{2.30}$$

Above, I is the moment of inertia of the waterplane area about the axis of inclination (remember, it is a barycentric axis), and I_{xy}, the **product of inertia** of the same area about the axes x and y. In German and some other languages I_{xy} is called *centrifugal moment of inertia*.

As we assumed that the angle ϕ is small, we can further write

$$x_B = \frac{I_{xy}}{\nabla} \phi$$

$$y_B = \frac{I}{\nabla} \phi \tag{2.31}$$

$$z_B = \frac{1}{2} \frac{I}{\nabla} \phi^2$$

The coordinate z_B is of second order and we can neglect it if ϕ is small. As to the x-coordinate let us remember that conventional ships in upright condition enjoy a port-to-starboard symmetry. This means that for such ships, in upright

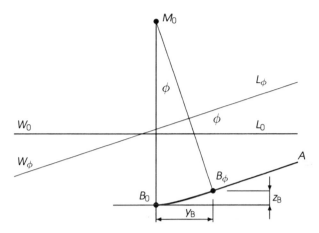

Figure 2.13 Calculation of metacentric radius

condition, the product of inertia is zero so that x_B is zero too. Then $\overline{B_0 B_\phi}$ in Figure 2.13 is essentially equal to y_B. For other floating bodies there is a three-dimensional theory that is beyond the scope of this book (see, for example, Appel, 1921; Hervieu, 1985). For our purposes it is sufficient to consider the projection of the curve of centres of buoyancy, B, on the plane that contains the initial centre of buoyancy, B_0, and is perpendicular to the axis of inclination. In this plane the length of the arc connecting B_0 to B_ϕ equals $\overline{BM}\phi$ (see Figure 2.13). As z_B is of second order, we can write

$$\frac{I}{\nabla}\phi = \overline{BM}\phi$$

and hence,

$$\overline{BM} = \frac{I}{\nabla} \qquad (2.32)$$

A statement of this important theorem is

> The magnitude of the metacentric radius, $\overline{BM}$, is equal to the ratio of the waterplane moment of inertia, about the axis of inclination, to the volume of displacement.

Returning to the third Eq. (2.31) we can see that z_B is always positive. This means that the curve of centres of buoyancy presents its concavity towards the waterline.

2.9 The curve of centres of buoyancy

Figure 2.14 shows a floating body inclined by some angle; the corresponding waterline is $W_1 L_1$ and the centre of buoyancy, B_1. Let us assume that the inclination increases by an additional, small angle, ϕ. Let the new waterline be $W_2 L_2$

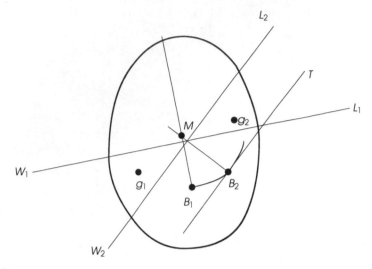

Figure 2.14 Properties of B and M curves

and the corresponding centre of buoyancy, B_2. For a small angle ϕ we can write the coordinates of the new centre of buoyancy as

$$y_B = \frac{I}{\nabla}\phi$$

$$z_B = \frac{1}{2}\frac{I}{\nabla}\phi^2$$

Differentiation of these equations yields

$$dy_B = \frac{I}{\nabla}d\phi$$

$$dz_B = \frac{I}{\nabla}\phi\,d\phi$$

which shows that the slope of the tangent to the B-curve in B_2 is

$$\left.\frac{dz_B}{dy_B}\right|_{B_2} = \phi$$

This is the assumed angle of inclination. We reach the important conclusion that the tangent to the B-curve, in a point B_ϕ, is parallel to the waterline corresponding to the centre of buoyancy B_ϕ.

We could reach the same conclusion by the following reasoning. In Figure 2.14 let the centre of volume of the emerged wedge be g_1 and that of the imersed wedge, g_2, and the volume of each one of them, v. Let the coordinates of g_1 be y_{g1}, z_{g1}, and those of g_2 be y_{g2}, z_{g2}. The coordinates of the initial centre of buoyancy, B_1, are y_{B1}, z_{B_1}, and those of B_2 are y_{B2}, z_{B_2}. Applying the lemma on moving volumes we write

$$y_{B2} - y_{B1} = (y_{g2} - y_{g1})\frac{v}{\nabla}$$

$$z_{B2} - z_{B1} = (z_{g2} - z_{g1})\frac{v}{\nabla}$$

or

$$\frac{z_{g2} - z_{g1}}{y_{g2} - y_{g1}} = \frac{z_{B2} - z_{B1}}{y_{B2} - y_{B1}} \tag{2.33}$$

which shows that $\overline{B_1 B_2}$ is parallel to $\overline{g_1 g_2}$. When ϕ tends to zero, $\overline{g_1 g_2}$ tends to the initial waterline and $\overline{B_1 B_2}$ to the tangent in B_1 to the B-curve.

2.10 The metacentric evolute

The buoyancy force is always normal to the waterline. As the tangent to the B-curve is parallel to the corresponding waterline, it follows that the buoyancy force is normal to the B-curve. In Figure 2.14 the normals to the B-curve in the points B_1 and B_2 intersect in a point M. In some languages this point is called *metacentric point*. When $B_1 \longrightarrow B_2$, the metacentric point tends to the metacentre.

Let the curve M be the locus of the metacentres corresponding to a given curve B. The curve M is the locus of centres of curvature of the curve B it is also the **envelope** of the normals to the curve B By definition, the curve M is the evolute of the curve B (see, for example Struik, 1961); it is called metacentric evolute. The term used in French is *développée métacentrique*, in German, *Metazentrische Evolute*, and in Italian, *evoluta metacentrica*.

Conversely, the curve B intersects at right angles the tangents to the meta-centric evolute. Then, by definition, the curve B is the *involute* of the curve M. The term used in French is *développante*; in German, *Evolvente*, and in Italian, *evolventa*.

The concepts of B and M curve are illustrated in Examples 2.5 and 2.6. Some readers may be familiar with another example of a pair of curves that stay one to the other in the relationship evolute–involute. The shape of the tooth flanks used today in most gears is that of an involute of circle.

2.11 Metacentres for various axes of inclination

In Eq. (2.32) the moment of inertia, I, is calculated about the axis of inclination. This axis passes through the centroid, F, of the waterplane and so does any other axis of inclination. It can be shown that there is a pair of orthogonal axes such that the moment of inertia about one of them is minimum and about the other maximum. Then, the metacentric radius corresponding to the former axis is minimum, and the moment about the latter axis is maximum. Correspondingly, one of the metacentric radii is minimum and the other maximum. In some European countries the smallest radius is denoted by r and is called *small metacentric radius*, while the largest radius is denoted R and is called *large metacentric radius*.

In the theory of moments of inertia the two axes for which we obtain the extreme values of moments of inertia are called **principal axes** and the corresponding moments, **principal moments of inertia**. When the waterplane area has an axis of symmetry, this axis is one of the principal axes; the other one is perpendicular to the first. The waterplane area of ships in upright condition has an axis of symmetry: the intersection of the waterplane and the centreline plane. The moment of inertia about this axis is the smallest one; it is used to calculate the **transverse metacentric radius**. The moment of inertia about the axis perpendicular in F to the centreline is the largest; it enters in the calculation of the **longitudinal metacentric radius**.

To give an idea of the relative orders of magnitude of the transverse and longitudinal metacentric radii, let us consider a parallelepipedic barge whose length is L, breadth, B, and draught, T. The volume of displacement equals $\nabla = LBT$. The transverse metacentric radius results from

$$\overline{BM} = \frac{LB^3/12}{LBT} = \frac{B^2}{12T}$$

The longitudinal metacentric radius is given by

$$\overline{BM_\mathrm{L}} = \frac{BL^3/12}{LBT} = \frac{L^2}{12T}$$

The ratio of the two metacentric radii is

$$\frac{\overline{BM_\mathrm{L}}}{\overline{BM}} = \left(\frac{L}{B}\right)^2$$

The length–breadth ratio ranges from 3.1, for some motor boats, to 10.5, for fast cruisers. Correspondingly, the ratio of the longitudinal to the transverse metacentric radius varies roughly between 10 and 110. As a rule of thumb, the longitudinal metacentric radius is of the same order of magnitude as the ship length.

2.12 Summary

A body submersed in a fluid is subjected to an upwards force equal to the weight of the displaced fluid. This is Archimedes' principle. The hydrostatic force predicted by this principle passes through the centre of volume of the displaced fluid; we call that point *centre of buoyancy* and denote it by the letter B.

For a floating body the weight of the displaced fluid equals the weight of that body. The symbol for the immersed volume is ∇; that for the displacement mass, Δ. If the density of the fluid is ρ, we can write

$$\Delta = \rho\nabla$$

Values of the density of water in different navigation ways are given in the Appendix of this chapter.

If a floating body is inclined by a small angle, the new waterplane intersects the initial one along a line that passes through its centroid, that is, through the centre of flotation.

If the floating body is a ship, using the notations described in Chapter 1 we write

$$\gamma C_{\mathrm{B}}LBT = \Sigma_{i=1}^{n}m_i$$

where m_i is the mass of the ith item aboard and n, the total number of ship items. Δ is called *displacement mass* and ∇, *displacement volume*. The above equation expresses the condition of equilibrium of forces. The condition of equilibrium of moments requires that the centre of gravity, G, of the floating body and its centre of buoyancy, B, lie on the same vertical. This condition is known as *Stevin's law*.

We say that a floating body is initially 'stable' if after a small perturbation of its position of equilibrium, that body returns to its initial position. To study initial stability, Bouguer introduced the notion of metacentre. Let the line of action of the buoyancy force in the initial position be R. If the floating body is inclined by a small angle, $\delta\phi$, the buoyancy force acts along a new line, say S. When $\delta\phi$ tends to zero, the intersection of the two lines, R and S, tends to a point, M, called metacentre.

The equilibrium of a floating body is stable if its metacentre lies above its centre of gravity. The distance from the centre of gravity to the metacentre, $\overline{GM}$, is called *metacentric height* and is considered positive upwards. The condition of initial stability can be expressed as

$$\overline{GM} > 0$$

The distance from the centre of buoyancy to the metacentre, $\overline{BM}$, is called *metacentric radius*. Its value is given by

$$\overline{BM} = \frac{I}{\nabla}$$

where I is the moment of inertia of the waterplane about the axis of inclination, a line that passes through the centroid of the area, that is through the centre of flotation, F. Let K be the origin of vertical coordinates. We can write

$$\overline{GM} = \overline{KB} + \overline{BM} - \overline{KG}$$

By its definition, the metacentre is the centre of curvature of the curve described by B for different angles of inclination. The curve described by the metacentre is the evolute of the curve of centres of buoyancy. The normals to the curve B are tangents to the curve M.

2.13 Examples

Example 2.1 – Melting ice cube
The following problem is sometimes presented as an intelligence quiz. We describe it here as a fine application of Archimedes' principle.

Let us suppose that somebody wants to cool a glass of water by putting in it a cube of ice made of the **same** water. Should he fear that when the cube melts the level of the water will rise?

Let the mass of the cube be M and the density of water, ρ. The volume of water displaced by the cube equals M/ρ. After meltdown the cube becomes a volume of water equal to M/ρ. *Conclusion*: The water volume in the glass does not change and neither does the water level.

Example 2.2 – Designing a buoy
This is a simple application of Archimedes' principle as the base of a design equation. Let us suppose that we want to design a spherical buoy for an instrument having a mass M. The buoy shall be made of 3 mm steel plate, of density ρ_S, and shall float so that the centre of the sphere lies in the waterplane (Figure 2.15).

To solve the problem we refer to Eq. (2.17),

$$\gamma \nabla = W$$

Instead of specific gravity and weight we can work with density and mass so that our design equation becomes

$$\rho_W \nabla = M$$

where ρ_W is the water density.

The volume of the submerged half-sphere is

$$\nabla = \frac{1}{2} \cdot \frac{4}{3} \pi d_o^3$$

where d_o is the outer diameter of the sphere. Let us measure d_o in metres.

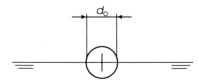

Figure 2.15 Designing a buoy

Then, the mass of the spherical shell is given by

$$M_{\text{steel}} = \rho_S \frac{4}{3} \pi \left[d_o^3 - (d_o - 0.003)^3 \right] \tag{2.34}$$

Putting things together we write

$$\rho_{\text{w}} \frac{1}{2} \cdot \frac{4}{3} \pi d_o^3 = \rho_{\text{s}} \frac{4}{3} \pi \left[d_o^3 - (d_o - 0.003)^3 \right] + M$$

which yields the design equation

$$\frac{\rho_{\text{w}}}{2} d_o^3 - 3 \times 0.003 \rho_S d_o^2 + 3 \times 0.003^2 \rho_S d_o - 0.003^3 \rho_S - \frac{3M}{4\pi} = 0 \tag{2.35}$$

This is a cubic equation. The general solution of cubic equations was found by Italian algebraists in the sixteenth century. Instead of calculating this solution we are going to numerically answer a particular example. Let our data be

water density	$1.025 \, \text{tm}^{-3}$
steel density	$7.850 \, \text{tm}^{-3}$
instrument mass	$0.010 \, \text{t}$

A MATLAB file, called buoy.m and that solves this equation can be found on the website of this book.

Running this file produces the results

```
do =
    0.2267
   -0.0444 +  0.1363i
   -0.0444 -  0.1363i
```

Obviously, only the first root is physically possible.

Another example of the use of Archimedes' principle in writing the design equation of a floating body can be found in Biran and Breiner (2002: 309–11).

Example 2.3
Figure 2.16(a) shows a cone floating top-down in water. The diameter of the base is D; the height, H, and the diameter of the waterplane area, d. Let the specific gravity of the cone material be γ_{c} and that of the water, γ_{w}. We want to find out under which conditions the cone can float as shown in the figure.

We begin by finding the draft, T. Archimedes' principle allows us to write

$$\gamma_{\text{c}} \frac{\pi D^2}{3 \cdot 4} H = \gamma_{\text{w}} \frac{\pi d^2}{3 \cdot 4} T \tag{2.36}$$

Geometrical similarity between the submerged cone and the whole cone yields

$$\frac{d}{D} = \frac{T}{H} \tag{2.37}$$

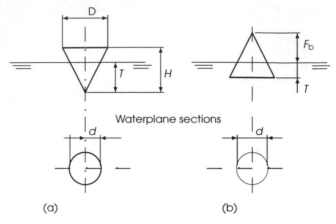

(a)   (b)

Figure 2.16 A floating cone

Substituting d from Eq. (2.37) into Eq. (2.36), and noting

$$\alpha = \left(\frac{\gamma_c}{\gamma_w}\right)^{1/3}$$

we obtain

$$T = \alpha H \qquad\qquad (2.38)$$

Other quantities necessary for checking the initial stability are

$$\overline{KG} = \frac{2}{3}H$$

$$\overline{KB} = \frac{2}{3}T = \frac{2}{3}\alpha H$$

$$\overline{BM} = \frac{I}{\nabla}$$

with

$$I = \frac{\pi d^4}{64}, \qquad \nabla = \frac{\pi d^2}{3 \cdot 4}T$$

the metacentric radius is

$$\overline{BM} = \frac{3d^2}{16T} = \frac{3\alpha}{16}\left(\frac{D}{H}\right)^2 H$$

and the metacentric height

$$\begin{aligned}
\overline{GM} &= \overline{KB} + \overline{BM} - \overline{KG} \\
&= \frac{2}{3}\alpha H + \frac{3\alpha}{16}\left(\frac{D}{H}\right)^2 H - \frac{2}{3}H
\end{aligned}$$

The cone is stable if

$$\frac{2}{3}\alpha + \frac{3\alpha}{16}\left(\frac{D}{H}\right)^2 - \frac{2}{3} > 0 \tag{2.39}$$

From Eq. (2.39) we can deduce a condition for the specific gravity of the cone material

$$\alpha > \frac{1}{1 + \frac{9}{32}\left(\frac{D}{H}\right)^2} \tag{2.40}$$

or, a condition for the D/H ratio:

$$\left(\frac{D}{H}\right)^2 > \frac{32}{9}\frac{1-\alpha}{\alpha} \tag{2.41}$$

Obviously, for the cone to float, the ratio α must be smaller than one. Thus, the complete condition for the cone material is

$$\frac{1}{1 + \frac{9}{32}\left(\frac{D}{H}\right)^2} < \alpha < 1 \tag{2.42}$$

Example 2.4

Figure 2.16(b) shows a cone floating top-up. Noting by F_b the freeboard, that is the difference $H - T$, Archimedes' principle yields the equation

$$\gamma_c \frac{\pi D^2}{3 \cdot 4}H = \gamma_w \frac{\pi}{3 \cdot 4}(D^2 H - d^2 F_b) \tag{2.43}$$

We obtain

$$F_b = \frac{\gamma_w - \gamma_c}{\gamma_w}\frac{D^2 H}{d^2} \tag{2.44}$$

Similarity gives us

$$d = \frac{D}{H}F_b \tag{2.45}$$

Combining Eqs. (2.44) and (2.45) we obtain

$$F_b = \left(\frac{\gamma_w - \gamma_c}{\gamma_w}\right)^{1/3} H$$

with

$$\beta = \left(\frac{\gamma_w - \gamma_c}{\gamma_w}\right)^{1/3}$$

we write for the freeboard

$$F_b = \beta H \tag{2.46}$$

and for the draught,

$$T = H - F_b = (1 - \beta)H \tag{2.47}$$

The diameter of the waterplane section is given by

$$d = \frac{D}{H} F_b = \beta D \tag{2.48}$$

To find the vertical coordinate of the centre of buoyancy we use the formula that gives the height of the centroid of a trapeze (see books on elementary geometry or engineering handbooks):

$$\overline{KB} = \frac{T}{3} \frac{D + 2d}{D + d} = \frac{(1 - \beta)}{3} \frac{1 + 2\beta}{1 + \beta} H \tag{2.49}$$

We calculate the metacentric radius as

$$\overline{BM} = \frac{I}{\nabla} \tag{2.50}$$

with

$$I = \frac{\pi \beta^4 D^4}{64}, \qquad \nabla = \frac{\pi (1 - \beta^3) D^2 H}{3 \cdot 4} \tag{2.51}$$

we obtain

$$\overline{BM} = \frac{3}{16} \frac{\beta^4}{1 - \beta^3} \frac{D^2}{H} \tag{2.52}$$

The height of the centre of gravity is

$$\overline{KG} = H/3 \tag{2.53}$$

and the resulting metacentric height is

$$\begin{aligned}
\overline{GM} &= \overline{KB} + \overline{BM} - \overline{KG} \\
&= \frac{1 - \beta}{3} \frac{1 + 2\beta}{1 + \beta} H + \frac{3}{16} \frac{\beta^4}{1 - \beta^3} \frac{D^2}{H} - \frac{H}{3}
\end{aligned} \tag{2.54}$$

The cone is stable if

$$\frac{(1 - \beta)(1 + 2\beta)}{1 + \beta} + \frac{9}{16} \frac{\beta^4}{1 - \beta^3} \left(\frac{D}{H}\right)^2 > 1 \tag{2.55}$$

We obtain a condition for the D to H ratio:

$$\left(\frac{D}{H}\right)^2 > \frac{32}{9} \frac{1 - \beta^3}{1 + \beta} \frac{1}{\beta^2} \tag{2.56}$$

The condition for the specific gravity of the cone material is

$$\frac{1-\beta^3}{1+\beta} \frac{1}{\beta^2} < \frac{9}{32}\left(\frac{D}{H}\right)^2 \qquad (2.57)$$

noting

$$C = \frac{9}{32}\left(\frac{D}{H}\right)^2$$

we can write

$$(C+1)\beta^3 + C\beta^2 - 1 > 0 \qquad (2.58)$$

In addition β must also fulfill the inequalities

$$0 < \beta < 1 \qquad (2.59)$$

Example 2.5 – A parallelepipedic barge

Let us consider a parallelepipedic barge; it has a constant, rectangular transverse section as shown in Figure 2.17. Let L be the length, B the breadth, H the depth and T the draught. For this simple body form we can calculate analytically the positions of the centre of buoyancy and of the metacentre. We shall do this in two ways:

1. Starting from known principles of mechanics and elementary results of differential geometry;
2. Using the theorems developed in this chapter.

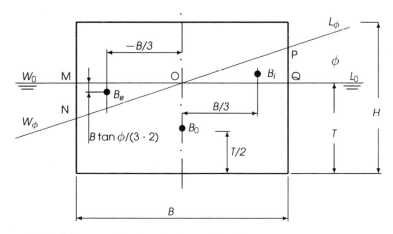

Figure 2.17 A barge with simple geometrical form

We begin this example by discussing the case in which the waterline reaches first the deck and later the bottom. Formally, this condition is expressed by

$$H - T < T \tag{2.60}$$

that is $H < 2T$. In upright condition the centre of buoyancy, B_0, is situated in the centreline plane and its height above the bottom equals $T/2$. As shown in Figure 2.18, we use a coordinate system with the origin in B_0, and measure y horizontally, positive rightwards, and z vertically, positive upwards.

In Figure 2.17 we consider that the barge heels to starboard by an angle ϕ and the new waterline is $W_\phi L_\phi$. We distinguish several phases:

1. The new waterline is situated between the original waterline, $W_0 L_0$, and the waterline passing through the corner of the deck. Formally, this case is defined by

$$0 \leq \phi \leq \arctan \frac{H - T}{B/2} \tag{2.61}$$

2. The waterline is situated between the waterline that passes through the starboard deck corner and the waterline that passes through the port-side bottom corner. Formally, this means

$$\arctan \frac{H - T}{B/2} < \phi \leq \arctan \frac{2T}{B} \tag{2.62}$$

3. As the angle ϕ increases, two other phases can be distinguished. However, it is easier to consider those phases as being symmetric to the first two.

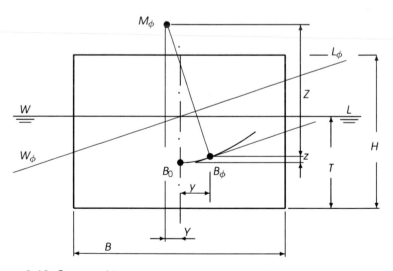

Figure 2.18 Centre of buoyancy and metacentre of simple barge

Phase 1

For the simple form considered in this example we can start from the principles of statics. We first observe that within the whole heel range defined by Eq. (2.61) the two waterlines $W_0 L_0$ and $W_\phi L_\phi$ intersect in the centreline plane. Indeed, the submerging and the emerging wedges thus defined are equal, that is the volume of displacement is constant (*isocarène* heeling). In other words we are dealing with a wall-sided barge.

To calculate the change of moment we multiply the volume of each wedge by the coordinate of its centroid measured from a convenient coordinate plane. Then, the coordinates of the centre of buoyancy, B_ϕ, are obtained by means of the lemma on moving volumes (Section 2.7). The calculations for the y-coordinate are shown in Table 2.2.

This is the place to stop for a short digression on this tabular form of calculations. Let us refer to Table 2.2. Column 2 contains the volumes of the initial hull, of the submerged wedge and of the emerged wedge. Column 3 contains the y-coordinates of the centres of volumes entered in column 2. As said, these coordinates are measured from the centreline plane; we call them *tcb*, an acronym for transverse centre of buoyancy. We use lower-case letters and reserve the upper-case notation, TCB, for the y-coordinate of the whole body. Column 4 contains the moments of the initial body and of the wedges, about the centreline plane. These moments are calculated as products of the terms in column 2, by those in column 3. The procedure is described symbolically by the expression $4 = 2 \times 3$ written in the subheading of column 4.

The sum of the terms in column 2 equals the total volume of the heeled barge; it is written in the cell identified by the entries volume and total. Similarly, the sum of the partial moments in column 4 is the moment of the heeled barge about the centreline plane; it appears in the cell corresponding to the entries moment and total. Dividing the moment of the heeled barge by its volume yields the y-coordinate of the centre of buoyancy of the heeled barge:

$$TCB = \frac{LB^3 \tan \phi/12}{LBT} = B^2 \tan \phi/(12T)$$

This result is written in the cell identified by the entries tcb and total.

Table 2.2 Calculating the transverse centre of buoyancy of the heeled barge

Solid	Volume	tcb	Moment
(1)	(2)	(3)	$(4) = (2) \times (3)$
Initial	LBT	0	0
Submerged wedge	$LB^2 \tan \phi/8$	$2B/(3 \cdot 2)$	$LB^3 \tan \phi/(3 \cdot 8)$
Emerged wedge	$-LB^2 \tan \phi/8$	$-2B/(3 \cdot 2)$	$LB^3 \tan \phi(3 \cdot 8)$
Total	LBT	$B^2 \tan \phi/(12T)$	$LB^3 \tan \phi /12$

A similar procedure is used to find the z-coordinate of the centre of buoyancy of the heeled barge; it is shown in Table 2.3. Calculations in tabular form are standard in Naval Architecture. More about them is written in Chapter 3 and we expect the reader to discover gradually the advantages of this way of solving problems. Obviously, Tables 2.2 and 2.3 can be consolidated. Then, the volumes are entered only once.

Tables 2.2 and 2.3 yield the **parametric equations** of the curve of centres of buoyancy:

$$y = \frac{1}{12}\frac{B^2}{T}\tan\phi$$

$$z = \frac{1}{24}\frac{B^2}{T}\tan^2\phi \tag{2.63}$$

We call the curve of centres of buoyancy **B curve**. From Eq. (2.63) we can derive

$$z = \frac{6T}{B^2}y^2 \tag{2.64}$$

This is the equation of a parabola.

The slope of the curves of centres of buoyancy is given by

$$\frac{dz}{dy} = \frac{dz/d\phi}{dy/d\phi} = \tan\phi \tag{2.65}$$

where

$$\frac{dy}{d\phi} = \frac{B^2}{12T}\frac{1}{\cos^2\phi} \tag{2.66}$$

and

$$\frac{dz}{d\phi} = \frac{B^2}{12T}\frac{\tan\phi}{\cos^2\phi} \tag{2.67}$$

Equation (2.65) shows that the tangent in B_ϕ has the slope ϕ, meaning that it is parallel to the corresponding waterline.

Table 2.3 Calculating the vertical centre of buoyancy of the heeled barge

Solid	Volume	vcb	Moment change
(1)	(2)	(3)	$(4) = 2 \times (3)$
Initial	LBT	0	0
Submerged wedge	$LB^2 \tan\phi/8$	$B\tan\phi/(3 \cdot 2)$	$LB^3 \tan^2\phi/(8 \cdot 3 \cdot 2)$
Emerged wedge	$-LB^2 \tan\phi/8$	$-B\tan\phi/(3 \cdot 2)$	$LB^3 \tan^2\phi/(8 \cdot 3 \cdot 2)$
Total	LBT	$B^2 \tan^2\phi/(24T)$	$LB^3 \tan^2\phi/(3 \cdot 8)$

To find the radius of curvature of the B curve we calculate

$$\frac{d^2 z}{dy^2} = \frac{1}{\cos^2 \phi} \frac{d\phi}{dy} = \frac{12T}{B^2} \tag{2.68}$$

and use a formula that can be found in many books on analysis or classic differential geometry (see, for example, Stoker, 1969: 26, or Gray, 1993: 11):

$$R = \frac{(1 + (dz/dy)^2)^{3/2}}{d^2 z/dy^2} = \frac{B^2}{12T} \frac{1}{\cos^3 \phi} \tag{2.69}$$

Now, let us use the theorems developed in this chapter. The volume of displacement of the barge is

$$\nabla = LBT$$

Equations (2.31) yield

$$
\begin{aligned}
x_B &= \frac{I_{xy}}{\nabla} \tan \phi = \frac{0}{LBT} = 0 \\
y_B &= \frac{I}{\nabla} \tan \phi = \frac{1}{12} \frac{B^2}{T} \tan \phi \\
z_B &= \frac{1}{2} \frac{I}{\nabla} \tan^2 \phi = \frac{1}{24} \frac{B^2}{T} \tan^2 \phi
\end{aligned}
\tag{2.70}
$$

These are exactly the results obtained in Tables 2.2 and 2.3. As to the metacentric radius, we calculate from Eq. (2.32)

$$\overline{BM_0} = \frac{I}{\nabla} = \frac{LB^3/12}{LBT} = \frac{1}{12} \frac{B^2}{T}$$

and, for any heel angle ϕ,

$$\overline{BM_\phi} = \frac{L(B/\cos \phi)^3/12}{LBT} = \frac{1}{12} \frac{B^3}{T} \frac{1}{\cos^3 \phi} \tag{2.71}$$

This is exactly the length of the radius of curvature obtained from Eq. (2.69).

Phase 2
In this phase the waterline passed the starboard deck corner and approaches the port-side bottom corner. If we consider the barge heeled by $90°$, so that the starboard side becomes the new bottom, the barge is again a wall-sided floating body. This observation allows us to continue the calculations in the same manner as for Phase 1. However, they would be more complex so that algebraic technicalities could obscure insight. To avoid this, we make a simplifying assumption (Hervieu, 1985): $T = H/2$. Then, the angle defining the limit between Phase 1 and Phase 2 is given by

$$\tan \phi = \frac{H}{B}$$

Substituting this value into Eq. (2.70) we find that at this angle the coordinates of the centre of buoyancy are

$$y_B = \frac{1}{12}\frac{B^2}{T}\tan\phi = \frac{B}{6}$$
$$z_B = \frac{1}{24}\frac{B^2}{T}\tan^2\phi = \frac{H}{12}$$

It is easy to see, in Figure 2.19 that these are the expected coordinates.

To continue the calculations in Phase 2, we use a new system of coordinates, η, ζ, with the origin in the centre of buoyancy, B_{90}, of the barge heeled by 90°. The relationships between the two systems of coordinates can be derived from Figure 2.20. We obtain thus

$$y_B = B/4 - \zeta_B$$
$$z_B = H/4 + \eta_B \tag{2.72}$$

The equations shown above are implemented in a MATLAB function called BARGE1 that can be found on the website provided for this book. The results of running the function with $B = 10$ and $H = 6$ are shown in Figure 2.21.

The reader is invited to experiment with various values of B and H and see how they influence the shape of the B and M curves. A more general treatment of the same problem can be found in Krappinger (1960).

Example 2.6 – B and M curves of Lido 9

Table 2.4 contains hydrostatic data of the vessel *Lido 9* for a volume of displacement equal to 44.16 m³ and the heel angles 0°, 15°, 30°,..., 90°. As shown

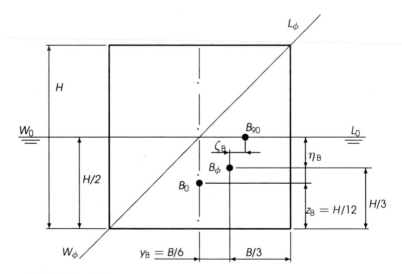

Figure 2.19 Simple barge

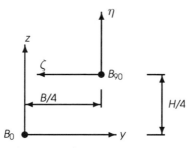

Figure 2.20 Coordinate systems for simple barge

in Figure 2.22, all the data are measured in a system of coordinates ξ, η, ζ. In this example, the axes $K\eta$ and $K\zeta$ rotate with an angle ϕ with respect to the axes Ky, Kz in which the hull surface is defined. The angle ϕ is the heel angle. The draft, T, is measured perpendicularly to the waterline; in our figure it is $T = \overline{KQ}$. As we see, $\overline{KN}$ is parallel to the waterline. The centre of buoyancy corresponding to the heel angle ϕ is marked B_ϕ and the metacentre, M_ϕ. In the table we dropped the subscripts ϕ. The height of the centre of buoyancy, $\overline{NB_\phi}$, is measured perpendicularly to the waterline and so is the height of the metacentre, $\overline{NM_\phi}$.

In this example we want to draw the curve of centres of buoyancy, B, and the metacentric evolute, M, at the given volume of displacement. With the data in Table 2.4 and the definitions shown in Figure 2.22 it is possible to draw manually these curves. Instead of this it is possible to use an M-file to draw the B- and M-curves for any ship we may want. The data is written in a convenient way, on an M-file named after the vessel we are studying. Thus, the contents of the file lido9.m can be found on the website of this book.

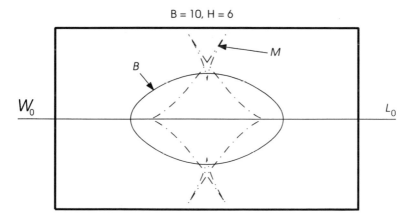

Figure 2.21 B and M curves of simple barge

Table 2.4 Data of vessel *Lido 9* at 44.16 m³ volume of displacement

Heel angle (°)	Draught (m)	$\overline{KN}$ (m)	$\overline{NB}$ (m)	$\overline{NM}$ (m)	LCB (m)	$\overline{NM}_L$ (m)
0	1.729	0	1.272	4.596	−1.735	23.371
15	1.575	1.122	1.121	3.711	−1.799	23.730
30	1.163	1.979	0.711	2.857	−1.932	23.154
45	0.600	2.595	0.107	1.830	−2.047	23.133
60	−0.012	2.945	−0.625	0.479	−2.072	17.473
75	−0.693	2.874	−1.393	−0.869	−2.008	14.298
90	−1.354	2.539	−2.108	−13.314	−1.970	12.792

Next, we project all points we are interested in on a transverse plane, that is a plane for which the longitudinal coordinate, x, is constant. We do this as in Figure 2.14. Let our plane be the midship section. For *Lido 9* this section is described by the points $P_1, P_2, \ldots, P_{15}$ whose coordinates are given in Table 2.5.

Let x_B, y_B be the coordinates of the centre of buoyancy, B, and x_M, y_M those of the metacentre, M. With the help of Figure 2.22 we can write

$$y_B = \overline{KN} \cos \phi - \overline{NB} \sin \phi$$
$$z_B = \overline{KN} \sin \phi + \overline{NB} \cos \phi$$

$$(2.73)$$

and

$$y_M = \overline{KN} \cos \phi - \overline{KM} \sin \phi$$
$$z_M = \overline{KN} \sin \phi + \overline{NM} \cos \phi$$

$$(2.74)$$

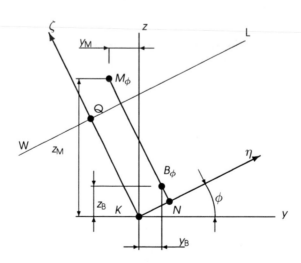

Figure 2.22 The coordinates of the points *B* and *M*

Table 2.5 Points defining the midship section of the Ship *Lido 9*

Point	y	z	Point	y	z
P_1	0.000	0.50	P_9	3.176	2.250
P_2	0.240	0.50	P_{10}	3.200	2.500
P_3	0.240	0.58	P_{11}	3.218	2.750
P_4	1.100	1.00	P_{12}	3.230	3.000
P_5	1.787	1.25	P_{13}	3.230	3.360
P_6	2.460	1.50	P_{14}	2.099	3.425
P_7	2.902	1.75	P_{15}	0.000	3.489
P_8	3.100	2.00			

Equations (2.73) can be rewritten in matrix form as

$$\begin{bmatrix} x_{\mathrm{B}} \\ y_{\mathrm{B}} \end{bmatrix} = \begin{bmatrix} \cos\phi & -\sin\phi \\ \sin\phi & \cos\phi \end{bmatrix} \begin{bmatrix} \overline{KN} \\ \overline{NB} \end{bmatrix} \tag{2.75}$$

Similarly, Eq. (2.74) can be written in matrix form as

$$\begin{bmatrix} x_{\mathrm{M}} \\ y_{\mathrm{M}} \end{bmatrix} = \begin{bmatrix} \cos\phi & -\sin\phi \\ \sin\phi & \cos\phi \end{bmatrix} \begin{bmatrix} \overline{KN} \\ \overline{NM} \end{bmatrix} \tag{2.76}$$

The **transformation matrix**

$$\begin{bmatrix} \cos\phi & -\sin\phi \\ \sin\phi & \cos\phi \end{bmatrix} \tag{2.77}$$

performs counter-clockwise rotation, around the origin, with the angle ϕ. In this example we need twice this rotation. As we may need it for more calculations in the future, it is worth programming a MATLAB function that evaluates the matrix and add this function to our toolbox. A possible listing of a file called `rotate.m` is given on the website of this book.

To draw the waterline we need a point on it. The easiest to calculate is the point Q shown in Figure 2.22. Here $\overline{KQ}$ corresponds to the draught T calculated by the program ARCHIMEDES. The equation of the waterline passing through this point is

$$z - T\cos\phi = \tan\phi(y - T\sin\phi) \tag{2.78}$$

The M-file, called `b_curve`, provided on the website of the book, performs all the calculations. The resulting plot is shown in Figure 2.23.

Table 2.4 contains a column that we did not use until now: the LCB values. We included these data to show that at finite angles of heel the centre of buoyancy can leave its initial transverse plane and move along the ship. This is the case of ships that do not have a fore-to-aft symmetry. Then, when the heel changes, the trim also changes until centre of gravity and centre of buoyancy lie again on the same vertical (Stevin's law).

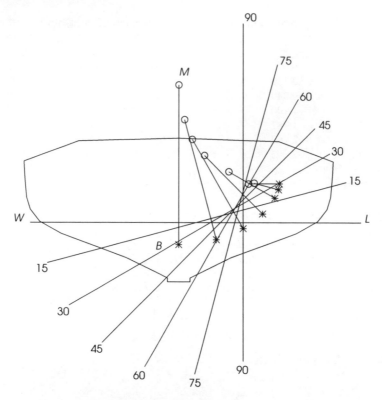

Figure 2.23 *B* and *M* curves of vessel *Lido 9*

Example 2.7 – Catamaran stability

Up to this point we have considered floating bodies whose buoyancy is provided by one submerged volume. If the floating body is a ship, we say that she is a **monohull ship**. In the example that follows we are going to show that stability can be greatly improved by distributing the buoyancy in two hulls. Then we talk about a *twin-hull ship*, but more often we use the term **catamaran**, a word derived from the Tamil 'kattumarum' composed of two words meaning 'to tie' and 'tree'. As the etymology indicates, catamarans have been in use for centuries in the Indian and Pacific Oceans. Today, many competition sailing boats and fast ferries are of the catamaran type.

Let us consider in Figure 2.24, a barge of breadth B and length L. Assuming the draught T, the displacement volume is

$$\nabla = LBT$$

and the metacentric radius,

$$\overline{BM} = \frac{B^3 L/12}{LBT} = \frac{1}{12}\frac{B^2}{T}$$

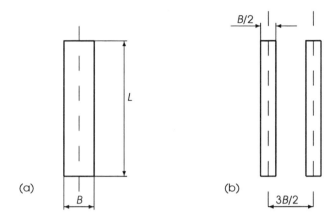

Figure 2.24 Monohull versus catamaran

We can obtain the same displacement volume with two hulls of breadth $B/2$, the same length, L, and the same draught, T. Assuming that the distance between the centrelines of the two hulls is $3B/2$, the resulting metacentric radius is

$$\overline{BM} = \frac{2}{LBT}\left[\frac{(B/2)^3}{12} + \frac{B}{2}L\left(\frac{3B}{2}\right)^2\right] = \frac{19B^2}{48T}$$

The first term between parantheses represents the sum of the moments of inertia of the waterlines about their own centrelines. The second term accounts for the parallel translation of the hulls from the plane of symmetry of the catamaran. The second term is visibly the greater. The ratio of the catamaran $\overline{BM}$ to that of the monohul is $\frac{19}{4}$. The improvement in stability is remarkable.

Catamarans offer also the advantage of larger deck areas and, under certain conditions, improved hydrodynamic performances. On the other hand, the weight of structures increases and the overall performance in waves must be carefully checked. It may be worth mentioning that also many vessels with three hulls, that is *trimarans*, have been built. Moreover, a company in Southampton developed a remarkable concept of a large ship with a main, slender hull, and four side hulls; that is a *pentamaran*.

Example 2.8 – Submerged bodies
Submerged bodies have no waterplane; therefore, their metacentric radii are equal to zero. Then the condition of initial stability is reduced to

$$\overline{GM} = \overline{KB} - \overline{KG} > 0$$

In simple words, the centre of gravity, G, must be situated under the centre of buoyancy, B. We invite the reader to draw a sketch showing the two mentioned points and derive the condition of stability by simple mechanical considerations. Submerged bodies do not develop hydrostatic moments that oppose inclinations, as they do not develop hydrostatic forces that oppose changes of depth.

Example 2.9 – An offshore platform

Figure 2.25 is a sketch of an *offshore platform* of the *semi-submersible* type. The buoyancy is provided by four pontoons, each of diameter b and length ℓ. The platform deck is supported by four columns. The depth of the platform is H and the draught, T.

Our problem is to find a condition for the height of the centre of gravity, $\overline{KG}$, for given platform dimensions. To do this, we calculate the limit value of $\overline{KG}$ for which the metacentric height, $\overline{GM}$, is zero. The metacentric radius is given by

$$\overline{BM} = \frac{I}{\nabla}$$

where the moment of inertia of the waterplane, I, and the volume of displacement, ∇, are

$$I = 4\left[\frac{\pi b^4}{64} + \left(\frac{\ell-b}{2}\right)^2 \frac{\pi b^2}{4}\right] = \frac{\pi b^2}{4}\left[5\left(\frac{b}{2}\right)^2 - 2\ell b + \ell^2\right]$$

$$\nabla \simeq 4\frac{\pi b^2}{4}\ell + 4\frac{\pi b^2}{4}T = \pi b^2(\ell + T)$$

In calculating the volume of displacement, ∇, we did not take into account the overlapping between column and pontoon ends. In conclusion

$$\overline{BM} = \frac{5\,(b/2)^2 - 2\ell b + \ell^2}{4(\ell + T)} \simeq \frac{\ell(\ell - 2b)}{4(\ell + T)} \tag{2.79}$$

where we neglected the term in b^2, usually small in comparison with other terms.

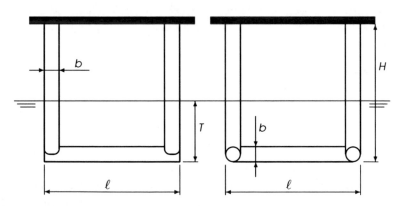

Figure 2.25 A semisubmersible platform

Table 2.6 Calculation of $\overline{KB}$

	Volume	Vertical arm	Moment
Pontoons	$\pi b^2 \ell$	$b/2$	$\frac{\pi b^3 \ell}{2}$
Columns	$\pi b^2 (T - b/2)$	$(T + b)/2$	$\frac{\pi b^2 (2T^2 + bT - b^2)}{4}$
Total	$\pi b^2 (\ell + T - b/2)$	$\frac{2T^2 + bT + 4b\ell}{4(\ell + T)}$	$\frac{\pi b^2 (2b\ell + 2T^2 + bT - b^2)}{4}$

The height of the centre of buoyancy above the base-line is calculated in Table 2.6. Neglecting the term in $-b^2$ we obtain

$$\overline{KB} = \frac{2T^2 + bT + 2b\ell}{4(\ell + T)} \tag{2.80}$$

The height of the metacentre above the baseline is given by

$$\overline{KM} = \overline{KB} + \overline{BM} = \frac{2T^2 + bT + \ell^2}{4(l + T)} \tag{2.81}$$

The condition for initial stability is

$$\overline{KG} < \frac{2T^2 + bT + \ell^2}{4(\ell + T)} \tag{2.82}$$

To rewrite Eq. (2.82) in non-dimensional form we define

$$\alpha = b/\ell, \qquad \beta = T/\ell$$

and obtain

$$\frac{\overline{KG}}{\ell} = \frac{2\beta^2 + \alpha\beta + 1}{4(1 + \beta)} \tag{2.83}$$

2.14 Exercises

Exercise 2.1 – Melting icebergs
In Example 2.1 we learnt that if an ice cube melts in a glass of water, the level of water does not change. Then, why do people fear that the meltdown of all icebergs would cause a water-level rise and therefore the flooding of lower coasts? Show that they are right.

Hint: Icebergs are formed on the continent and they are made of fresh water, while oceans consist of salt water. The density of salt water is greater than that of fresh water.

Exercise 2.2 – The tip of the iceberg

Icebergs are formed from compressed snow; their average density is 0.89 tm^{-3}. The density of ocean water can be assumed equal to 1.025 tm^{-3}. Calculate what part of an iceberg's volume can be seen above the water and explain the meaning of the expression 'The tip of the iceberg'.

Hint: See Exercise 2.1.

Exercise 2.3 – Draughts of a parallelepipedic barge

Consider a parallelepipedic (or, with another term, a box-shaped) barge characterized by the following data:

Length, L	10 m
Breadth, B	3 m
Mass, Δ/g	30 t

Find the draught, T_1, in fresh water, and the draught, T_2, in ocean water. See the Appendix of this chapter for various water densities.

Exercise 2.4 – Whisky on the rocks

Instead of considering a cube of ice floating in a glass of water, as in Example 2.1, let us think of a cube of ice floating in a glass of whisky. What happens when the cube melts?

Exercise 2.5 – A lemma about moving masses in three-dimensional

Prove the lemma in Section 2.7 for a three-dimensional system of masses and a three-dimensional displacement of one of the masses.

Exercise 2.6 – A wooden parallelepiped

The floating condition of a wooden, homogeneous block of square cross-section depends on its specific gravity. Three possible positions are shown in Figure 2.26.

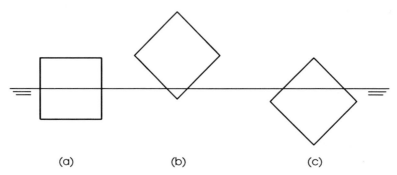

(a)   (b)   (c)

Figure 2.26 Different floating conditions of a wooden, parallelepipedic block

1. Find the ranges of specific gravity enabling each position.
2. For each range find a suitable kind of wood. To do this look through tables of wood properties.
3. Can you imagine other floating positions? In the affirmative, calculate the corresponding specific-gravity range.

Hint: A floating position is possible if the corresponding metacentric height is positive.

Exercise 2.7 – B and M curves – variable heel
Table 2.7 contains the same data items as Table 2.4, but calculated at 5-degree intervals. With this 'resolution' it is possible to plot smooth B and M curves. First, write the data on a file lido9a similar to file lido9. Next, modify the programme cited in Example 2.6 to plot only the B and M curves of the vessel whose data are called from the keyboard. Run the program with the data given at 5-degree intervals and print a hardcopy of the resulting plot.

Exercise 2.8 – B and M curves – variable trim
Table 2.8 contains data of the vessel *Lido 9* for constant volume of displacement equal to 44.16 m³, upright condition, and trim varying between −0.3 and 1.1 m.

Table 2.7 Data of vessel *Lido 9* at 44.16 m³ volume of displacement and 5-degree heel intervals, trim = −0.325 m

Heel angle (°)	Draught (m)	$\overline{KN}$ (m)	$\overline{NB}$ (m)	$\overline{NM}$ (m)	LCB (m)	$\overline{NM}_L$ (m)
0	1.729	0	1.272	4.596	−1.735	23.371
5	1.711	0.399	1.255	4.438	−1.740	23.693
10	1.659	0.776	1.204	4.119	−1.761	24.008
15	1.575	1.122	1.121	3.711	−1.799	23.730
20	1.462	1.432	1.009	3.341	−1.841	23.813
25	1.324	1.716	0.872	3.073	−1.887	23.464
30	1.163	1.979	0.711	2.857	−1.932	23.154
35	0.985	2.215	0.528	2.464	−1.971	22.822
40	0.796	2.419	0.326	2.105	−2.002	22.810
45	0.600	2.595	0.107	1.830	−2.047	23.133
50	0.402	2.749	−0.126	1.537	−2.106	21.837
55	0.198	2.870	−0.372	1.082	−2.113	19.757
60	−0.012	2.945	−0.625	0.479	−2.072	17.473
65	−0.235	2.960	−0.883	−0.185	−2.041	16.162
70	−0.464	2.931	−1.140	−0.543	−2.025	15.117
75	−0.693	2.874	−1.393	−0.869	−2.008	14.298
80	−0.919	2.788	−1.640	−1.171	−1.994	13.633
85	−1.140	2.678	−1.878	−1.446	−1.981	13.121
90	−1.354	2.539	−2.108	−13.314	−1.970	12.792

Table 2.8 Data of vessel *Lido 9* at 44.16 m^3 volume of displacement, 0.1 m trim intervals, upright condition

Trim (m)	Draught (m)	$\overline{NB}$ (m)	$\overline{NM}$ (m)	LCB (m)	$\overline{NM}_L$
−1.000	1.673	1.174	4.536	−2.777	23.681
−0.900	1.653	1.192	4.550	−2.623	23.904
−0.800	1.668	1.208	4.564	−2.468	24.069
−0.700	1.683	1.224	4.577	−2.313	24.163
−0.600	1.697	1.238	4.585	−2.157	24.145
−0.500	1.709	1.251	4.589	−2.001	23.954
−0.400	1.721	1.24	4.592	−1.848	23.584
−0.300	1.732	1.276	4.598	−1.697	23.293
−0.200	1.742	1.286	4.604	−1.548	22.951
−0.100	1.750	1.295	4.610	−1.401	22.556
0.000	1.758	1.304	4.614	−1.257	22.108
0.100	1.765	1.311	4.615	−1.114	22.137
0.200	1.772	1.319	4.614	−0.971	22.115
0.300	1.777	1.324	4.612	−0.829	22.046
0.400	1.782	1.329	4.610	−0.690	21.910
0.500	1.786	1.333	4.606	−0.546	21.707
0.600	1.789	1.336	4.603	−0.407	21.431
0.700	1.792	1.338	4.599	−0.270	21.116
0.800	1.793	1.339	4.599	−0.135	20.895
0.900	1.794	1.340	4.597	0.000	20.871
1.000	1.795	1.340	4.594	0.135	20.834
1.100	1.795	1.338	4.590	0.269	20.829

The LCB values in column 5 are equivalent to the $\overline{KN}$ values in Example 2.6, Figure 2.22.

Write the data on an M-file, lido9b.m, and use the program b_curve to plot the B- and M-curves. Here the M-curve is the locus of the longitudinal metacentre.

2.15 Appendix – Water densities

	Density (t m^{-3})
Fresh water	1.000
Eastern Baltic Sea	1.003
Western Baltic Sea	1.015
Black Sea	1.018
Oceans	1.025
Red Sea	1.044
Caspian Sea	1.060
Dead Sea	1.278

3

Numerical integration in naval architecture

3.1 Introduction

In Chapter 2, we have learnt that the evaluation of ship properties, such as displacement and stability, requires the calculation of areas, centroids and moments of plane figures, and of volumes and centres of volumes. Such properties are calculated by integration. In the absence of an explicit definition of the hull surface, in terms of calculable mathematical functions, the integrations cannot be carried out by analytic methods. The established practice has been to describe the hull surface by tabulated data, as shown in Chapter 1, and to use these data in numerical calculations.

Two methods for numerical integration are described in this chapter: the **trapezoidal** and **Simpson's rules**. The treatment is based on Biran and Breiner (2002). The rules are exemplified on integrands defined by explicit mathematical expressions; this is done to convince the reader that the two methods of numerical integration are efficient, and to allow an evaluation of errors. The first examples are followed in Chapter 4 by Naval-Architectural applications to real ship data presented in tabular form.

Many Naval-Architectural problems require the calculation of the definite integral

$$\int_a^b f(x)\,\mathrm{d}x$$

of a function bounded in the finite interval $[a, b]$. We approximate the definite integral by the weighted sum of a set of function values, $f(x_1), f(x_2), \ldots, f(x_n)$, evaluated, or measured, at n points $x_i \in [a, b]$, $i = 1, 2, \ldots, n$, i.e.

$$\int_a^b f(x)\,\mathrm{d}x \approx \sum_{i=1}^{n} a_i f(x_i) \tag{3.1}$$

In Naval Architecture, the coefficients a_i are called **multipliers**; in some books on Numerical Methods they are called **weights**.

There are several ways of deriving formulae for numerical integration – also called **quadrature formulae** – of the form shown in Eq. (3.1); three of them are mentioned below:

1. By geometrical reasoning, considering $\int_a^b f(x)\,dx$ as the area under the curve $f(x)$, between $x = a$ and $x = b$.
2. By approximating the function $f(x)$ by an interpolating polynomial, $P(x)$, and integrating the latter instead of the given function, so that

$$\int_a^b f(x)\,dx \approx \int_a^b P(x)\,dx$$

3. By developing the given function into a Taylor or MacLaurin series and integrating the first terms of the series.

The first approach yields a simple intuitive interpretation of the rules for numerical quadrature and of the errors involved. This interpretation enables the user to derive the rules whenever required, and to adapt them to particular situations, for instance, when changing the subintervals of integration. On the other hand, each rule must be derived separately. The advantages of the other approaches are:

- The derivation is common to a group of rules which thus appears as particular cases of a more general method.
- The derivation yields an expression of the error involved.

In the next two sections, we shall use the geometrical approach to derive the two most popular rules, namely the **trapezoidal** and **Simpson's rules**. These two methods are sufficient for solving most problems encountered in Naval Architecture. The error terms will be given without derivation; however, interpretations of the error expressions will follow their presentation.

3.2 The trapezoidal rule

Let us consider the function $f(x)$ represented in Figure 3.1. We assume that we know the values $f(x_1), f(x_2), \ldots, f(x_5)$ and we want to calculate the definite integral

$$I = \int_{x_1}^{x_5} f(x)\,dx \tag{3.2}$$

The integral in Eq. (3.2) represents the area under the curve $f(x)$. Let us connect the points $f(x_1), f(x_2), \ldots, f(x_5)$ by straight line segments (the dashed-dotted

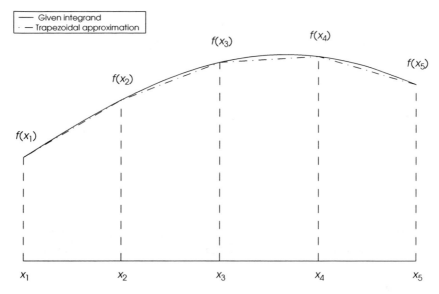

Figure 3.1 The derivation of the trapezoidal rule

lines in the figure). We approximate the area under the curve by the sum of the areas of four trapezoids, i.e. the area of the trapezoid with base $\overline{x_1 x_2}$ and heights $f(x_1), f(x_2)$, plus the area of the trapezoid with base $\overline{x_2 x_3}$ and heights $f(x_2), f(x_3)$, and so on. We obtain

$$I \approx (x_2 - x_1)\frac{f(x_1) + f(x_2)}{2} + (x_3 - x_2)\frac{f(x_2) + f(x_3)}{2} + \cdots \qquad (3.3)$$

For constant x-spacing, $x_2 - x_1 = x_3 - x_2 = \cdots = h$, Eq. (3.3) can be reduced to a simpler form:

$$I \approx h\left[\frac{1}{2}f(x_1) + f(x_2) + f(x_3) + \cdots + f(x_{n-1}) + \frac{1}{2}f(x_n)\right] \qquad (3.4)$$

We call the intervals $[x_1, x_2]$, $[x_2, x_3]$, and so on, *subintervals*.

As an example, let us calculate

$$\int_{0°}^{90°} \sin x \, dx$$

The calculation presented in tabular form is as follows:

Angle (°)	$\sin x$	Multiplier	Product
0	0	1/2	0
15	0.2588	1	0.2588
30	0.5000	1	0.5000
45	0.7071	1	0.7071
60	0.8660	1	0.8660
75	0.9659	1	0.9659
90	1.0000	1/2	0.5000
Sum	–	–	3.7979

The calculations were performed with MATLAB and the precision of the display in the *short format*, i.e. four decimal digits, was retained. To obtain the approximation of the integral, we multiply the sum in column 4 by the constant subinterval, h:

$$\left(\pi \times \frac{15}{180} \right) \times 3.7979 = 0.9943$$

Above we measured the interval in radians, as we should do in such calculations. Equation (3.3) in matrix form yields

$$I \approx \frac{1}{2} \left[\begin{array}{c} (x_2 - x_1)(x_3 - x_2)(x_4 - x_3)(x_5 - x_4) \end{array} \right] \left[\begin{array}{c} y_1 + y_2 \\ y_2 + y_3 \\ y_3 + y_4 \\ y_4 + y_5 \end{array} \right] \right] \tag{3.5}$$

The generalized form of Eq. (3.5) is implemented in MATLAB by the `trapz` function that can be called with two arguments:

1. the column vector x,
2. the column vector y, of the same length as x, or a matrix y, with the same number of rows as x.

If the points $x_1, x_2, \ldots, x_n$ are equally spaced, i.e., if

$$x_2 - x_1 = x_3 - x_2 = \cdots = h$$

the `trapz` function can be called with one argument, namely the column vector (or matrix) y. In this case, the result must be multiplied by the common x-interval, h.

3.2.1 Error of integration by the trapezoidal rule

In any subinterval $[x_i, x_{i+1}]$, the error of the approximation I_i, obtained by the trapezoidal rule equals

$$I_i - \int_{x_i}^{x_{i+1}} f(x)\,\mathrm{d}x = \frac{1}{12}h^3\frac{\mathrm{d}^2 f(\xi_i)}{\mathrm{d}x^2} \tag{3.6}$$

where ξ_i is a point in the subinterval (x_i, x_{i+1}) and $h = x_{i+1} - x_i$. Usually, the interval of integration $[x_1, x_m]$, is divided into several subintervals; if we assume that they are equal, and note by I the trapezoidal approximation over the whole interval, we can write

$$\left| I - \int_{x_1}^{x_m} f(x)\,\mathrm{d}x \right| = \left| \sum_{i=1}^{m-1} \frac{h^3}{12}\frac{\mathrm{d}^2 f(\xi)}{\mathrm{d}x^2} \right|$$

$$\leq \frac{x_m - x_1}{12}h^2 \max_{\xi \in [x_1, x_m]} \left| \frac{\mathrm{d}^2 f(\xi)}{\mathrm{d}x^2} \right| \tag{3.7}$$

We do not know the maximum value of the derivative in Eq. (3.7); otherwise, we would have been able to calculate the exact value of the integral. We can, however, say the following:

- By substituting in Eq. (3.7) the maximum value of $\mathrm{d}^2 f(x)/\mathrm{d}x^2$ in the interval $[x_1, x_m]$, we can calculate an upper boundary of the error.
- The error is proportional to the square of h; if we halve the subinterval, the error is reduced approximately in the ratio $1/4$.
- The method is exact if $\mathrm{d}^2 f(x)/\mathrm{d}x^2 = 0$. This is the case for linear functions. As a matter of fact, the derivation of the trapezoidal rule was based on a linear approximation of $f(x)$.

Example 3.1
In this example, we consider the integral

$$\int_0^{\pi/2} \{1 + \sin(x)\}\,\mathrm{d}x = [x - \cos(x)]_0^{\pi/2}$$

$$= \pi/2 + 1 = 2.570\ 796\ 326\ 794\ 90$$

To calculate the same integral numerically by means of the trapezoidal rule, we begin by dividing the interval $[0, \pi/2]$ into two subintervals and obtain the value 2.518 855 775 763 42. The error equals -2.02% of the correct value. We can reduce the error by halving the subinterval h. Experimenting with subintervals equal to $\pi/8, \pi/16, \ldots, \pi/128$, we obtain the results shown in Table 3.1 where they are compared with the results yielded by Simpson's rule (see Section 3.3). For $h = \pi/8$, Figure 3.1 shows the error as the sum of the small areas contained between the dashed-dotted line (the trapezoids) and the solid line (the given

Table 3.1 Results by trapezoidal and by Simpson's rule

Subinterval	Integral	
	Trapezoidal rule	Simpson's rule
$\pi/4$	2.51885577576342	2.57307620428711
$\pi/8$	2.55791212776767	2.57093091176909
$\pi/16$	2.56758149868107	2.57080462231886
$\pi/32$	2.56999300727997	2.57079684347960
$\pi/64$	2.57059552111492	2.57079635905990
$\pi/128$	2.57074612688700	2.57079632881103

curve). This area looks really small. The errors in per cent of the true values are shown in Table 3.2. As predicted by Eq. (3.7), each time we divide the subinterval h by 2, the error is divided approximately by 4. It is easy to see that as $h \to 0$, the trapezoidal approximation of the integral tends to the true value.

In this example, by reducing the size of the subinterval h we could make the error negligible. This was easy because we had an explicit expression for $f(x)$, and we could evaluate as many values of $f(x)$ as we wanted. When there is no explicit mathematical definition, as it happens when the ship lines are defined only by drawings or tables of offsets, the number of function values that can be measured, or evaluated, is restricted by practical limitations. In such cases, we must be satisfied if the precision of the integration is consistent with the precision of the measurements, or of calculations involving the same constants and variables. To understand this point better, let us suppose that we want to calculate the ship displacement mass as $\Delta = \rho\nabla$, where ρ is the density of the surrounding water. It makes no sense to be very precise in the calculation of the displacement volume ∇, if we multiply it afterwards by a conventional value of the density ρ. The density varies from sea to sea (see table in Appendix A of Chapter 2), and in the same sea it varies with temperature. In most calculations, it would be impossible to take into account these variations, and the Naval Architect or the ship Master has to use the value prescribed by the regulations relevant to

Table 3.2 Per cent error by trapezoidal and by Simpson's rule

Subinterval	Per cent error	
	Trapezoidal rule	Simpson's rule
$\pi/4$	−2.0204	0.08868371
$\pi/8$	−0.5012	0.00523515
$\pi/16$	−0.1251	0.00032268
$\pi/32$	−0.0312	0.00002010
$\pi/64$	−0.0078	0.00000126
$\pi/128$	−0.0195	0.00000008

the ship under consideration. For example, for oceans and the Mediterranean sea, various regulations specify the value $1.025 \, t \, m^{-3}$. An exception is the *inclining experiment*, a case in which the actual density must be measured. But, even in that case the precision of the measurement is limited and not better than that of the ∇-value calculated with the rules described in this chapter.

3.3 Simpson's rule

In Figure 3.2, the solid line passing through the points B, C and D represents the integrand $f(x)$. We want to calculate the integral of $f(x)$ between $x = A$ and $x = E$, i.e. the area ABCDEFA. This time we shall approximate $f(x)$ by a parabola whose equation has the form

$$f(x) = a_0 + a_1 x + a_2 x^2 \tag{3.8}$$

The parabola is represented by a dashed-dotted line in Figure 3.2. We need three points to define this curve; therefore, in addition to the values of $f(x)$ calculated at the two extremities, i.e. at the points B and D, we shall also evaluate $f(x)$ at the half-interval, obtaining the point C. Let

$$\overline{AB} = f(x_1), \qquad \overline{FC} = f(x_2), \qquad \overline{ED} = f(x_3)$$

$$h = \frac{\overline{AE}}{2} = \frac{(x_3 - x_1)}{2}$$

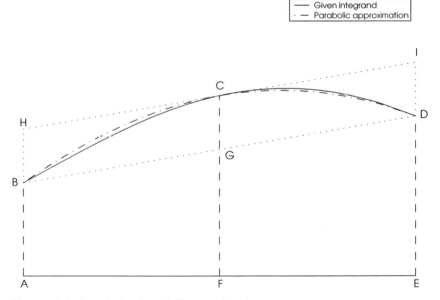

Figure 3.2 The derivation of Simpson's rule

We divide the total area under $f(x)$ into two partial areas:

1. the trapezoid ABDEA,
2. the parabolic segment BCDGB.

The first area equals

$$\overline{AE} \cdot \frac{\overline{AB} + \overline{ED}}{2} = 2h \cdot \frac{f(x_1) + f(x_3)}{2}$$

For the second area, we use a result from geometry that says that the area of a parabolic segment equals two-thirds of the area of the circumscribed parallelogram. Correspondingly, we calculate the second area as two-thirds of the circumscribed parallelogram BHID, i.e.

$$\frac{2}{3} \cdot \overline{AE} \cdot \overline{CG} = \frac{2}{3} \cdot 2h \left(f(x_2) - \frac{f(x_1) + f(x_3)}{2} \right)$$

Adding the two partial sums yields

$$\int_{x_1}^{x_3} f(x)\, dx \approx \frac{h}{3} [f(x_1) + 4f(x_2) + f(x_3)] \tag{3.9}$$

which is the elementary form of **Simpson's rule**.

Usually we have to integrate the function $f(x)$ over a larger interval $[a, b]$. Then, we achieve a better approximation by dividing the given interval into more subintervals. From the way we derived Eq. (3.9) we see that the number of subintervals must be even, say $n = 2k$, where k is a natural number. Let

$$h = \frac{a - b}{n} = x_2 - x_1 = x_3 - x_2 = \cdots = x_{n+1} - x_n$$

Applying Eq. (3.9) for each pair of subintervals, and adding all partial sums, we get

$$\int_{x_1}^{x_{n+1}} f(x)\, dx = \frac{h}{3} [f(x_1) + 4f(x_2) + 2f(x_3)$$
$$+ 4f(x_4) + \cdots + 4f(x_n) + f(x_{n+1})] \tag{3.10}$$

which is the extended form of Simpson's rule, for equal subintervals. This form is very helpful when calculations are carried out manually. As an example, let us calculate

$$\int_{0°}^{90°} \sin x\, dx$$

In tabular form, the calculation is

Angle (°)	$\sin x$	Multiplier	Product
0	0	1	0
15	0.2588	4	1.0353
30	0.5000	2	1.0000
45	0.7071	4	2.8284
60	0.8660	2	1.7321
75	0.9659	4	3.8637
90	1.0000	1	1.0000
Sum	–	–	11.4595

To obtain the approximation of the integral, we multiply the sum in column 4 by the constant subinterval:

$$\left(\pi \times \frac{15}{180} \right) \times \frac{11.4595}{3} = 1.0000$$

When a computer is used, there is no need to have all subintervals equal and it is sufficient to have **pairs of equal intervals**. A MATLAB function called `simp` that implements Eq. (3.9) is described in Biran and Breiner (2002, Chapter 10).

As an example, let us calculate by Simpson's rule the same integral that we exemplified in Section 3.2. As shown in Tables 3.1 and 3.2, the results are much better than those obtained with the trapezoidal rule.

3.3.1 Error of integration by Simpson's rule

Denoting by I_i the approximation obtained by Simpson's rule in the subinterval $[x_1, x_3]$, the error equals

$$I_i - \int_{x_1}^{x_3} f(x)\, dx = h^5 \frac{1}{90} \frac{d^4 f(\xi)}{dx^4} \tag{3.11}$$

where $x_1 \leq \xi \leq x_3$. Summing up the errors in all pairs of subintervals, and denoting by I the approximation calculated with Simpson's rule, we obtain

$$\left| I - \int_{x_1}^{x_n} f(x)\, dx \right| = \left| \sum_{1}^{n/2} \frac{d^4 f(\xi)}{dx^4} \cdot \frac{h^5}{90} \right| \tag{3.12}$$

$$\left| I - \int_{x_1}^{x_n} f(x)\, dx \right| \leq \frac{x_n - x_1}{180} h^4 \max_{\xi \in [x_1, x_n]} \left| \frac{d^4 f(\xi_m)}{dx^4} \right| \tag{3.13}$$

At this point we can say the following about Simpson's rule:

1. If we divide h by 2, the error decreases approximately in the ratio $1/16$.
2. Simpson's rule yields the exact result if $d^4 f/dx^4 = 0$. This is certainly true for second-degree parabolas, which is not surprising because we assumed such a curve when we developed the rule. It is interesting that the method is also exact for cubics (third-degree curves).
3. For an equal number of subintervals, Simpson's rule yields better results than the trapezoidal rule. On the other hand, Simpson's rule imposes a serious constraint: the number of subintervals must be even, or, equivalently, $f(x)$ must be evaluated at an uneven number of equally spaced points, or, in other words, an uneven number of ordinates. If, for example, we calculate the area of a waterline, we need an uneven number of equally spaced stations.

Example 3.2
We refer again to Example 3.1 using this time Simpson's rule. We can experiment with decreasing subintervals and obtain the results shown in Table 3.1, where they are compared with the results yielded by the trapezoidal rule. The convergence is considerably faster than that obtained in the case of the trapezoidal rule. The per cent errors are shown in Table 3.2. As predicted by Eq. (3.13), each time we divide the subinterval h by 2, the error decreases approximately in the ratio $1/16$. Note also that only two subintervals yield better results with Simpson's rule than eight with the trapezoidal rule.

3.4 Calculating points on the integral curve

The trapezoidal and Simpson's rules produce one number for an interval of ordinates, i.e.

$$I(a, b) = \int_a^b f(x)\,dx$$

Sometimes we are interested not only in one number, but also in a sequence of numbers that describe the integral as a function within the given interval,

$$I(x) = \int_a^x f(x)\,dx, \quad a \le x \le b \tag{3.14}$$

Thus, in certain hydrostatic calculations we may need to know the areas of transverse sections (stations) as functions of draught (see Chapter 4).

Another example is that of calculations of dynamic stability which require the knowledge of the area under the curve of the righting arm as function of the heel angle. The latter subject is discussed in Chapters 6 and 8. An appropriate name for a procedure that yields such an integral is **integral with variable upper limit**.

Let us consider a sequence of points, x_1, x_2, ..., x_n, and a sequence of values $f(x_1)$, $f(x_2)$, ..., $f(x_n)$. In the first example above, the values of the independent variable, x_i, represent draught, the functions $f(x_i)$, half-breadth and the integral, the area of the station up to that draught.

In the second example, x_i is a heel angle, $f(x_i)$, the righting arm, $\overline{GZ}$ and the integral, the area under the righting-arm curve up to the respective angle. We could calculate the integral in Eq. (3.14) by applying one of the integration rules over the interval $[x_1, x_2]$, then over $[x_1, x_3]$, and so on. This procedure would be awkward. Table 3.3 illustrates an algorithm that yields the integral with variable upper limit in a 'continuous' calculation. Let us detail the algorithm.

In column 1, we write the current numbers of the points at which we know the values of the function to be integrated. In column 2, we write the x_i values, i.e. the draughts in the first example given above, or the heel angles in the second example. In column 3, we write the values of the functions $f(x_i)$ at the points x_i shown in column 2. For columns 3 and 4, the algorithm is

Write 0 in column 4, line 1
For $i = 1 : (n - 1)$

- Pick up the value in column 4, line i
- Go left and add the value in column 3, line i
- Go down and add the value in column 3, line $i + 1$
- Write the result in column 4, line $i + 1$

End

In column 5, line i, we write the result of the product of the content of column 4, line i, by half of the subinterval of integration. Visual inspection of column 5 shows that the expressions appearing there are exactly those yielded by the trapezoidal rule over the intervals $[x_1, x_2]$, $[x_1, x_2]$, ..., $[x_1, x_n]$.

Table 3.3 The algorithm for integration with variable upper limit

No.	Position	Function	Sums	Integrals
1	x_1	$f(x_1)$	0	0
		$\downarrow$	$\hookleftarrow$	
2	x_2	$f(x_2)$		
		$\hookrightarrow$	$f(x_1) + f(x_2)$	$\frac{f(x_1)}{2} + \frac{f(x_2)}{2}$
		$\downarrow$	$\hookleftarrow$	
3	x_3	$f(x_3)$		
		$\hookrightarrow$	$f(x_1) + 2f(x_2) + f(x_3)$	$\frac{f(x_1)}{2} + f(x_2) + \frac{f(x_3)}{2}$
		$\downarrow$	$\hookleftarrow$	
...	...	...	...	...
n	x_n	$f(x_n)$	$f(x_1) + 2f(x_2) + \cdots + 2f(x_{n-1}) + f(x_n)$	$\frac{f(x_1)}{2} + f(x_2) + \cdots + f(x_{n-1}) + \frac{f(x_n)}{2}$

Let us illustrate the above procedure with the example of $\int_0^x \sin x \, \mathrm{d}x$ in the interval $[0, \, 2\pi]$. As known

$$\int_0^x \sin \xi \, \mathrm{d}\xi = -\cos \xi|_0^x = 1 - \cos x \qquad (3.15)$$

Figure 3.3 shows the implementation of the algorithm in Microsoft's Excel (MS Excel). A MATLAB function based on the same algorithm, called `intvar`, can be found in the website of this book. The algorithm is used in the following MATLAB lines that solve the particular problem exemplified above:

```
x   = 0: 10: 180;      % angles in degrees
phi = pi*x/180;        % angles in radians
y   = sin(phi);        % function to be integrated
l   = length(y);
y1  = y(1:(1-1));
s1  = [ 0 cumsum(y1) ];
y2  = y(2:1);
s2  = [ 0 cumsum(y2) ];
% now multiply by half subinterval
S   = (pi*10/(2*180))*(s1 +s2);
```

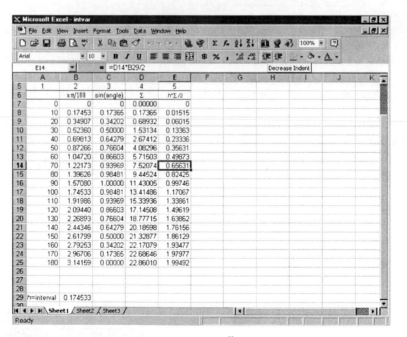

Figure 3.3 An Excel spreadsheet for $\int_0^x \sin x \, dx$ in the interval $[\, 0, 2\pi]$

Table 3.4 Integral with variable upper limit – comparing
the analytic result with that obtained in MATLAB

Angle (°)	Analytic result	Numerical result
0	0	0
10	0.0152	0.0152
20	0.0603	0.0602
30	0.1340	0.1336
40	0.2340	0.2334
50	0.3572	0.3563
60	0.5000	0.4987
70	0.6580	0.6563
80	0.8264	0.8243
90	1.0000	0.9975
100	1.1736	1.1707
110	1.3420	1.3386
120	1.5000	1.4962
130	1.6428	1.6386
140	1.7660	1.7616
150	1.8660	1.8613
160	1.9397	1.9348
170	1.9848	1.9798
180	2.0000	1.9949

In our experience, the MATLAB procedure is slightly more exact than the
Excel spreadsheet. Table 3.4 compares the result yielded by Eq. (3.15) with
those obtained with the MATLAB function. The agreement between the results
obtained, analytically in Excel, and in MATLAB is remarkable.

3.5 Intermediate ordinates

The integration rules developed in Sections 3.2 and 3.3 were based on a subdivision into equal subintervals. This procedure is not always the best one. Let
us consider, for example, the waterline shown in Figure 3.4. We may appreciate
that the shape of the curve between Stations 0 and 1 suits neither the trapezoidal
nor Simpson's rule; applying either of them would yield large errors. We learnt
that reducing the intervals would also reduce the errors. Therefore, let us introduce an **intermediate** station between Stations 0 and 1 and appropriately call it
Station $\frac{1}{2}$. We introduce another intermediate station between Stations 9 and 10
and call it $9\frac{1}{2}$. We invite the reader to check that the corresponding sequence of
trapezoidal multipliers is now

$$1/4,\ 2/4,\ 3/4,\ 4/4,\ 1,\ \ldots,\ 1,\ 4/4, 3/4,\ 2/4,\ 1/4$$
$$= 1/4,\ 1/2,\ 3/4,\ 1,\ \ldots,\ 1,\ 3/4,\ 1/2,\ 1/4$$

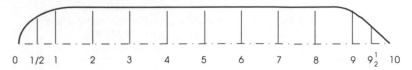

0   1/2 1   2   3   4   5   6   7   8   9 $9\frac{1}{2}$ 10

Figure 3.4 Intermediate ordinates at Station $\frac{1}{2}$ and Station $9\frac{1}{2}$

The subdivision illustrated in Figure 3.4 suits Simpson's rule too, because we have a pair of equal subintervals $\delta L/2$, four pairs of equal subintervals δL and a pair of equal subintervals $\delta L/2$.

3.6 Reduced ordinates

We present in this section another way of overcoming the problem described in the preceding section. In continuation, we show how the same method can be adapted for a more difficult case.

Let us consider the thick, solid-line curve shown in the left-hand side of Figure 3.5; it may be, for example, the after part of a waterline. If we calculate the area under the curve by the trapezoidal rule, and enter 0 for the half-breadth at Station 0 and the actual half-breadth at Station 1, we miss the whole shaded area. If we use Simpson's rule with the same values, plus the actual half-breadth at Station 2 (remember, for Simpson's rule we must take two equal subintervals), we obtain, in fact, the area under the dashed line, and this can be again less than the actual area.

The right-hand side of Figure 3.5 shows a simple way of improving the result. Let us draw the line BC so that the two shaded areas look equal. Our intention is to rely upon visual appreciation because we are looking for a quick procedure. Then, we take the length of the segment $\overline{AC}$ as the **reduced ordinate** at Station 0.

Above, the curve we are interested in begins exactly at one station. Frequently it happens that the curve begins or ends between stations. Such a case is illustrated in Figure 3.6, which may represent the forward part of a waterline.

To obtain a reduced ordinate, we begin by applying the procedure described above, and substitute the given curve arc by the straight line segment $\overline{AB}$. Next, we connect the point A to the point C and draw $\overline{BE}$ parallel to $\overline{AC}$. The reduced

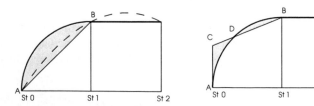

Figure 3.5 Reduced ordinates – a simple case

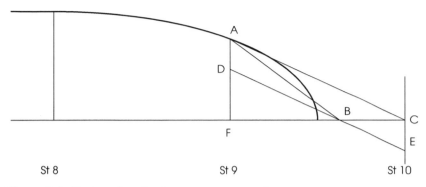

Figure 3.6 Reduced ordinates – a more complex case

ordinate is $\overline{CE}$ and we use it with a minus sign. To prove that the proposed procedure yields the correct result, we extend the segment $\overline{BE}$ until it intercepts Station 9 at point D. We are looking for the area of the triangle ABF, but this area equals the area of the triangle ACF minus that of the triangle ABC. Now, the area of the triangle ABC is half the area of the parallelogram ACED. Noting $\overline{AF} = y_9$, the half-breadth at Station 9 and FC $= \delta L$, we can write

$$\text{Area} = \frac{y_9 \cdot \delta L}{2} - \frac{\overline{CE} \cdot \delta L}{2} \qquad (3.16)$$

This is exactly the result we would obtain by applying the trapezoidal rule with the value y_9 for Station 9 and the length of the segment $\overline{CE}$ taken with the minus sign.

3.7 Other procedures of numerical integration

We described in this section two rules for numerical integration: the trapezoidal and Simpson's rules. Additional methods of integration have been developed and employed. For example, a third rule popular in English-language literature is *Simpson's second rule* in which the given integrand is approximated by a third-degree parabola. This rule is applied on sets of three equal subintervals, or, in other words, sets of four equally spaced ordinates. This is a very serious constraint. As shown in Chapter 13, CAD programs used today in Naval Architecture describe the hull surface by *piecewise polynomials*, i.e. they fit polynomials and combination of polynomials to curve segments and surface patches. Then, it is possible to use the polynomial coefficients to obtain the integrals by simple algebraic formulae. For example, if a segment of a waterline is described by the equation

$$y = c_1 x^2 + c_2 x + c_3 \qquad (3.17)$$

then the area enclosed between the curve segment, the centreline, and the stations $x = a$, $x = b$ is

$$\int_a^b y \, \mathrm{d}x = \frac{c_1}{3}x^3 + \frac{c_2}{2}x^2 + c_3x \Big|_a^b = \frac{c_1}{3}(b^3 - a^3) + \frac{c_2}{2}(b^2 - a^2) + c_3(b - a)$$

(3.18)

Similar equations can be derived for other properties, namely moments and moments of inertia.

3.8 Summary

Naval Architecture requires the calculation of areas, moments of areas, moments of inertia of areas, volumes and moments of volumes. Such calculations involve definite integrals. Usually, the hull surface is defined by line drawings or tables of offsets, and not by explicit mathematical expressions. Then, the integrals can be obtained only by numerical methods. In a numerical method, we approximate the integral by a weighted sum of a finite set of function values, i.e.

$$\int_a^b f(x) \, \mathrm{d}x \approx \sum_{i=1}^n a_i f(x_i)$$

(3.19)

Two methods that implement such approximations are introduced in this chapter: the trapezoidal rule and Simpson's rule. The trapezoidal rule approximates the given curve by straight line segments, while Simpson's rule approximates it by a parabola. The rules are exemplified on integrands for which we know the exact solutions. Thus, it is possible to show convincingly that the approximations yield satisfactory results. Also, it is possible to see that, as the number of ordinates – i.e. the number of points at which the integral is evaluated – increases, the error decreases. The number of ordinates must be limited for practical reasons. This is possible because it is sufficient to maintain a precision consistent with measurements or other calculations. Simpson's rule yields, on one hand, results closer to the exact value. On the other hand, it imposes a serious constraint: the number of subintervals must be even.

By applying the rule of integration over one interval we obtain one number. In Naval Architecture, it is sometimes necessary to have a set of numbers that describe the integral curve as a function of the independent variable, i.e.

$$I(x) = \int_a^x f(x) \mathrm{d}x, \quad a \le x \le b$$

This integral with variable upper limit can be obtained with the aid of an elegant algorithm described in this chapter.

The shape of curves encountered in Naval Architecture can be such that over certain intervals, generally towards their ends, it may be necessary to use smaller subintervals of integration. We then use intermediate ordinates. In the case of a waterline, these ordinates are intermediate stations.

In the lines plan, some lines can terminate within a subinterval, and not at the end of the subinterval. For example, by construction the design waterline usually begins at the aft perpendicular AP, and ends at the forward perpendicular FP. Most other waterlines can begin and end between stations. For good approximations of the areas under such curves, while using the initially given subdivision into subintervals, the lines must be corrected yielding reduced ordinates that will be used in the integration.

3.9 Examples

Example 3.1
Calculate the integral

$$\int_0^{45} x^3 \, \mathrm{d}x$$

by the following methods:
(a) analytic, (b) trapezoidal rule, five ordinates, (c) trapezoidal rule, nine ordinates, (d) Simpson's rule, five ordinates and (e) Simpson's rule, nine ordinates.

Solution

(a)

$$\int_0^{45} x^3 \, \mathrm{d}x = \left. \frac{x^4}{4} \right|_0^{45} = 1\,025\,156.25$$

(b) The following values were calculated in MS Excel:

No. of ordinate	Trapezoidal multiplier	x	$f(x)$	Products
(1)	(2)	(3)	(4)	$(5 = 2 \times 4)$
1	1/2	0.00	0.00	0.00
2	1	11.25	1423.83	1423.83
3	1	22.50	11390.63	11390.63
4	1	33.75	38443.36	38443.36
5	1/2	45.00	91125.00	45562.50
Sum	–	–	–	96820.31
Integral			$(45/4)\text{Sum}/3 =$	1089228.52

The error is

$$E = 1\,025\,156.25 - 1\,089\,228.52 = -64\,072.27$$

and the relative error is

$$E_\mathrm{r} = 100 \times \frac{E}{1\,025\,156.25} = -6.25\%$$

(c) The following values were calculated in MS Excel:

No. of ordinate	Trapezoidal multiplier	x	$f(x)$	Products
(1)	(2)	(3)	(4)	$(5 = 2 \times 4)$
1	1/2	0.00	0.00	0.00
2	1	5.63	177.98	177.98
3	1	11.25	1423.83	1423.83
4	1	16.88	4805.42	4805.42
5	1	22.50	11390.63	11390.63
6	1	28.13	22247.31	22247.31
7	1	33.75	38443.36	38443.36
8	1	39.38	61046.63	61046.63
9	1/2	45.00	91125.00	45562.50
Sum	–	–	–	185097.66
Integral			(45/8)Sum/3 $=$	1041174.32

The error is

$$E = 1\,025\,156.25 - 1\,041\,174.32 = -16\,018.07$$

and the relative error is

$$E_\mathrm{r} = 100 \times \frac{E}{1\,025\,156.25} = -1.56\%$$

(d) The following values were calculated in MS Excel:

No. of ordinate	Simpson's multiplier	x	$f(x)$	Products
(1)	(2)	(3)	(4)	$(5 = 2 \times 4)$
1	1	0.00	0.00	0.00
2	4	11.25	1423.83	5695.31
3	2	22.50	11390.63	22781.25
4	4	33.75	38443.36	153773.44
5	1	45.00	91125.00	91125.00
Sum	–	–	–	273375.00
Integral			(45/4)Sum/3 $=$	1025156.25

The error is

$$E = 1\,025\,156.25 - 1\,025\,156.25 = 0$$

and the relative error is

$$E_r = 100 \times \frac{E}{1\,025\,156.25} = 0\%$$

(e) The following values were calculated in MS Excel:

No. of ordinate	Simpson's multiplier	x	$f(x)$	Products
(1)	(2)	(3)	(4)	$(5 = 2 \times 4)$
1	1	0.00	0.00	0.00
2	4	5.63	177.98	711.91
3	2	11.25	1423.83	2847.66
4	4	16.88	4805.42	19221.68
5	2	22.50	11390.63	22781.25
6	4	28.13	22247.31	88989.26
7	2	33.75	38443.36	76886.72
8	4	39.38	61046.63	244186.52
9	1	45.00	91125.00	91125.00
Sum	–	–	–	546750.00
Integral			(45/8)Sum/3 $=$	1025156.25

The error is

$$E = 1\,025\,156.25 - 1\,025\,156.25 = 0$$

and the relative error is

$$E_r = 100 \times \frac{E}{1\,025\,156.25} = 0\%$$

***MATLAB** solution*

(a) Analytic:

```
format long
a = 45^4/4 = 1.025156250000000e+006
```

(b) Trapezoidal rule, five ordinates:

```
x = 0: 45/4: 45;
y = x.^3;
b = trapz(x, y) = 1.089228515625000e+006
error = a - b = -6.4072e+004
percent_error = 100*(a - b)/a = -6.2500 %
```

(c) Trapezoidal rule, nine ordinates:

```
x = 0: 45/8: 45;
y = x.^3;
c = trapz(x, y) = 1.041174316406250e+006
error = a - c = -1.6018e+004
percent_error = 100*(a - c)/a = -1.5625 %
```

(d) Simpson's rule, five ordinates:

```
x = 0: 45/4: 45;
y = x.^3;
d = simp(x', y') = 1.025156250000000e+006
error = a - d = 0
percent_error = 100*(a - d)/a = 0 %
```

(e) Simpson's rule, nine ordinates:

```
x = 0: 45/8: 45;
y = x.^3;
e = simp(x', y') = 1.025156250000000e+006
error = a - e = 0
percent_error = 100*(a - e)/a = 0 %
```

3.10 Exercises

Exercise 3.1
Calculate the integral

$$\int_{-\pi/2}^{\pi/2} \sin x \, dx$$

by the following methods:

(a) analytic, (b) trapezoidal rule, five ordinates, (c) trapezoidal rule, nine ordinates, (d) Simpson's rule, five ordinates and (e) Simpson's rule, nine ordinates. Analyze the errors and explain your results.

Exercise 3.2
Find the trapezoidal multipliers corresponding to integration over the set of stations

$$0, \frac{1}{2}, 1, 1\frac{1}{2}, 2, 3, \ldots, 8, 8\frac{1}{2}, 9, 9\frac{1}{2}, 10$$

Exercise 3.3
Find the Simpson's multipliers corresponding to integration over the set of stations

$$0, \frac{1}{2}, 1, 2, 3, \ldots, 8, 9, 9\frac{1}{2}, 10$$

4
Hydrostatic curves

4.1 Introduction

In the preceding chapter we learnt several methods of numerical integration used in Naval Architecture. In this chapter we are going to apply them to the calculation of areas, centroids, moments of inertia of areas, volumes, and centres of volume. We call these properties **hydrostatic data** and show how to plot them, as functions of draught, in curves that allow further calculations.

Another set of plots consists of **Bonjean curves**; they enable the user to calculate the displacement and the centres of buoyancy for a given waterline, in an upright condition. The waterline can be not only a straight line, as is the case in still water, but also a curve. The latter case can arise when the hull is deflected because of a longitudinal bending moment or thermal expansion, or when the vessel floats in waves. The vessel is said to be in a **hogging condition** if the keel is concave downwards, and in a **sagging condition** if the keel is concave upwards.

All the properties mentioned above are represented as functions of draught. Certain functional relationships exist between some of those curves. Three such properties are described in this chapter.

Another subject dealt with in this chapter is that of **affine hulls**, i.e. hulls obtained from given ship lines by multiplying by the same **scale factor** all dimensions parallel to an axis of coordinates. The properties of an affine hull can be derived by simple formulae from the properties of the parent hull.

Within this chapter we use the following notations:

i station number, as in the lines drawing;

j station number defined such that the distance from the origin of x-coordinates is $j \, \delta L$;

x_i x-coordinate of station i;

y_i half-breadth of station i on a given waterline;

α_i integration multiplier for station i; for Simpson's rule we assume that the common factor $1/3$ is included in α_i;

δL subinterval of integration along the x-axis;

δT subinterval of integration along the z-axis;

For the above definitions we have, obviously, $j = 0$ in the origin of coordinates.

4.2 The calculation of hydrostatic data

4.2.1 Waterline properties

In this section, we refer to Figure 4.1 and assume that all waterlines are symmetric about the centreline. This assumption is true for almost all ships in upright condition.

We calculate the waterplane area, of a given waterline, as

$$A_{\mathrm{W}} = 2 \int_a^b y \, \mathrm{d}x \approx 2 \left(\sum_{i=n_1}^{n_2} \alpha_i y_i \right) \delta L \qquad (4.1)$$

where the waterline begins at station n_1, with $x = a$, and ends at station n_2, with $x = b$.

The **moment of the waterplane area about a transverse axis** passing through the origin of coordinates is

$$M_x = 2 \int_a^b xy \, \mathrm{d}x \approx 2 \left(\sum_{i=n_1}^{n_2} \alpha_i x_i y_i \right) \delta L = 2 \left(\sum_{i=n_1}^{n_2} \alpha_i j_i y_i \right) \delta L^2 \qquad (4.2)$$

Leaving the indexes n_1 and n_2, we write the **x-coordinate of the centre of flotation** of the given line as

$$x_{\mathrm{F}} = \frac{M_x}{A_{\mathrm{W}}} = \frac{2 \left(\sum \alpha_i j_i y_i \right) \delta L^2}{2 \left(\sum \alpha_i y_i \right) \delta L} = \frac{\left(\sum \alpha_i j_i y_i \right)}{\left(\sum \alpha_i y_i \right)} \delta L \qquad (4.3)$$

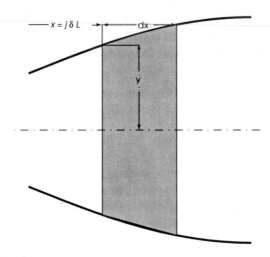

Figure 4.1 An element of waterline area

The notation x_F corresponds to the DIN 81209 standard. The notation used in English-language texts is **LCF**, an acronym for **longitudinal centre of flotation**. The corresponding curve is shown in Figure 4.2. To calculate the transverse **moment of inertia of the waterplane area**, i.e. the moment of inertia about the centreline, we first write the moment of inertia of the elemental area shown in grey in Figure 4.1:

$$dI_T = \frac{(2y)^3 \, dx}{12} = \frac{2}{3} y^3 \, dx \tag{4.4}$$

Then, the moment of inertia of the whole waterplane equals

$$I_T = \int_a^b \frac{2}{3} y^3 \, dx \approx \frac{2}{3} \left(\sum_{i=0}^n \alpha_i y_i^3 \right) \delta L \tag{4.5}$$

The **moment of inertia of the waterplane area about a transverse axis** passing through the origin of coordinates is calculated as

$$I_y = 2 \int_a^b x^2 y \, dx \approx 2 \left(\sum_{i=0}^n \alpha_i x_i^2 y_i \right) \delta L = 2 \left(\sum_{i=0}^n \alpha_i j_i^2 y_i \right) \delta L^3 \tag{4.6}$$

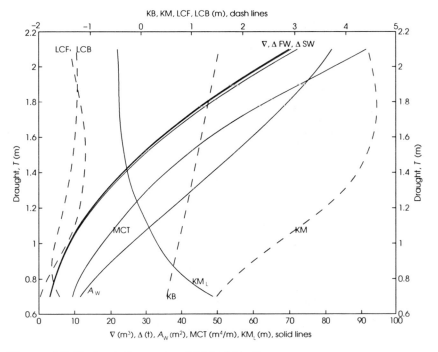

Figure 4.2 Hydrostatic curves of Ship *Lido 9*

In Subsection 2.8.1, we learnt that, for small angles of inclination, the initial and the inclined waterlines intersect themselves along a line passing through the centre of flotation (barycentric axis). For longitudinal inclinations, that is trim, in intact condition, this is almost always true. Therefore, we are interested in finding the moment of inertia of the waterplane area about the transverse barycentric axis. We find this moment, called **longitudinal moment of inertia**, by using a theorem on the parallel translation of the axes of coordinates

$$I_{\mathrm{L}} = I_y - x_{\mathrm{F}}^2 A_{\mathrm{W}} \tag{4.7}$$

The geometrical properties of the waterplane area can be conveniently calculated in a spreadsheet such as that shown in Table 4.1. The table contains the data of the lowest waterline in Figure 1.11 and it was calculated in MS Excel.

The final results are obtained by using the sums in Table 4.1 as follows:

$$
\begin{aligned}
\delta L \;&=\; 0.893\,\mathrm{m} \\
A_{\mathrm{W}} \;&=\; 2 \times 0.893 \times 9.507 = 16.98\,\mathrm{m}^2 \\
LCF \;&=\; \frac{-4.124}{9.507} \times 0.893 = -0.387\,\mathrm{m} \\
I_{\mathrm{T}} \;&=\; \frac{2}{3} \times 13.091 \times 0.893^2 = 7.79\,\mathrm{m}^4 \\
I_y \;&=\; 2 \times 50.058 \times 0.893^3 = 71.29\,\mathrm{m}^4 \\
I_{\mathrm{L}} \;&=\; 71.29 - (-0.387)^2 \times 16.98 = 68.75\,\mathrm{m}^4
\end{aligned}
$$

Table 4.1 A waterline sheet

Station No.	Trapezoidal multiplier, α_i	Half-breadth, y_i	Levers, j_i	Functions of area, $\alpha_i y_i$	Functions of moments, $\alpha_i j_i y_i$	Functions of I_x, $\alpha_i j_i^2 y_i$	Cubes of half-breadth, y_i^3	Functions of I_T, $\alpha_i y_i^3$
1	2	3	4	$5 = 2 \times 3$	$6 = 5 \times 4$	$7 = 6 \times 4$	$8 = 3^3$	$9 = 2 \times 8$
0	1/2	0.000	−5	0.000	0.000	0.000	0.000	0.000
1	1	0.900	−4	0.900	−3.600	14.400	0.729	0.729
2	1	1.189	−3	1.189	−3.567	10.701	1.681	1.681
3	1	1.325	−2	1.325	−2.650	5.300	2.326	2.326
4	1	1.377	−1	1.377	−1.377	1.377	2.611	2.611
5	1	1.335	0	1.335	0.000	0.000	2.379	2.379
6	1	1.219	1	1.219	1.219	1.219	1.811	1.811
7	1	1.024	2	1.024	2.048	4.096	1.074	1.074
8	1	0.749	3	0.749	2.247	6.741	0.420	0.420
9	1	0.389	4	0.389	1.556	6.224	0.059	0.059
10	1/2	0.000	5	0.000	0.000	0.000	0.000	0.000
Sums	–	–	–	9.507	−4.124	50.058	–	13.091

We recommend the reader to check the plausibility of the results by comparing them with the data of the circumscribed rectangle. For example, the area of this rectangle is

$$2 \times 1.377 \times 8.928 = 24.559 \, \mathrm{m}^2$$

that is greater than the waterplane area, and so it should be.

Table 4.1 requires a few explanations. The fourth line contains column numbers. An expression like $5 = 2 \times 3$ means that the numbers in column 5 are the products of numbers in column 2 by numbers in column 3. Similarly, the numbers in column 7 are the products of numbers in column 6 by numbers in column 4. This means that the number 14.400, for example, is obtained by one multiplication, namely $-3.600 \times (-4)$, and not by two multiplications and a squaring operation, $1 \times (-4)^2 \times (0.900)$. Proceeding in this way we spare computer resources and reduce the possibilities of errors. Another simplification results from the use of the factors j, most of them integers. Multiplication by integers is easier when carried out manually, and it is not affected by numerical errors. Thus, instead of multiplying the products in column 5 by x-distances that are 'real' numbers (fractional values), and introduce numerical errors at each station, we multiply by integers. Then, the sums of the products in columns 6 and 7 are multiplied only once by the length, and the square of the length of the subinterval of integration, δL, which can be a real number.

Let us make a final comment on the use of electronic spreadsheets for calculations such as those in Table 4.1. The values of half-breadths, y_i, are entered only once, in column 3, although they are repeatedly used in all calculations. In this way, we reduce the possibilities of errors that can occur when entering a number. Moreover, if we must change the value of a half-breadth, we do it in one place only, and the change spreads automatically over the whole table.

Instead of using an electronic spreadsheet, such as MS Excel, one can write a programme in a suitable language, for example, MATLAB. Such a programme can be useful if the calculations are chained with other computer operations. For the reasons explained above, we recommend to write the programme following the principles used in the waterline sheet shown in Table 4.1.

4.2.2 Volume properties

We can obtain the displacement volume corresponding to a given draught, T_0, by integrating 'vertically' the waterplane areas from the lowest hull point to the given draught:

$$\nabla = \int_0^{T_0} A_W \, \mathrm{d}z \approx \left(\sum_{i=1}^{i_T} \alpha_i A_{Wi} \right) \delta T \tag{4.8}$$

The moment of the displacement volume above the base line can also be obtained by 'vertical' integration:

$$M_B = \int_0^{T_0} T A_W \, dz \approx \left(\sum_{i=1}^{i_T} \alpha_i z_i A_{Wi} \right) \delta T = \left(\sum_{i=1}^{i_T} \alpha_i j_i A_{Wi} \right) \delta^2 T \quad (4.9)$$

where z_i is the z-coordinate of the ith waterline and j_i the number of the waterline counted from the baseline.

From Eqs. (4.8) and (4.9), we calculate the vertical coordinate of the centre of buoyancy, z_B, as

$$z_B = \frac{M_B}{\nabla} \approx \frac{\left(\sum_{i=1}^{i_{T_0}} \alpha_i j_i A_{Wi} \right) \delta^2 T}{\left(\sum_{i=1}^{i_{T_0}} \alpha_i A_{Wi} \right) \delta T} = \frac{\left(\sum_{i=1}^{i_{T_0}} \alpha_i j_i A_{Wi} \right)}{\left(\sum_{i=1}^{i_{T_0}} \alpha_i A_{Wi} \right)} \delta T \quad (4.10)$$

The notation z_B is that prescribed in the DIN 81209 standard. The notations common in English-language books are $\overline{KB}$, or VCB, the latter being the acronym of **vertical centre of buoyancy**. The procedure used with Eq. (4.10) yields bad approximations for the lowest waterlines. Therefore, we recommend to neglect the results for the first waterlines. As shown in Section 4.4, we can also calculate the displacement and the vertical centre of buoyancy by 'longitudinal' integration of values read in Bonjean curves.

4.2.3 Derived data

Let us suppose that we know the displacement, Δ_0, corresponding to a given draught, T_0, and we want to find by how many tons that displacement will change if the draught changes by δT cm. Let the waterplane area be A_W m^2 and the water density ρ_W t m^{-3}. For a small draught change, we may neglect the slope of the shell (in other words we assume a wall-sided hull) and we write

$$\delta\Delta = \rho_W A_W \, \delta T$$

If we measure Δ in tons, and δT in centimetres, we obtain

$$\delta\Delta = \rho_W \frac{A_W}{\delta T} \times 100 \quad (4.11)$$

We call the quantity $\rho_W A_W / 100$ **tons per centimetre immersion** and use for it the notation **TPC**. In older, English-language books, we find the notation TPI as an acronym for **tons per inch**. This quantity is calculated from an expression similar to Eq. (4.11), but adapted for English and American units. For SI units

$$TPC = \frac{A_W}{100} \times \rho_W \quad (4.12)$$

where ρ_W should be taken from the Appendix of Chapter 2. The problem posed above can be inverted: find the change in draught, δT, corresponding to a change of displacement, $\delta\Delta$. The obvious answer is

$$\delta T = \frac{\delta\Delta}{TPC}$$

The above calculations yield good approximations as long as the changes $\delta\Delta$ and δT are small. In fact, Eq. (4.11) is a linearization of the relationship between displacement volume and waterplane area.

Trim calculations will be discussed in more detail in Chapter 7. However, as one quantity required for those calculations is derived from hydrostatic data and is usually presented with the latter, we introduce this quantity here. Let us calculate the moment necessary to change the trim by 1 m. If the length between perpendiculars is L_{pp} and is measured in m, the corresponding angle of trim is defined by

$$\arctan\theta = \frac{1}{L_{pp}} \tag{4.13}$$

The notation θ for the angle of trim corresponds to the standards ISO 7463 and DIN 81209-1. At the angle of trim given by Eq. (4.13), the displacement and buoyancy forces are separated by a distance $\overline{GM_L}\sin\theta$, where $\overline{GM_L}$ is the **longitudinal metacentric height** calculated as

$$\overline{GM_L} = \overline{KB} + \overline{BM_L} - \overline{KG}$$

The couple formed by the displacement and buoyancy forces is

$$\Delta\overline{GM_L}\sin\theta$$

For small angles of trim, we assume $\tan\theta \approx \sin\theta$ and then the **moment to change trim by 1 m** is equal to

$$M_{CT} = \frac{\Delta\overline{GM_L}}{L_{pp}} \tag{4.14}$$

where M_{CT} is measured in t m/m, Δ in t, and $\overline{GM_L}$ and L_{pp}, in m. Although the SI unit is the metre, some design offices use the 'moment to change trim by 1 cm'. Then, the value of M_{CT} given by Eq. (4.14) should be divided by 100.

In the first design stages $\overline{KG}$ is not known. As $\overline{BM_L} \gg \overline{KB} - \overline{KG}$, we can assume the approximation $\overline{GM_L} \approx \overline{BM_L}$.

In Table 4.2, calculated with the ARCHIMEDES programme, the moment to change trim is based on the displacement volume, ∇, and is measured in m^4/m. Let us check, for example, the value corresponding to the draught 1.9 m. We rewrite Eq. (4.14) as

$$M_{CT} = \frac{\nabla\overline{BM_L}}{L_{pp}} \tag{4.15}$$

Table 4.2 Hydrostatic data of ship *Lido 9*

Data	Units	Draught							
	m	0.700	0.900	1.100	1.300	1.500	1.700	1.900	2.100
Trim difference by head > 0)	m	0.000	0.000	0.000	0.000	0.000	0.000	0.000	0.000
Volume of displacement	m³	2.998	6.090	11.212	18.669	28.379	40.314	54.197	69.825
LCB Fwd of midship	m	−1.599	−1.747	−1.600	−1.446	−1.329	−1.268	−1.246	−1.266
KB	m	0.506	0.660	0.819	0.973	1.120	1.263	1.401	1.536
Waterline area	m²	11.529	20.221	31.449	42.998	54.183	64.708	74.088	81.810
LCF	m	−1.973	−1.648	−1.298	−1.150	−1.092	−1.137	−1.259	−1.388
Long moment of inertia	m⁴	144.830	218.207	334.093	469.420	642.827	857.657	1129.524	1416.003
Moment to change trim	m⁴/m	9.344	14.078	21.554	30.285	41.473	55.333	72.872	91.355
Transverse moment of inertia	m⁴	2.950	9.364	25.814	55.665	93.061	134.428	171.925	201.990
Longitudinal, KM	m	48.813	36.491	30.615	26.117	23.772	22.538	22.242	21.815
Transverse, KM	m	1.490	2.198	3.121	3.955	4.400	4.598	4.574	4.429
Block coefficient, CB	–	0.110	0.126	0.149	0.177	0.216	0.261	0.301	0.342
Waterline coefficient, CW	–	0.296	0.377	0.461	0.531	0.620	0.712	0.783	0.841
Midship coefficient, CM	–	0.069	0.124	0.172	0.220	0.280	0.344	0.398	0.444
Prismatic coefficient, CP	–	–	–	0.870	0.807	0.773	0.758	0.758	0.770

and calculate

$$M_{\mathrm{CT}} = \frac{54.197(22.242 - 1.401)}{15.5} = 72.872\,\mathrm{m^4/m}$$

This is exactly the value appearing in Table 4.2.

4.2.4 Wetted surface area

We call **wetted surface** area the hull area in contact with the surrounding water. When we speak about a certain value of the wetted surface area we mean the value corresponding to a given draught. We need this quantity when we calculate the **ship resistance**, i.e. the force by which the water opposes the forward motion of the ship. Besides this, the protection against corrosion, be it active or passive, depends on the value of the wetted surface area. The methods used to calculate the wetted surface area can be extended to the evaluation of the shell area up to any given height. The total shell area is needed for a preliminary estimation of the weight of shell plates and the weight of paint.

In the past, the wetted surface area was calculated as the area of the hull **expansion**. In simple terms, to do this one has to 'open' the hull surface and lay it flat on a plane. This operation can be done exactly for certain surfaces called *developable* (see Chapter 13) such as the surfaces of cubes, cylinders or cones. Many hull surfaces are not developable, for some only the **middlebody** is developable. Then, the Naval Architect must be satisfied with an approximation, such as described in Comstock (1967, pp. 39–41). Recent computer programmes for Naval Architecture calculate the wetted surface area by methods of differential geometry. Approximate formulae for calculating the wetted surface area of many ship types can be found in the literature of speciality. If the chosen hull belongs to a series of models tested in a towing tank, the wetted surface area is usually included in the data supplied by the experimenting institution.

4.3 Hydrostatic curves

Table 4.2 shows the hydrostatic data of the Ship *Lido 9*, for draughts between 0.7 and 2.1 m, as calculated by the ARCHIMEDES programme. The data appear at discrete draught intervals. It is usual to represent those data also as **hydrostatic curves** that allow interpolation at any required draught. Such curves are part of the documentation that must be onboard, for use by deck officers in calculations required for the operation of the vessel. Many ships are provided today with board computers that store the input data of the vessel and enable the officers to calculate immediately any data they need. Even in those cases the hydrostatic curves and the knowledge to use them should be present for emergency cases in which the computer fails.

There are no universally accepted standards for plotting hydrostatic data and we can find a wide variety of 'styles'. For our purposes we choose a simple model that can be accommodated in the space of a textbook page, but still shows the major features common to all representations. The curves are plots of functions of the draught, T, at constant trim and heel. In general, the trim equals zero (ship on even keel), but it is possible to plot hydrostatic curves for any given, non-zero trim. The heel is almost always zero. The hydrostatic curves represent data calculated for parallel waterplanes. Romance languages use a short, elegant term for this situation. For instance, in French one talks about 'carènes isoclines', while Italian uses the term 'carene isocline'.

Let us refer to Figure 4.2. The draught axis is vertical, positive upwards. The various properties are measured horizontally, each at its own scale, so that all curves can be contained in the same paper format. In our example, the curves of volume of displacement, ∇, displacement in fresh water, ΔFW, displacement in salt water, ΔSW, waterplane area, A_{W}, moment to change trim by one metre, MCT, and longitudinal metacentre above keel, KM_{L}, are measured along the lower scale that is to be read as 0–100 m^3, 0–100 t, 0–100 m^2, 0–100 m^4/m, or 0–100 m, respectively. The vertical centre of buoyancy, $\overline{KB}$, the transverse metacentre above keel, $\overline{KM}$, the longitudinal centre of flotation, LCF, and the longitudinal centre of buoyancy, LCB, are measured along the upper scale graduated from -2 to 5 m. To simplify things, we plot the coefficients of form, C_{B}, C_{M}, C_{P}, and C_{WL} in another graph shown in Figure 4.3.

Let us return to the volume and displacement values represented in hydrostatic curves. The displacement volume, ∇, is usually the volume of the moulded hull.

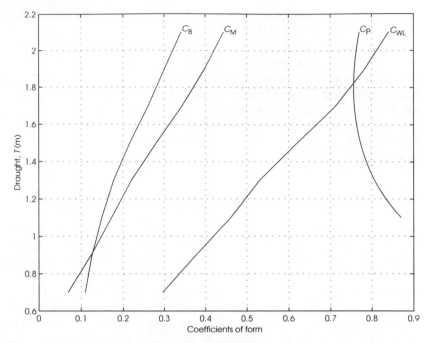

Figure 4.3 Coefficients of form of Ship *Lido 9*

The displacements in fresh and in salt water should be **total displacements** that include the displacements of shell plates and appendages. Appendages found in all kinds of ships include rudders, propellers, propeller shafts and struts, bilge keels, and roll fins. The sonar domes of warships are also appendages if they do not appear in the lines drawing and are not directly taken into account in hydrostatic calculations. The volumes of tunnels that accommodate bow thrusters should be subtracted from the volume of the moulded, submerged hull when calculating total displacements.

American literature recommends to calculate separately the volumes and moments of shell plates and appendages, and to add them to those of the moulded hull. This procedure requires detailed knowledge of all appendages and shell plates, an information not available in early design stages. An approximate, simple method consists in adding a certain percentage to the moulded displacement volume. This amounts to multiplying the moulded volume by a **displacement factor** that is the sum of surrounding-water density and the relative part of appendages and shell plates. Examples of values found in European projects are

$$\Delta_{FW} = (1.000 + 0.0008)\nabla = 1.008\nabla$$

for a vessel displacing a few hundred tons, and

$$\Delta_{FW} = (1.000 + 0.0005)\nabla = 1.005\nabla$$

for larger vessels. The corresponding displacements in salt water of density $1.025\,\mathrm{t\,m^{-1}}$ are

$$\Delta_{\mathrm{FW}} = (1.025 + 0.0008)\nabla = 1.033\nabla$$
$$\Delta_{\mathrm{FW}} = (1.025 + 0.0005)\nabla = 1.030\nabla$$

To understand why the additional percentage decreases with increasing volume, let us remember that volumes increase like the cubes of dimensions, while surfaces, such as those of plates and rudders, increase like the square of dimensions.

4.4 Bonjean curves and their use

Figure 4.4 shows the midship section of the Ship *Lido 9* in solid, thick line. Its equation is of the form

$$z = f(y)$$

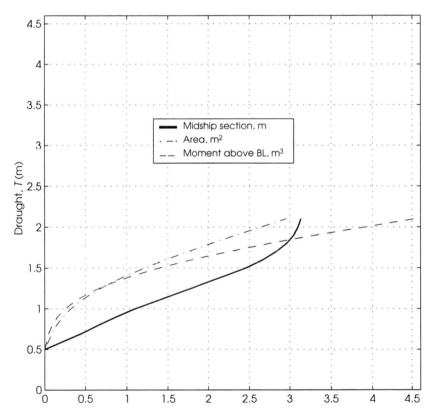

Figure 4.4 The meaning of Bonjean curves

The Bonjean curves are defined by the equations

$$A = \int_{\text{keel}}^{T} y \, dz \qquad (4.16)$$

$$M = \int_{\text{keel}}^{T} zy \, dz \qquad (4.17)$$

The first integral yields the **sectional area** as function of draught, while the second integral is the moment of the sectional area about the base line, also as function of draught.

Figure 4.5 shows the Bonjean curves of the Ship *Lido 9*. The ship outline appears in solid line. The scales along the x-axis and the T-axis are different, otherwise the drawing format would be too long. The waterline appearing in the figure corresponds to the mean draught 2 m and the trim 0.5 m. The data corresponding to this line are written in Table 4.3; they are read along horizontal lines starting from the intersection of the waterline with the corresponding station. For example, the midship station is intersected by the waterline a small

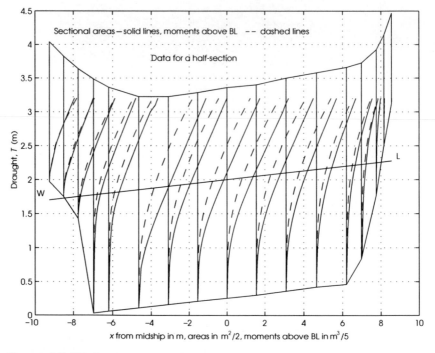

Figure 4.5 Bonjean curves of Ship *Lido 9*

Table 4.3 A Bonjean sheet

Station No.	Trapezoidal multiplier, α_i	Lever arm, j_i	Sectional area, A_i	Functions of area, $\alpha_i A_i$	Moment from MS, $\alpha_i j_i A_i$	Moment above BL, M_i	Functions of moment, $\alpha_i M_i$
1	2	3	4	$5 = 2 \times 4$	$6 = 3 \times 5$	7	$8 = 2 \times 7$
0	1/4	−5	0.23	0.06	−0.29	0.37	0.09
$\frac{1}{2}$	1/2	−4.5	0.68	0.34	−1.53	0.93	0.47
1	3/4	−4	1.04	0.78	−3.12	1.45	1.09
2	1	−3	2.99	2.99	−8.98	3.83	3.83
3	1	−2	2.21	2.21	−4.41	3.11	3.11
4	1	−1	2.62	2.62	−2.62	3.76	3.76
5	1	0	2.68	2.68	0.00	3.93	3.93
6	1	1	2.42	2.42	2.42	3.68	3.68
7	1	2	2.09	2.09	4.17	3.29	3.29
8	1	3	1.51	1.51	4.54	2.47	2.47
9	3/4	4	0.87	0.65	2.60	1.45	1.09
$9\frac{1}{2}$	1/2	4.5	0.43	0.21	0.97	0.77	0.38
10	1/4	5	0.03	0.01	0.04	0.06	0.01
Sum	–	–	–	18.57	−6.21	–	27.20

distance below 2 m. On the horizontal corresponding to that draught, we read the sectional area

$$A = 2 \times 1.34 = 2.68 \, \text{m}^2$$

and the moment about BL

$$M = 5 \times 0.79 = 3.95 \, \text{m}^3$$

To simplify the example, we neglect the data corresponding to the ship volumes aft of Station 0 and forward of Station 10. The respective values are indeed very small and by not including them we can integrate by either trapezoidal or Simpson's rule without having to correct multipliers.

The final results are calculated as follows:

$$\delta L = 1.55 \, \text{m}$$
$$\nabla = 2 \times 1.55 \times 18.57 = 57.57 \, \text{m}^3$$
$$LCB = \frac{-6.21}{18.57} \times 1.55 = -0.518$$
$$\overline{KB} = \frac{27.20}{18.57} = 1.465 \, \text{m}$$

4.5 Some properties of hydrostatic curves

In Section 4.2, we have learnt how to calculate hydrostatic data and represent them as functions of draught, for constant trim and heel. In addition to the functional dependence of each variable on draught, certain relationships between various curves hold true. In this section, we are going to show three of them. Relationships between various hydrostatic curves have been used to check visually the correctness of hydrostatic calculations. Such checks were obviously very useful when calculations were carried out by tedious manual procedures, even if with the help of mechanical integrating devices. Today we rely on the correctness and accuracy of computer programmes, but errors can still occur when plotting the output of the programmes by means of procedures that are not part of the hydrostatic programme. Besides this, reading this section is a good exercise in understanding the meaning of hydrostatic data.

In Figure 4.6, we consider a floating body with the waterline WL. The centre of buoyancy is B, the displacement volume is ∇, and the waterplane area A_W. The moment of the submerged volume about the plane zOy is $x_B \nabla$, the moment of the submerged volume about the plane xOz equals $y_B \nabla$, and the moment of the submerged volume about the plane yOx is $z_B \nabla$.

Let us assume that the waterline rises by a draught change equal to δT. Then, the submerged volume increases by $\delta V = A_W \, \delta T$. Let the centre of the additional volume be F. When δT tends to zero, F tends to the centroid of the waterline, that is to the *centre of flotation*. The moments of the submerged volume change by

$$
\begin{aligned}
\delta(x_B \nabla) &= x_F A_W \, \delta T \\
\delta(y_B \nabla) &= y_F A_W \, \delta T \\
\delta(z_B \nabla) &= z_F A_W \, \delta T
\end{aligned}
\tag{4.18}
$$

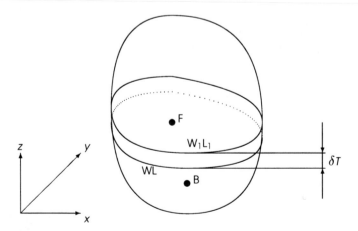

Figure 4.6 Properties of 'isocline' floating bodies

Expanding the left-hand side of Eq. (4.18), we obtain

$$
\begin{aligned}
\nabla \delta x_{\mathrm{B}} + x_{\mathrm{B}}\,\delta\nabla &= x_{\mathrm{F}}A_{\mathrm{W}}\,\delta T \\
\nabla \delta y_{\mathrm{B}} + y_{\mathrm{B}}\,\delta\nabla &= y_{\mathrm{F}}A_{\mathrm{W}}\,\delta T \\
\nabla \delta z_{\mathrm{B}} + z_{\mathrm{B}}\,\delta\nabla &= z_{\mathrm{F}}A_{\mathrm{W}}\,\delta T
\end{aligned}
\tag{4.19}
$$

Dividing by $\delta\nabla = A_{\mathrm{W}}\,\delta T$, rearranging terms and passing to infinitesimal quantities we rewrite Eq. (4.19) as

$$
\begin{aligned}
x_{\mathrm{F}} - x_{\mathrm{B}} &= \frac{\mathrm{d}(x_{\mathrm{B}})}{\mathrm{d}T}\frac{\nabla}{A_{\mathrm{W}}} \\
y_{\mathrm{F}} - y_{\mathrm{B}} &= \frac{\mathrm{d}(y_{\mathrm{B}})}{\mathrm{d}T}\frac{\nabla}{A_{\mathrm{W}}} \\
z_{\mathrm{F}} - z_{\mathrm{B}} &= \frac{\mathrm{d}(z_{\mathrm{B}})}{\mathrm{d}T}\frac{\nabla}{A_{\mathrm{W}}}
\end{aligned}
\tag{4.20}
$$

Let us consider the first of Eq. (4.20) and assume $x_{\mathrm{F}} = x_{\mathrm{B}}$. The left-hand side becomes zero and so must be the right-hand side. The displacement volume, ∇, can equal zero only at the lowest point of the hull, where A_{W} is also zero. For any other point for which $x_{\mathrm{F}} = x_{\mathrm{B}}$ we must have $\mathrm{d}(x_{\mathrm{B}})/\mathrm{d}T = 0$. In the hydrostatic curves this means

> Where the curve of the longitudinal centre of flotation, *LCF*, intersects the curve of the longitudinal centre of buoyancy, *LCB*, the tangent to the latter curve is vertical.

We can easily verify this result on the curves shown in Figure 4.2. It may happen that for some ship forms the two curves do not intersect. We turn now to the third part in Eq. (4.20). Except at the lowest point of the hull, z_{F} can never equal z_{B}. It results that $\mathrm{d}(z_{\mathrm{B}})/\mathrm{d}T$ can never be zero in any other place than the lowest point of the hull. In other words, the $\overline{KB}$ curve can have a vertical tangent only in its origin. This result, which can be checked in Figure 4.2, corresponds to our intuition. Indeed, as the draught increases, so must do the z-coordinate of the centre of buoyancy. Finally, let us divide, side by side, the first part in Eq. (4.20) by the last. We obtain

$$
\frac{x_{\mathrm{F}} - x_{\mathrm{B}}}{z_{\mathrm{F}} - z_{\mathrm{B}}} = \frac{\mathrm{d}x_{\mathrm{B}}}{\mathrm{d}z_{\mathrm{B}}}
\tag{4.21}
$$

and remark that $z_{\mathrm{F}} = T$. To discover the geometric significance of Eq. (4.21) let us examine Figure 4.7 built with data of the Ship *Lido 9*; it contains a plot of z_{B} as function of x_{B}, or, with alternative notations, $\overline{KB}$ values as function of

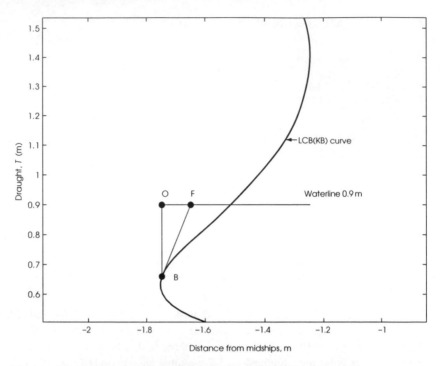

Figure 4.7 Relationship between centre of flotation and centre of buoyancy

LCB. The point F is the centre of flotation corresponding to a draught of 0.9 m, and the point B, the centre of flotation for the same draught. We can write

$$\tan(\angle OBF) = \frac{\mathrm{d}x_B}{\mathrm{d}z_B} = \frac{\overline{OF}}{\overline{BO}} = \frac{x_F - x_B}{z_F - z_B} \tag{4.22}$$

which proves Eq. (4.21).

Conventional ships are symmetric about their centrelines. Then, $y_F = y_B = 0$ and so is $\mathrm{d}(y_B)/\mathrm{d}T$. For floating bodies that have no port-to-starboard symmetry, it makes sense to divide the second part in Eq. (4.20) by the third and obtain

$$\frac{y_F - y_B}{z_F - z_B} = \frac{\mathrm{d}y_B}{\mathrm{d}z_B} \tag{4.23}$$

Then, a property similar to that derived for the $z_B(x_B)$-curve can be found for the $z_B(y_B)$-curve. Examples of floating bodies that have no port-to-starboard symmetry are ships with permanent list caused by unsymmetrical loading, by negative metacentric height or by flooding.

4.6 Hydrostatic properties of affine hulls

One way of obtaining new ship lines is to derive them by a transformation, or mapping, of some suitable, given lines. The simplest transformation is that in which all dimensions parallel to one of the coordinate axes are multiplied by the same scale factor. Thus, let all dimensions parallel to the x-axis be multiplied by r_x, all dimensions parallel to the y-axis be multiplied by r_y, and those parallel to the z-axis, by r_z. We say then that we obtain a hull **affine** to the **parent** hull, or that we obtain the new hull forms by an **affine transformation**. In fact, the transformations we are talking about are a subset of what is known in geometry as **affine mappings**, more specifically **scaling**.

The case $r_x = r_y = r_z = r$ is particularly important; it yields a hull that is **geometrically similar** to the parent hull. For example, the lines of a ship and those of her model used in basin tests are geometrically similar. The results of basin tests can be extrapolated to the actual ship size by the laws of **dimensional analysis**. When designing a new ship with the hull geometrically similar to that of a successful ship one spares the costs of basin tests.

Modern computer programmes for hydrostatic calculations can find the properties of affine hulls by changing only the scale factors, r_x, r_y, r_z, and not all the input, that is the offsets. However, it is possible to derive the hydrostatic properties of affine hulls by simple explicit expressions based on geometric considerations. This possibility is important because it permits a straightforward calculation of the scale factors that would yield the desired properties. In this section, we are going to show with a few examples how to proceed. The reader may continue by solving the exercises proposed at the end of the chapter.

Let us begin by calculating the displacement volume, ∇_1, of a new hull affine to a parent hull having the displacement volume

$$\nabla_0 = \int \int \int \mathrm{d}x \mathrm{d}y \mathrm{d}z \qquad (4.24)$$

The dimensions of the new hull change as $x_1 = r_x x$, $y_1 = r_y y$, $z_1 = r_z z$ so that the new displacement volume is

$$\nabla_1 = \int \int \int \mathrm{d}x_1 \, \mathrm{d}y_1 \, \mathrm{d}z_1 = \int \int \int r_x \, \mathrm{d}x \cdot r_y \, \mathrm{d}y \cdot r_z \, \mathrm{d}z = r_x r_y r_z \nabla_0$$

$$(4.25)$$

For geometrically similar hulls, we obtain $\nabla_1 = r^3 \nabla_0$.

With a similar reasoning, we can find that for scale factors r_x, r_y, r_z, the new longitudinal centre of buoyancy is $LCB_1 = r_x LCB_0$, the new longitudinal centre of flotation is $LCF_1 = r_x LCF_0$, and the new vertical centre of buoyancy, $\overline{KB_1} = r_z \overline{KB_0}$.

4.7 Summary

The methods of numerical integration learnt in Chapter 3 can be applied to the calculation of hydrostatic data. The properties of waterplanes are the area, A_W, the longitudinal coordinate of the centre of flotation, LCF, the transverse moment of inertia, I_T, and the longitudinal moment of inertia, I_L. These properties can be conveniently calculated in an electronic spreadsheet. The input data, i.e. the half-breadths, are entered only once, but are used repeatedly in all calculations. The various quantities are calculated each in a separate column. In the same line, corresponding to one station, the calculations are chained in a way that reduces the number of required arithmetic operations.

The hydrostatic data are calculated at discrete intervals, as functions of draught, for constant trim and heel. These data are plotted in hydrostatic curves that allow interpolation. These curves are part of the documentation that must be present aboard the ship and are used in calculations related to the operation of the vessel. A summary of the data yielded by hydrostatic calculations is given in Table 4.4.

The Bonjean curves represent the areas of transverse sections, and the moments of these areas above the baseline, as functions of draught. Bonjean curves are used in the processing of the results of inclining experiment (see Chapter 7).

Certain relationships exist between some hydrostatic curves. They can be used for visual checks of the hydrostatic curves.

One method of deriving new ship lines consists in multiplying by the same scale factor all dimensions parallel to an axis of coordinates. Such transformations are called affine transformations. The properties of a new hull, affine to a

Table 4.4 A summary of hydrostatic calculations

Quantity	Notation	How to calculate it
Waterplane area	A_W	Eq. (4.1)
Moment of waperplane area about a transverse axis	M_x	Eq. (4.2)
Longitudinal centre of flotation	x_F, LCF	Eq. (4.3)
Transverse moment of inertia of waterplane area	I_T	Eq. (4.5)
Moment of inertia of waterplane about a transverse axis	I_y	Eq. (4.6)
Longitudinal moment of inertia of waterplane area	I_L	Eq. (4.7)
Displacement volume	∇	Eq. (4.8), Table 4.3
Moment of displacement volume above base line	M_B	Eq. (4.9)
Vertical centre of buoyancy	$z_B, \overline{KB}, VCB$	Eq. (4.10), Table 4.3
Longitudinal centre of buoyancy	x_B, LCB	Table 4.3
Tons per centimetre immersion	TCP	Eq. (4.12)
Moment to change trim by one metre	M_{CT}	Eqs. (4.14) and (4.15)

parent hull, can be derived from the properties of the parent hull by simple algebraic expressions. An important case of affine transformation is that in which the three scale factors are equal. Two hulls related in this way are geometrically similar. Affine transformations do not change the coefficients of form.

4.8 Example

Example 4.1 – the displacement of geometrically similar hulls
Let us assume, for example, that we derive a geometrically similar hull by increasing the linear dimensions with the scale factor 10%. The displacement volume increases by the factor $1.1^3 = 1.331$. For a quick estimate, let us write

$$\nabla_1 = r^3 \nabla_0 \tag{4.26}$$

Taking natural logarithms of both sides yields

$$\ln \nabla_1 = 3 \ln r + \ln \nabla_0 \tag{4.27}$$

We differentiate both sides considering ∇_0 constant and obtain

$$\frac{d\nabla_1}{\nabla_1} = 3\frac{dr}{r} \tag{4.28}$$

We have now a rule for simple and quick approximation: the percent change of the displacement volume equals three times the percent ratio change.

4.9 Exercises

Exercise 4.1
Modify Table 4.1 for a coordinate origin in AP and repeat the calculation. Check the results with those shown in the original table.

Exercise 4.2
Modify Table 4.1 for use with Simpson's rule and repeat the calculations.

Exercise 4.3
Verify the values of M_{CT} in Table 4.2, for the draughts 1.8 and 2.1 m, using the displacement–volume, $\overline{KB}$ and $\overline{KM_L}$ values shown there.

Exercise 4.4
Modify Table 4.3 for a coordinate origin in AP and repeat the calculation. Check the results with those shown in the original table.

Exercise 4.5
Modify Table 4.3 for use with Simpson's rule and repeat the calculations.

Exercise 4.6
Using the data of Ship *Lido 9* plot a figure in which you can verify the property described by Eq. (4.21) for the draught values 1.7, 1.9 and 2.1 m.

Exercise 4.7
Show that affine transformations leave the coefficients of form unchanged. In mathematical terminology, **the coefficients of form are invariants of affine transformations**.

Exercise 4.8
Show that for affine hulls, the metacentric radius, $\overline{BM}$, behaves like B^2/T.

6

Simple models of stability

6.1 Introduction

In Chapter 5 we learnt how to calculate and how to plot the righting arm in the curve of statical stability. It may be surprising that for a very long period the metacentric height and the curve of righting arms were considered sufficient for appreciating the ship stability. We do not proceed so in other engineering fields. As pointed out by Wendel (1965), one first finds out the resistance to ship advance and only afterwards dimensions the engine. Also, we first calculate the load on a beam and only afterwards we dimension it. Similarly, we should determine the **heeling moments** and then compare them with the righting moment. It was only at the beginning of the twentieth century that Middendorf proposed such a procedure for large sailing ships. His book, *Bemastung und Takelung der Schiffe*, was first published in Berlin, in 1903, and it contained the first proposal for a ship-stability criterion. In 1933, Pierrottet wrote in a publication of the test basin in Rome that the stability of a ship must be assessed by comparing the heeling moments with the righting moment. He detailed his proposal in 1935, in a meeting of INA, but had no immediate followers. Thus, in 1939 Rahola published in Helsinki his doctoral thesis; it was based on extensive statistics and a very profound analysis of the qualities of stable and unstable vessels. Rahola proposed then a stability criterion that considered only the metacentric height and the curve of the righting arm. The Naval-Architectural community appreciated Rahola's work and his proposal was used, indeed, as a stability standard and stood at the basis of stability regulations issued later by national and international authorities.

It was only after the Second World War that the issue of comparing heeling and righting arms was brought up again. German researchers used then a very appropriate term: *Lever arm balance* (Hebelarm Bilanz). Eventually, newer stability regulations made compulsory the comparison of lever arms and we show in this chapter how to do it.

Heeling moments can be caused by wind, by the centrifugal force developed in turning, by transverse displacements of masses, by towing or by the lateral pull developed in cables that connect two vessels during the transfer of loads at sea. In Chapter 5 we have shown that, when the ship heels at constant displacement, it is sufficient to consider the righting arm as an indicator of stability. Then, to assess the ship stability it is necessary to compare the righting arm with a **heeling**

arm. According to the DIN-ISO standard, we note the heeling arm by the letter ℓ and indicate the nature of the righting arm by a subscript. To obtain a generic heeling arm, ℓ_g, corresponding to a heeling moment, M_g, we divide that moment by the ship weight

$$\ell_g = \frac{M_g}{g\Delta} \tag{6.1}$$

where Δ is the displacement mass and g, the acceleration due to gravity. In older practice it has been usual to measure the displacement in unit of force. Then, instead of Eq. (6.1) one had to use

$$\ell_g = \frac{M_g}{\Delta}$$

Much attention should be paid to the system of units used in calculation. From now on we constantly use the displacement mass in calculations. At this point it may seem that we defined the heeling arm as above just to be able to compare the righting arm with a quantity having the same physical dimensions (and units!). In Section 6.7, we prove that this definition is mathematically justified.

In Figure 6.1, we superimposed the curve of a generic heeling arm, ℓ_g, over the curve of the righting arm, $\overline{GZ}$. For almost all positive heeling angles shown in the plot the righting arm is positive. We define the righting arm as positive if when the ship is heeled to starboard, the righting moment tends to return it

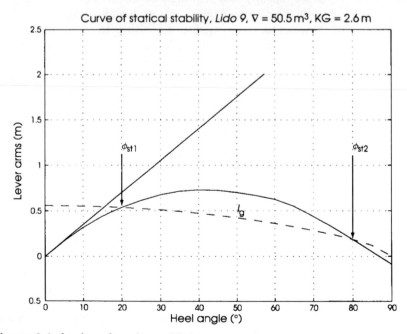

Figure 6.1 Angles of static equilibrium

towards port. In the same figure the heeling arm is also positive, meaning that the corresponding heeling moment tends to incline the ship towards starboard. What happens if the ship heels in the other direction, i.e. with the port side down? Let us extend the curve of statical stability by including negative heel angles, as in Figure 6.2. The righting arms corresponding to negative heel angles are negative. For a ship heeled towards port, the righting moment tends, indeed, to return the vessel towards starboard, therefore it has another sign than in the region of positive heel angles. The heeling moment, however, tends in general to heel the ship in the same direction as when the starboard is down and, therefore, it is positive. Summarizing, the righting-arm curve is symmetric about the origin, while the heeling-arm curves are symmetrical about the lever-arm axis.

In this chapter we present simplified models of various heeling arms, models that allow reasonably fast calculations. Approximate as they may be, those models stand at the basis of regulations that specify the stability requirements for various categories of ships. In most cases, practice has shown that ships complying with the regulations were safe. The requirements themselves are explained in Chapters 8 and 10. By the end of this chapter, we briefly explain why the simplifying assumptions are necessary in Naval-Architectural practice.

We can appreciate the stability of a vessel by comparing the righting arm with the heeling arm as long as the heeling moment is applied gradually and inertia forces and moments can be neglected. When the heeling moment appears suddenly, as caused, for example, by a gust of wind, one has to compare the

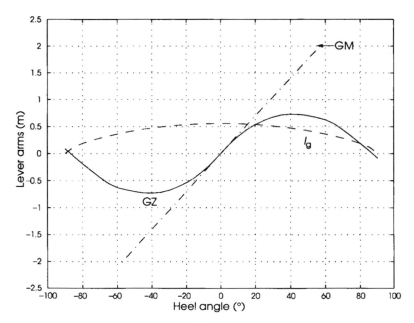

Figure 6.2 Curve of statical stability extended for heeling towards both ship sides

heeling energy with the work done by the righting moment. Such situations are discussed in the section on dynamical stability. In continuation we show how moving loads, solid or liquid, endanger the ship stability, and we develop formulae for calculating the reduction of stability. Other situations in which the stability is endangered are those of grounding or positioning in dock. We show how to predict the moment in which those situations may become critical. This chapter also discusses the situations in which a ship sails with a negative metacentric height.

6.2 Angles of statical equilibrium

Figure 6.1 shows the curve of a heeling arm, ℓ_g, superimposed on the curve of the righting arm, $\overline{GZ}$. In general, those curves intersect at two points; they are noted here as ϕ_{st1} and ϕ_{st2}. Both points correspond to positions of statical equilibrium because at both points the righting arm and the heeling arm are equal, and, therefore, the righting moment and the heeling moment are also equal. Only the first point corresponds to a position of stable equilibrium, while the second point corresponds to a situation of unstable equilibrium. In this section, we give an intuitive proof of this statement; for a rigorous proof, see Section 6.7.

Let us first consider the equilibrium in the first static angle, ϕ_{st1}, and assume that some perturbation causes the ship to heel further to starboard by a small angle, $\delta\phi$. When the perturbation ceases at the angle $\phi_{st1} + \delta\phi$, the righting arm is larger than the heeling arm, returning thus the ship towards its initial position, at the angle ϕ_{st1}. Conversely, if the perturbation causes the ship to heel towards port, to an angle $\phi_{st1} - \delta\phi$, when the perturbation ceases the righting arm is smaller than the heeling arm, so that the ship returns towards the initial position, ϕ_{st1}. This situation corresponds to the definition of stable equilibrium given in Section 2.4.

Let us see now what happens at the second angle of equilibrium, ϕ_{st2}. If some perturbation causes the ship to incline further to starboard, the heeling arm will be larger than the righting arm and the ship will capsize. If the perturbation inclines the ship towards port, after its disappearance the righting arm will be larger than the heeling arm and the ship will incline towards port regaining equilibrium at the first static angle, ϕ_{st1}. We conclude that the second static angle, ϕ_{st2}, corresponds to a position of unstable equilibrium.

6.3 The wind heeling arm

We use Figure 6.3 to develop a simple model of the heeling moment caused by a beam wind, i.e. a wind perpendicular to the centreline plane. In this situation the wind heeling arm is maximal. In the simplest possible assumption the wind generates a force, F_V, that acts in the centroid of the lateral projection of the

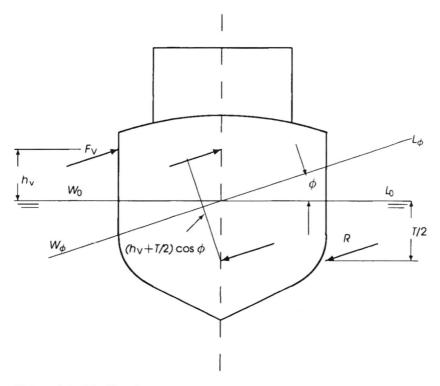

Figure 6.3 Wind heeling arm

above-water ship surface, and has a magnitude equal to

$$F_V = p_V A_V$$

where p_V is the wind pressure and A_V is the area of the above-mentioned projection of the ship surface. Let us call A_V **sail area**.

Under the influence of the force F_V the ship tends to drift, a motion opposed by the water with a force, R, equal in magnitude to F_V. To simplify calculations we assume that R acts at half-draught, $T/2$. The two forces, F_V and R, form a torque that inclines the ship until the heeling moment equals the righting moment. The value of the heeling moment in the upright condition is $p_V A_V(h_V + T/2)$, where h_V is the height of the sail-area centroid above $W_0 L_0$. The heeling arm in upright condition is

$$\ell_V(0) = \frac{p_V A_V(h_V + T/2)}{g\Delta}$$

How does the heeling arm change with the heeling angle? In the case of a 'flat' ship, i.e. for $B = 0$, the area exposed to the wind varies proportionally to $\cos \phi$. In Figure 6.3, we show that for a flat ship the forces F_V and R would act in the centreline plane, both horizontally, i.e. parallel to the inclined waterline $W_\phi L_\phi$.

Then, the lever arm of the torque would be proportional to $\cos \phi$. Summing up, the wind heeling arm equals

$$\ell_V(\phi) = \frac{p_V A_V \cos \phi}{g\Delta} \left(h_V + \frac{T}{2} \right) \cos \phi = \frac{p_V A_V (h_V + T/2)}{g\Delta} \cos^2 \phi$$

(6.2)

This is the equation proposed by Middendorf and that prescribed by the stability regulations of the US Navy; it can be found in more than one textbook on Naval Architecture where it is recommended for all vessels. The reader may feel some doubts about the strong assumptions accepted above. In fact, other regulatory bodies than the US Navy adopted wind-heeling-arm curves that do not behave like $\cos^2 \phi$. The respective equations are described in Chapters 8 and 10. Our own critique of the above model, and a justification of some of its underlying assumptions, are presented in Section 6.12.

The wind pressure, p_V, is related to the wind speed, V_W, by

$$p_V = \frac{1}{2} c_w \rho V_W^2$$

(6.3)

where c_w is an aerodynamic resistance coefficient and ρ is the air density. The coefficient c_w depends on the form and configuration of the sail area. An average value for c_w is 1.2. Wegner (1965) quotes a research that yielded $1.00 \le c_w \le 1.36$, and two Japanese researchers, Kinohita and Okada, who measured c_w values ranging between 0.95 and 1.24. Equation (6.3) shows that the wind heeling arm is proportional to the square of the wind speed. In this section, we considered the wind speed as constant over all the sail area. This assumption is acceptable for a fast estimation of the wind heeling arm. However, we may know from our own experience that wind speed increases with height above the water surface. Some stability regulations recognize this phenomenon and we show in Chapters 8 and 10 how to take it into account. Calculations with variable wind speed, i.e. considering the **wind gradient**, yield lower, more realistic heeling arms for small vessels whose sail area lies mainly in the low-wind-speed region. It may be worth mentioning that engineers take into account the wind gradient in the design of tall buildings and tall cranes.

6.4 Heeling arm in turning

When a ship turns with a linear speed V, in a circle of radius R_{TC}, a centrifugal force, F_{TC}, develops; it acts in the centre of gravity, G, at a height $\overline{KG}$ above the baseline. From mechanics we know that

$$F_{TC} = \Delta \frac{V^2}{R_{TC}}$$

Under the influence of the force F_{TC} the ship tends to drift, a motion opposed by the water with a reaction R. To simplify calculations, we assume again that the water reaction acts at half-draught, i.e. at a height $T/2$ above the baseline. The two forces, F_{TC} and R, form a torque whose lever arm in upright condition is $(KG - T/2)$. For a heeling, flat ship this lever arm is proportional to $\cos \phi$. Dividing by the displacement force, we obtain the **heeling lever of the centrifugal force in turning circle**:

$$\ell_{TC} = \frac{1}{g} \frac{V^2}{R_{TC}} \left(\overline{KG} - \frac{T}{2} \right) \cos \phi \tag{6.4}$$

The speed V to be used in Eq. (6.4) is the speed in turning, smaller than the speed achieved when sailing on a straight line path. The turning radius, R_{TC}, and the speed in turning, V, are not known in the first stages of ship design. If results of basin tests on a ship model, or of sea trials of the ship, or of a sister ship, are available, they should be substituted in Eq. (6.4). The stability regulations of the German Navy, BV 1033, provide formulae for approximations to be used in the early design stages of naval ships (see Chapter 10). A discussion of this subject can be found in Wegner (1965). This author uses a non-dimensional factor

$$C_D = \left(\frac{V_D}{V_0} \right)^2 \frac{L_{pp}}{R_{TC}} \tag{6.5}$$

where V_D is the ship speed in turning and V_0, the speed on a straight line path. Substituting into Eq. (6.4) yields

$$\ell_{TC} = C_D \frac{V_0^2}{gL_{pp}} \left(\overline{KG} - \frac{T}{2} \right) \cos \phi \tag{6.6}$$

Quoting *Handbuch der Werften*, Vol. VII, Wegner shows that for 95% of 80 cargo ships the values of C_D ranged between 0.19 and 0.25. For a few trawlers the values ranged between 0.30 and 0.35.

6.5 Other heeling arms

A dangerous situation can arise if many passengers crowd on one side of the ship. There are two cases when passengers can do this: when attracted by a beautiful seascape or when scared by some dangerous event. In the latter case, passengers can also be tempted to go to upper decks. The resulting heeling arm can be calculated from

$$\ell_P = \frac{np}{\Delta} (y \cos \phi + z \sin \phi) \tag{6.7}$$

where n is the number of passengers, p, the average person mass, y, the horizontal coordinate of the centre of gravity of the crowd and z, the vertical translation of said centre. The second term between parentheses accounts for the virtual metacentric-height reduction. Wegner (1965) recommends to assume that up to seven passengers can crowd on a square metre, that the average mass of a passenger plus some luggage is 80 kg, and that the height of a passenger's centre of gravity above deck is 1.1 m. Smaller values are prescribed by the regulations described in Chapters 8 and 10. Wegner recommends to include in the deck area all areas that can be occupied by panicking people, e.g. tables, benches and skylights. Other heeling moments can occur when a tug tows a barge. The barge can drift and then the tension in the towing cable can be decomposed into two components, one parallel to the tug centreline and the other perpendicular to the first. The latter component can cause capsizing of the tug. The process is very fast and there may be no survivors. To avoid this situation tugs must be provided with quick-release mechanisms that free instantly the towing cable. Lateral forces also appear when fishing vessels tow nets or when two vessels are connected by cables during replenishment-at-sea operations. Special provisions are made in stability regulations for the situations mentioned above; they are presented in Chapters 8 and 10. **Icing** is a phenomenon known to ship crews sailing in very cold zones. The accumulation of ice has a double destabilizing effect: it raises the centre of gravity and it increases the sail area. The importance of ice formation should not be underestimated. For example, Arndt (1960a) cites cases in which blocks of ice 1 m thick developed on a poop deck, or walls of 60 cm of ice formed on the front surface of a bridge. Therefore, stability regulations take into account the effect of ice.

6.6 Dynamical stability

Until now we assumed that the heeling moments are applied gradually and that inertial moments can be neglected. Shortly, we studied **statical stability**. Heeling moments, however, can appear, or increase suddenly. For example, wind speed is usually not constant, but fluctuates. Occasionally, sudden bursts of high intensity can occur; they are called *gusts*. As another example, loosing a weight on one side of a ship can cause a sudden heeling moment that sends down the other side. In the latter cases we are interested in **dynamical stability**. It is no more sufficient to compare righting with heeling arms; we must compare the energy of the heeling moment with the work done by the opposing righting moment. It can be easily shown that the energy of the heeling moment is proportional to the area under the heeling-arm curve, and the work done by the righting moment is proportional to the area under the righting-arm curve. To prove this, let us remember that the work done by a force, F, which produces a motion from x_1 to x_2 is equal to

$$W = \int_{x_1}^{x_2} F \, \mathrm{d}x \tag{6.8}$$

If the path of the force F is an arc of circle of radius r, the length of the arc that subtends an angle $d\phi$ is $dx = r\, d\phi$. Substituting into Eq. (6.8), we get

$$W = \int_{\phi_1}^{\phi_2} Fr\, d\phi = \int_{\phi_1}^{\phi_2} M\, d\phi \qquad (6.9)$$

where M is a moment.

A ship subjected to a sudden heeling moment M_h, applied when the roll angle is ϕ_1, will reach for an instant an angle ϕ_2 up to which the energy of the heeling moment equals the work done by the righting moment, so that

$$W = \int_{\phi_1}^{\phi_2} \frac{M_h}{g}\, d\phi = \int_{\phi_1}^{\phi_2} \overline{\Delta GZ}\, d\phi \qquad (6.10)$$

or

$$\int_{\phi_1}^{\phi_2} \frac{M_h}{g\Delta}\, d\phi = \int_{\phi_1}^{\phi_2} \overline{GZ}\, d\phi \qquad (6.11)$$

This condition is fulfilled in Figure 6.4 where the area under the heeling-arm curve is $A_2 + A_3$, and the area under the righting-arm curve is $A_1 + A_3$. As A_3 is common to both areas, the condition is reduced to $A_1 = A_2$. Moseley is quoted

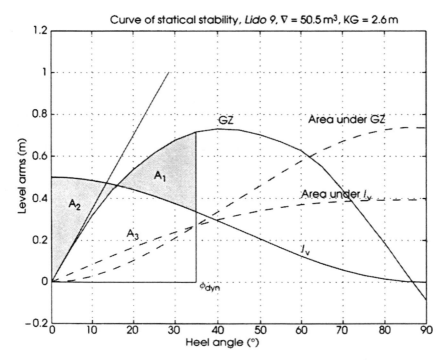

Figure 6.4 Dynamical stability

for having proposed the calculation of dynamical stability as early as 1850. It took several marine disasters and many years until the idea was accepted by the Naval-Architectural community.

In Figure 6.4, we marked with ϕ_{dyn} the maximum angle reached by the ship after being subjected to a gust of wind. An elegant way to find this angle is to calculate the areas under the curves as functions of the heel angle, ϕ, plot the resulting curves and find their points of intersection. The algorithm for calculating the integrals with variable upper limit is described in Section 3.4.

In Figure 6.4, we assumed that the gust of wind appeared when the ship was in an upright condition, i.e. $\phi_1 = 0$. As shown in Figure 6.5, the situation is less severe if $\phi_1 > 0$, and more dangerous if $\phi_1 < 0$. In both graphs the maximum dynamical angle is found by plotting the curve

$$\int_{\phi_1}^{\phi} \overline{GZ}\, d\phi - \int_{\phi_1}^{\phi} l_v\, d\phi$$

and looking for the point where it crosses zero. An analogy with a swing (or a pendulum) is illustrated in Figure 6.6. Many readers may have tried to accelerate a swing by pushing it periodically. Thus, they may know that a push given in position (a) sends the swing to an angle that is much larger than the angle achieved by pushing at position (b). Moreover, pushing the swing while it is in position

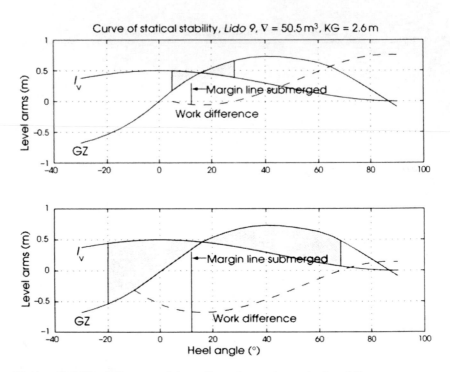

Figure 6.5 The influence of the roll angle on dynamical stability

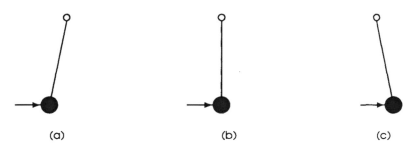

Figure 6.6 Swing analogy

(c) proves very difficult. The physical explanation is simple. In position (a), the energy transferred from the push is added to the potential energy accumulated by the swing, the latter energy acting to return the swing rightwards. In position (c), the potential energy accumulated by the swing tends to return it to position (b), opposing thus the energy impacted by the push. The influence of the roll angle on dynamical stability is taken into consideration by some stability regulations (see Chapter 8).

6.7 Stability conditions – a more rigorous derivation

We describe the dynamics of heeling by Newton's equation for rotational motion

$$J\frac{\mathrm{d}^2\phi}{\mathrm{d}t^2} + g\Delta\overline{GZ} = M_{\mathrm{H}} \tag{6.12}$$

where J is the **mass moment of inertia** of the ship, Δ, the mass displacement and M_{H}, a heeling moment. The mass moment of inertia is calculated as the sum of the products of masses by the square of their distance from the axis of roll

$$J = \sum_{i=1}^{n} \left(y_i^2 + z_i^2\right) m_i \tag{6.13}$$

where y_i is the transverse and z_i is the height coordinate of the mass i. In the SI system, we measure J in $\mathrm{m}^2\,\mathrm{t}$. In Eq. (6.12) we neglected damping and added mass, terms briefly introduced in Section 6.12 and used in Chapter 12. We also neglect the **coupling** of heeling with other ship motions.

Let us multiply by $\mathrm{d}\phi$ on both sides of Eq. (6.12), we obtain

$$J\frac{\mathrm{d}^2\phi}{\mathrm{d}t^2}\mathrm{d}\phi + g\Delta\overline{GZ}\,\mathrm{d}\phi = M_{\mathrm{H}}\,\mathrm{d}\phi \tag{6.14}$$

We transform the factor that multiplies J as follows:

$$\frac{\mathrm{d}^2\phi}{\mathrm{d}t^2}\mathrm{d}\phi = \frac{\mathrm{d}\dot\phi}{\mathrm{d}t}\mathrm{d}\phi = \frac{\mathrm{d}\dot\phi}{\mathrm{d}\phi}\cdot\frac{\mathrm{d}\phi}{\mathrm{d}t}\cdot\mathrm{d}\phi = \dot\phi\,\mathrm{d}\dot\phi \tag{6.15}$$

and integrate between an initial angle, ϕ_0, and a final angle, ϕ_f,

$$J \int_{\phi_0}^{\phi_f} \dot{\phi} \, d\dot{\phi} + g\Delta \int_{\phi_0}^{\phi_f} \overline{GZ} \, d\phi = \int_{\phi_0}^{\phi_f} M_H \, d\phi \tag{6.16}$$

The result is

$$\frac{1}{2} J(\dot{\phi}^2(\phi_f) - \dot{\phi}^2(\phi_0)) = \int_{\phi_0}^{\phi_f} M_H \, d\phi - g\Delta \int_{\phi_0}^{\phi_f} \overline{GZ} \, d\phi \tag{6.17}$$

The left-hand side of the above equation represents kinetic energy, K. In the position of stable equilibrium the potential energy has a minimum. As the sum of potential and kinetic energies is constant in a system such as that under consideration (it is a **conservative** system), the kinetic energy has a maximum in the position of statical equilibrium. The conditions for maximum are

$$\frac{dK}{d\phi} = 0, \qquad \frac{d^2 K}{d\phi^2} < 0 \tag{6.18}$$

Substituting K by the right-hand side of Eq. (6.17) and differentiating, we obtain

$$\frac{M_H}{g\Delta} = \overline{GZ}$$
$$\frac{d(M_H/g\Delta)}{d\phi} < \frac{d\overline{GZ}}{d\phi} \tag{6.19}$$

The first part of Eq. (6.19) shows that at the point of statical equilibrium the righting arm equals the heeling arm. The second part of the equation shows that at the point of stable statical equilibrium the slope of the righting arm must be greater than that of the heeling arm. This is a rigorous proof that the first static angle corresponds to a position of stable equilibrium, while the second does not.

Until now we looked for the angles of statical equilibrium. Let us examine the dynamical phenomenon, i.e. the behaviour of the heeling angle, ϕ, as function of time. The conditions for maximum dynamic angle are

$$\dot{\phi} = 0, \qquad \ddot{\phi} < 0 \tag{6.20}$$

Substituting the first part of Eq. (6.20) in Eq. (6.16), we obtain

$$\int_{\phi_0}^{\phi_f} \overline{GZ} \, d\phi = \int_{\phi_0}^{\phi_f} \frac{M_H}{g\Delta} \, d\phi \tag{6.21}$$

Equation (6.21) represents the condition of equality of the areas under the righting and the heeling arms. The second part of Eq. (6.20) when applied to Eq. (6.12) yields the condition

$$\overline{GZ} > \frac{M_H}{g\Delta} \tag{6.22}$$

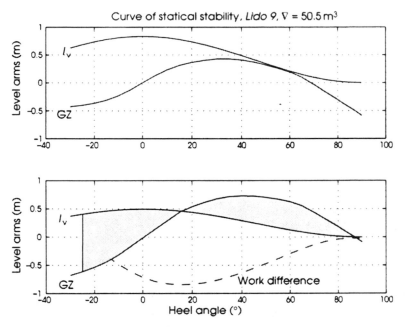

Figure 6.7 Two limiting cases of instability

Figure 6.7 shows two limiting cases. In the upper plot the first part of condition (6.19) is fulfilled, while the second is not. Therefore, in this case there is no angle of stable statical equilibrium and the ship is lost. In the lower Figure 6.7 the areas under the righting-arm and the heeling-arm curves are equal, but condition (6.22) is not fulfilled. Therefore, under the shown gust of wind the ship will capsize.

6.8 Roll period

For small angles of heel, and assuming $M_{\mathrm{H}} = 0$, we rewrite Eq. (6.12) as

$$J\frac{\mathrm{d}^2\phi}{\mathrm{d}t^2} + g\Delta\overline{GM}\phi = 0 \qquad (6.23)$$

We say that this equation describes **unresisted roll**. We define the **mass radius of gyration**, i_{m}, by

$$J = i_{\mathrm{m}}^2\Delta \qquad (6.24)$$

Substituting the above expression into Eq. (6.23) and rearranging yields

$$\frac{\mathrm{d}^2\phi}{\mathrm{d}t^2} + \frac{g\overline{GM}}{i_{\mathrm{m}}^2}\phi = 0 \qquad (6.25)$$

With the notation

$$\omega_0 = \sqrt{\frac{g\overline{GM}}{i_{\mathrm{m}}^2}} \tag{6.26}$$

the solution of this equation is of the form $\phi = \Phi \sin(\omega_0 t + \epsilon)$, where ω_0 is the **natural angular frequency** of roll and ϵ, the **phase**. The **natural period of roll** is the inverse of the roll frequency, f_0, defined by

$$\omega_0 = 2\pi f_0$$

Using algebra, we obtain

$$T_0 = 2\pi \frac{i_{\mathrm{m}}}{\sqrt{g\overline{GM}}} \ (\mathrm{s}) \tag{6.27}$$

We conclude that the larger the metacentric height, $\overline{GM}$, the shorter the roll period, T_0. If the roll period is too short, the oscillations may become unpleasant for crew and passengers, and can induce large forces in the transported cargo. Tangential forces developed in rolling are proportional to the angular acceleration, i.e. to

$$\frac{\mathrm{d}^2\phi}{\mathrm{d}t^2} = -\Phi\omega_0^2 \sin(\omega_0 t + \epsilon)$$

a quantity directly proportional to $\overline{GM}$.

Thus, while a large metacentric height is good for stability, it may be necessary to impose certain limits on it. IMO (1995), for example, referring to ships carrying timber on deck, recommends to limit the metacentric height to maximum 3% of the ship breadth (Paragraph 4.1.5.5). Norby (1962) quotes researches carried out by Kempf, in Germany, in the 1930s. Kempf defined a non-dimensional rolling factor, $T\sqrt{g/B}$, and, on the basis of extensive statistics found that:

- for values of Kempf's factor under 8 the ship motions are **stiff**;
- for values between 8 and 14 the roll is **comfortable**;
- for factor values above 14 the motions are **tender**.

When the motions become too tender the ship master will worry because the metacentric height may be too low.

The exact value of the radius of gyration, i_{m}, can be calculated from Eq. (6.24) and requires the knowledge of all masses and their positions. This knowledge is not always available, certainly not in the first phases of ship design. Therefore, it is usual to assume that the radius of gyration, i_{m}, is proportional to the ship breadth, B, i.e.

$$i_{\mathrm{m}} = aB$$

Let us define

$$c = 2a = \frac{2i_m}{B}$$

Substituting into Eq. (6.27), we obtain

$$T_0 = \frac{\pi c B}{\sqrt{g\overline{GM}}} \tag{6.28}$$

As $\pi \approx \sqrt{g}$, we can rewrite Eq. (6.28) as

$$T_0 \approx \frac{cB}{\sqrt{\overline{GM}}} \tag{6.29}$$

Rose (1952) quotes the following c values: large cargo and passenger vessels, 0.85; small cargo and passenger vessels, 0.77; loaded ore carriers, 0.81; tugs, 0.76; wide barges, 0.79. These values are based on old-type vessels. More recently, Costaguta (1981) recommends to take $i_m = B/3$ for merchant ships, and $c = 0.8$–0.9 for round-bilge, motor yachts. Some shipyards use $i_m = 0.35B$.

For actual ships, i_m can be obtained experimentally by measuring the roll period. When i_m is known, Eq. (6.27) can be used to control the metacentric height by measuring the roll period. This can be done automatically and on-line with the help of modern technology. Wendel (1960b) describes an instrument that did the job many years ago. The use of the roll period as a stability indicator is discussed, for example, by Norby (1962) and Jons (1987).

Normally, the roll period is measured in the still water of a harbour, and the ship is tied by the stern and by the aft to minimize other motions than roll. When measuring the roll period in a seaway it is necessary to distinguish between the ship own period and the period of encounter with the waves (see Jons, 1987 and Chapter 9).

6.9 Loads that adversely affect stability

6.9.1 Loads displaced transversely

In Figure 6.8, we consider that a mass m, belonging to the ship displacement Δ, is moved transversely a distance d. A heeling moment appears and its value, for any heeling angle ϕ is $dm \cos \phi$. As a result, the ship centre of gravity G moves to a new position, G_1, the distance $\overline{GG_1}$ being equal to

$$\overline{GG_1} = \frac{dm}{\Delta} \tag{6.30}$$

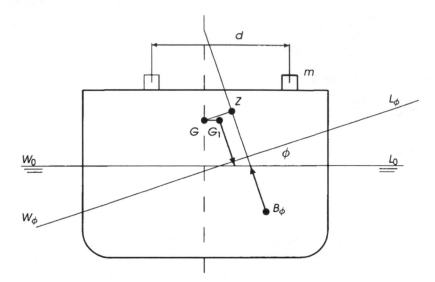

Figure 6.8 The destabilizing effect of a mass moved transversely

and the righting arm is reduced to an **effective** value

$$\overline{GZ}_{\text{eff}} = \overline{GZ} - \frac{dm}{\Delta} \cos \phi \qquad (6.31)$$

We invite the reader to check that the above reduction occurs when the vessel is inclined towards the side to which the mass m was moved, while the righting arm increases if the ship is inclined towards the other side.

6.9.2 Hanging loads

In Figure 6.9, we consider a mass m suspended at the end of a rope of length h. When an external moment causes the ship to heel by an angle ϕ, the hanging mass moves transversely a distance $h \tan \phi$, and the ship's centre of gravity moves in the same direction a distance

$$\overline{GG_1} = \frac{hm}{\Delta} \tan \phi \qquad (6.32)$$

In Figure 6.10, we see that the righting arm is reduced from $\overline{GZ}$ to $\overline{G_1 Z_1} = \overline{GZ}_{\text{eff}}$. The effect is the same as if the centre of gravity, G, moved to a higher position, G_V, given by

$$\overline{GG_V} = \frac{\overline{GG_1}}{\tan \phi} = \frac{hm}{\Delta} \qquad (6.33)$$

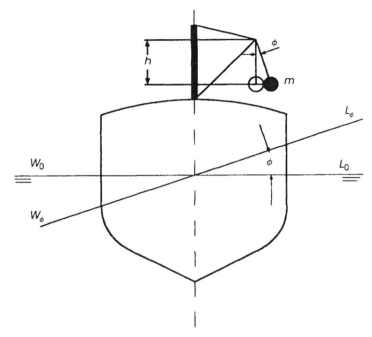

Figure 6.9 Hanging load

As a result, we use for initial-stability calculations a corrected, or **effective metacentric height**

$$\overline{GM}_{\text{eff}} = \overline{GM} - \frac{hm}{\Delta} \tag{6.34}$$

The destabilizing effect appears immediately after raising the load sufficiently to let it move freely. Looking at Eq. (6.34) we see that the metacentric height is reduced by the same amount that would result from raising the load by a distance h. In other words, we can consider that the mass acts in the hanging point.

6.9.3 Free surfaces of liquids

Liquids with free surfaces are a very common kind of moving load. Any engine-propelled vessel needs fuel and lubricating-oil tanks. Tanks are needed for carrying fresh water. The cargo can be liquid; then tanks occupy a large part of the vessel. Tanks cannot be filled to the top. Liquids can have large thermal expansion coefficients and space must be provided to accommodate for their expansion, otherwise unbearable pressure forces may develop. In conclusion, almost all vessels carry liquids that can move to a certain extent endangering thus the ship stability. A partially filled tank is known as a **slack tank**.

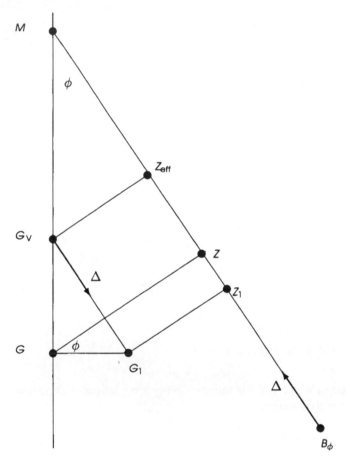

Figure 6.10 Effective metacentric height

Figure 6.11(a) shows a tank containing a liquid whose surface is free to move within a large range of heeling angles without touching the tank top or bottom. Let us consider that the liquid volume behaves like a ship hull and consider the free surface a waterplane. Then, the centre of gravity of the liquid is the buoyancy centre of the liquid hull. Therefore, we use for it the notation b_0. While the ship heels, the centre of gravity of the liquid moves along the curve of the centre of the buoyancy, 'around' the metacentre, m. The horizontal distance between the initial position b_0 and the inclined position b_ϕ is

$$\overline{b_0 m} \tan \phi$$

If v is the volume occupied by the liquid, i_B, the moment of inertia of the liquid surface with respect to the barycentric axis parallel to the axis of heeling and ρ,

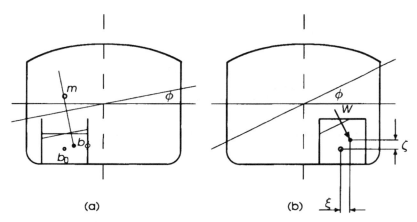

Figure 6.11 The free-surface effect

the liquid density, the heeling moment produced by the inclination of the liquid surface is

$$M_l = \rho v \frac{i_B}{v} \tan \phi = \rho i_B \tan \phi$$

where M_l has the dimensions of mass times length.

As a result, the ship centre of gravity moves transversely a distance equal to

$$\overline{GG_1} = \frac{\rho i_B}{\Delta} \tan \phi \qquad (6.35)$$

By comparison with the preceding section, we conclude that the effective metacentric height is

$$\overline{GM_{\text{eff}}} = \overline{GM} - \frac{\rho i_B}{\Delta} \qquad (6.36)$$

and the **effective righting arm,**

$$\overline{GZ_{\text{eff}}} = \overline{GZ} - \frac{\rho i_B}{\Delta} \sin \phi \qquad (6.37)$$

Instead of considering the free-surface effect as a virtual reduction of the metacentric height and of the righting lever, we can take it into account as the heeling lever of free movable liquids. Its value is

$$\ell_F = \frac{\rho i_B}{\Delta} \qquad (6.38)$$

and the respective curve is proportional to $\sin \phi$. The latter approach is that adopted in the stability regulations of the German Navy.

The reduction of stability caused by the liquids in slack tanks is known as **free-surface effect**. Two of its features must be emphasized:

- the mass of the liquid plays no role, only the moment of inertia of the free surface appears in equations;
- the effect does not depend on the position of the tank.

In general, ships have more than one tank, and different tanks can contain different liquids. The destabilizing effects of all tanks must be summed up when calculating the effective metacentric height

$$\overline{GM}_{\text{eff}} = \overline{GM} - \frac{\sum_{k=1}^{n} \rho_k i_{Bk}}{\Delta} \tag{6.39}$$

and the effective righting arm,

$$\overline{GZ}_{\text{eff}} = \overline{GZ} - \frac{\sum_{k=1}^{n} \rho_k i_{Bk}}{\Delta} \sin \phi \tag{6.40}$$

where n is the total number of tanks.

Often the liquid surface is not free to behave as in Figure 6.11(a) and its shape changes when it reaches the tank top or bottom. Then, we cannot use the equations shown above. The same happens when the heeling angle is large and the forms of the tank such that the shape of the free surface changes in a way that cannot be neglected. In such cases the exact trajectory of the centre of gravity must be calculated. As shown in Figure 6.11(b), the resulting heeling moment is

$$M_\ell = W(\xi \cos \phi + \zeta \sin \phi) \tag{6.41}$$

where W is the liquid mass, ξ is the horizontal distance and ζ is the vertical distance travelled by the centre of gravity.

Some books and articles on Naval Architecture contain tables and curves that allow the calculation of the free-surface effect for various tank proportions. Present-day computer programmes can calculate exactly and quickly the position of the centre of gravity for any heel angle. For example, one can describe the tank form as a hull surface and run the option for cross-curves calculations. Therefore, correction tables and curves are not included in this book.

The free-surface effect can endanger the ship, or even lead to a negative metacentric height. Therefore, it is necessary to reduce the free-surface effect. The usual way to do this is to subdivide tanks by *longitudinal bulkheads*, such as shown in Figure 6.12. If the left-hand figure would refer to a parallelepipedic hull, the moment of inertia of the liquid surface in each tank would be $1/2^3 = 1/8$ that of the undivided tank. Having two tanks, the total moment of inertia, and the corresponding free-surface effect, are reduced in the ratio 1/4. An usual arrangement in tankers is shown in Figure 6.12(b).

Some materials that are not really liquid can behave like liquids. Writes Price (1980), 'Whole fish when carried in bulk in a vessel's hold behave like liquid', and should be considered as such in stability calculations.

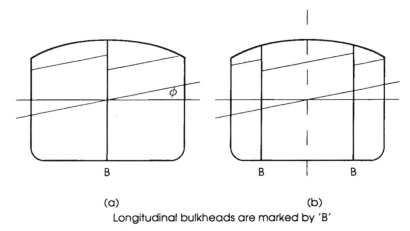

(a)   (b)

Longitudinal bulkheads are marked by 'B'

Figure 6.12 Reducing the free-surface effect

We end this section by noting that transverse bulkheads do not reduce the free-surface effect of slack tanks.

6.9.4 Shifting loads

Shifting loads, also called **sliding loads**, such as grain, coal and sand are a very dangerous type of moving loads. Arndt (1968) lists 31 incidents due to sliding loads, 13 of them leading to sinking, one to abandoning the ship. Those accidents occurred between July 1954 and November 1966. More cases are cited in the literature of speciality. Unlike liquid loads, materials considered in this section do not move continuously during the ship roll. Shifting loads stay in place until a certain roll angle is reached and then they slide suddenly.

The sides of a mass of **granular** materials, like those cited above, are inclined. The angle between the side and the horizontal is called **angle of repose** and is an important characteristic of the material. The angle of repose of most grain loads ranges between 20° and 22°, but for barley it reaches 46° (see Price, 1980). The angles of repose of ores range between 34° for copper from Norway, and 60° for copper from Peru.

Let the angle of repose be ρ_R. During roll, the mass of the granular material stays in place until the heel angle exceeds the angle of repose, i.e. $\phi > \rho_R$. Then, the granular load slides suddenly and its centre of gravity moves horizontally a distance ξ, and up a distance ζ. By analogy with Figure 6.11(b), we can calculate a reduction of the metacentric height equal to

$$\overline{GG_V} = \frac{m_L(\xi \cos \phi + \zeta \sin \phi)}{\Delta}$$

While the ship rolls back, the load does not move until its angle exceeds the angle of repose. Wendel (1960b) describes this process and shows how the reduction of metacentric height can be represented by a loop that reminds the phenomenon of hysteresis known mainly from the theory of magnetism. The accelerations induced by ship motions can cause load shifting at angles that are smaller than the angle of repose. The behaviour of granular materials is further complicated by settling and by variations of humidity. For a detailed discussion, see Arndt (1968).

6.9.5 Moving loads as a case of positive feedback

In all cases of moving loads we can assume that an external moment m_h, caused the ship to heel by an angle ϕ. Consequently, the load moved to the same side producing another heeling moment m_a that is added to the external moment. This process is illustrated in Figure 6.13. Control engineers will recognize in this process an example of **positive feedback**. Following Birbănescu-Biran (1979), we can, indeed, use simple block-diagram techniques to retrieve some of the relationships found above. A simplified development follows; a more rigorous one can be found in the cited reference. Readers familiar with the elements of Control Engineering can understand this section without difficulty; other readers may skip it. However, making a little effort to understand the block diagram in Figure 6.13 can provide more insight into the moving-load effect.

In Figure 6.13, $G(s)$ is the **ship transfer function** and $H(s)$ is the moving-load transfer function. In the forward branch of the ship-load system, the **Laplace transform** of the heel angle $\Phi(s)$ is related to the Laplace transform of the effective heeling moment $M_e(s)$ by

$$\Phi(s) = G(s)M_e(s) \tag{6.42}$$

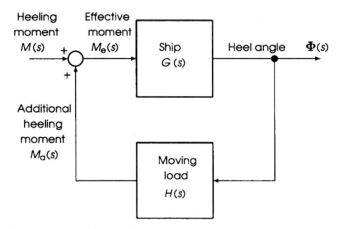

Figure 6.13 Moving loads as a case of positive feedback

The Laplace transform of the additional heeling moment $M_a(s)$, induced by the moving load, is related to the Laplace transform of the heel angle by

$$M_a(s) = H(s)\Phi(s) \tag{6.43}$$

Substituting $M_e(s)$ in Eq. (6.42) by the sum of the moments $M(s)$ and $M_a(s)$ yields

$$\Phi(s) = G(s)(M(s) + H(s)\Phi(s)) \tag{6.44}$$

Finally, the transfer function of the ship-load system is given by

$$\frac{\Phi(s)}{M(s)} = \frac{G(s)}{1 - G(s)H(s)} \tag{6.45}$$

To find the transfer function of the ship, we refer to Eq. (6.25) to which we add a heeling moment, m_e, in the right-hand side

$$\frac{d^2\phi}{dt^2} + \frac{g\overline{GM}}{i_m^2}\phi = \frac{g}{i_m^2}\frac{M_e}{\Delta} \tag{6.46}$$

Applying the Laplace transform, with zero initial conditions and rearranging, we obtain the ship transfer function

$$\frac{\Phi(s)}{M_e(s)} = \frac{g/i_m^2\Delta}{s^2 + (g/i_m^2)\overline{GM}} \tag{6.47}$$

Substitution of the above transfer function into Eq. (6.45) yields

$$\frac{\Phi(s)}{M(s)} = \frac{g/i_m^2\Delta}{s^2 + (g/i_m^2)\left(\overline{GM} - H(s)/\Delta\right)} \tag{6.48}$$

The factor

$$\left(\overline{GM} - \frac{H(s)}{\Delta}\right)$$

is the effective metacentric height.

From the preceding sections, it can be found that the transfer function of a hanging load is $H(s) = mh$, and the transfer function of a free liquid surface is $H(s) = \rho i_B$. Equation (6.48) yields the condition for bounded response:

$$\overline{GM} - \frac{H(s)}{\Delta} > 0$$

Indeed, if this condition is fulfilled, $\phi(t)$ is a sinusoidal function of time with bounded amplitude. If the condition is not fulfilled, the heel angle is given by a hyperbolic sine, a function whose amplitude is not bounded. We retrieved thus, by other means, the famous condition of initial stability. A diagram such as that in Figure 6.13 can be the basis of a SIMULINK© programme for simulating the roll of a ship with moving loads aboard.

6.10 The stability of grounded or docked ships

6.10.1 Grounding on the whole length of the keel

Figure 6.14 shows a ship grounded on the whole length of the keel. If local tide lowers the sea level, at a certain draught the ship will loose stability and capsize. To plan the necessary actions, the ship master must know how much time remains until reaching the critical draught. A similar situation occurs when a ship is laid in a floating dock. While ballast water is pumped out of the dock, the draught of the ship decreases. Props must be fully in place before the critical draught is reached.

In Figure 6.14, we consider that the draught T descended below the value T_0 corresponding to the ship displacement mass Δ. Then, the ship weight is supported partly by the buoyancy force $g\rho\nabla_T$ and partly by the reaction R:

$$g\rho\nabla_T + R = g\Delta \tag{6.49}$$

where ∇_T is the submerged volume at the actual draught T. The ship heels and for a small angle ϕ, the condition of stability is

$$g\rho\nabla_T\overline{KM}\sin\phi > g\Delta\overline{KG}\sin\phi \tag{6.50}$$

or

$$\overline{KM} > \frac{\Delta\overline{KG}}{\rho}\frac{1}{\nabla_T} \tag{6.51}$$

Simplifying we obtain

$$\overline{KM} > \frac{\nabla}{\nabla_T}\overline{KG} \tag{6.52}$$

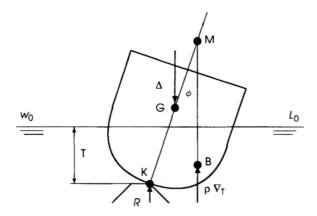

Figure 6.14 Ship grounded on the whole keel length

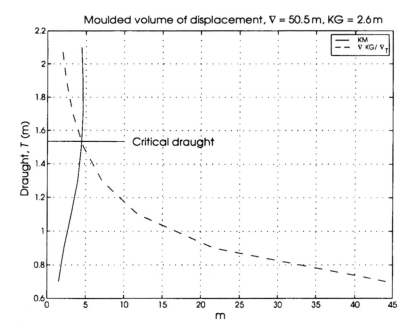

Figure 6.15 Finding the critical draught of a ship grounded on the whole keel length

where ∇ is the displacement volume corresponding to the ship mass Δ. As an example, Figure 6.15 shows the curves $\overline{KM}$ and $\nabla\overline{KG}/\nabla_T$ as functions of draught, i.e. local depth T, for the ship *Lido 9*. The critical draught in this case is 1.53 m.

6.10.2 Grounding on one point of the keel

Figure 6.16 shows a ship grounded on one point of the keel; let this point be P_0. We draw a horizontal line through P_0; let P_1 be its intersection with the vertical passing through the centre of gravity G, and P_3 is the intersection with the vertical passing through the centre of buoyancy B and the metacentre M. Taking moments about the line P_0P_3 we write

$$\rho\nabla_T\overline{P_3M}\sin\phi > \Delta\overline{P_1G}\sin\phi \tag{6.53}$$

or

$$\overline{P_3M} > \frac{\nabla}{\nabla_T}\overline{P_1G} \tag{6.54}$$

The similarity of the triangles P_0MP_3 and $P_0P_2P_1$ lets us write

$$\frac{\overline{P_3M}}{\overline{P_1P_2}} = \frac{\overline{P_3P_0}}{\overline{P_1P_0}} \tag{6.55}$$

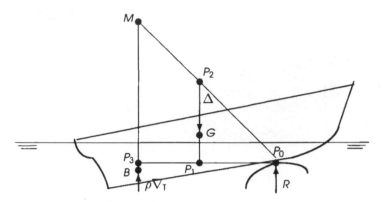

Figure 6.16 Ship grounded on one point of the keel

Taking moments of forces about the point P_0 gives

$$\rho \nabla_T \overline{P_3 P_0} = \Delta \overline{P_1 P_0} \tag{6.56}$$

Combining Eqs. (6.54)–(6.56) yields the condition

$$\overline{P_1 P_2} > \overline{P_1 G} \tag{6.57}$$

In other words, the point P_2 plays the role of metacentre. From Figure 6.16 and Eq. (6.57), we see that pulling the ship to the left increases the distance $\overline{GP_2}$, while pulling the ship to the right reduces it.

6.11 Negative metacentric height

The metacentric height $\overline{GM}$ can become negative if the centre of gravity is too high, or if the influence of moving loads is important. Even with a negative metacentric height, ships with certain forms can still find a position of stable equilibrium at an angle of heel that does not endanger them immediately. An example is shown in Figure 6.17 where the $\overline{GZ}$ curve is based on the data of a small cargo ship built in 1958. The solid line represents the righting-arm curve in ballast, departure condition. Let us assume that for some reason the centre of gravity G moves upwards a distance $\delta \overline{KG} = 0.75\,\text{m}$. The dotted-dashed curve represents the quantity $\delta \overline{KG} \sin \phi$ that must be subtracted from the initial righting-arm curve. The two curves intersect at approximately $10°$ and $55°$. The resulting righting-arm is shown in Figure 6.18. The ship finds a position of stable equilibrium at $\phi_1 \approx 10°$; she sails permanently heeled at this angle called **angle of loll**. Looking again at Figure 6.17, we see that the first intersection of the two

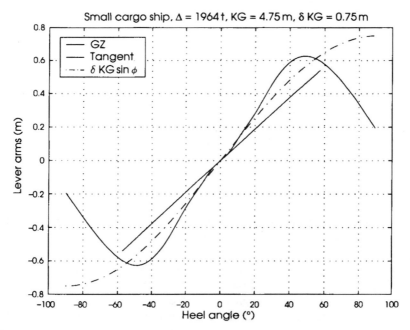

Figure 6.17 Stability with negative metacentric height

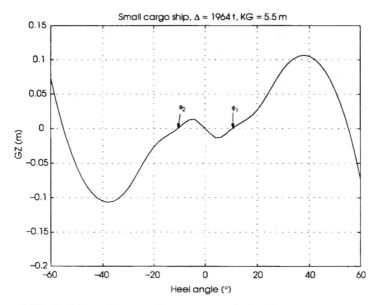

Figure 6.18 Stability with negative metacentric height

curves is possible because the first part of the $\overline{GZ}$ curve lies above the tangent in the origin. It can be shown that the corresponding metacentric evolute has ascending branches at $\phi = 0$.

In Figure 6.18 we can see that, if a ship sailing with a positive angle of loll receives a wave, a small gust of wind, or some other perturbation coming from the starboard, she will incline to the port-side and stay there at a negative angle of loll, $\phi_2 = -\phi_1$. In a seaway, such a ship can oscillate between ϕ_1 and ϕ_2. This kind of abrupt oscillation, different from a continuous roll, is characteristic for negative metacentric heights.

An angle of loll can be corrected only by lowering the centre of gravity, not by moving loads transversely, or by filling ballast tanks on the higher side. Hervieu (1985) proves this in two ways, first by considering the metacentric evolute, next by examining the righting-arm curve. We adopt here the second approach.

We first assume that the ship master tries to correct the loll by moving a mass $m = 2\,t$. As the breadth of the ship is $11.9\,m$, we can assume that the mass is moved a distance $d = 6\,m$ towards port. The correcting arm, $dm \cos \phi/\Delta$, is shown as a dotted line in Figure 6.19. Subtracting this correcting arm from the initial righting-arm curve, we obtain the dashed line. The starboard angle of loll ϕ_3 is smaller than the initial angle ϕ_1, but the port-side angle of loll increases from ϕ_2 to ϕ_4. Also, we see that the area A under the $\overline{GZ}$ curve is somewhat reduced.

Next, we assume that, unsatisfied by the first result, the ship master moves more masses towards port, until $m = 4.25\,t$. Figure 6.20 shows now the limit situation in which the correcting-arm curve is tangent to the initial $\overline{GZ}$ curve. The starboard angle of loll ϕ_3 is smaller than in the previous case, but still not zero. On the other hand, the port-side angle of loll ϕ_4 is sensibly larger than the uncorrected one, and the area A is smaller.

Finally, we consider in Figure 6.21 a very grave case with a still higher centre of gravity ($\overline{KG} = 5.55\,m$) and assume that the ship master decides to move more masses until $m = 6.5\,t$. There is no position of equilibrium at starboard and the ship can find one only with the port-side down, at an angle of loll ϕ_4 sensibly larger than the initial angle ϕ_2. The area A under the righting-arm curve is small and a not-too-large moment tending to incline the vessel towards port can cause capsizing.

Ships whose righting-arm curves do not present inflexions like that shown in Figure 6.17 cannot find an angle of loll. The reader is invited to examine such a case in Exercise 6.4.

Once, it was not unusual to see that a ship carrying timber on deck sailed out of harbour with an angle of loll. Today, Paragraph 4.1.3 of IMO (1995) specifies for such vessels that, "At all times during a voyage, the metacentric height GM_0 should be positive after correction for free surface effects...", and even requires that in the departure condition the metacentric height be not less than $0.10\,m$.

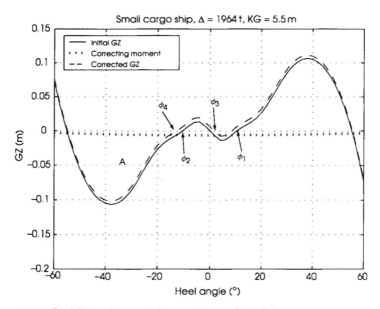

Figure 6.19 Stability with negative metacentric height

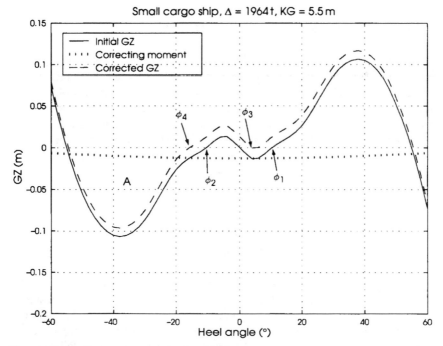

Figure 6.20 Correcting an angle of loll

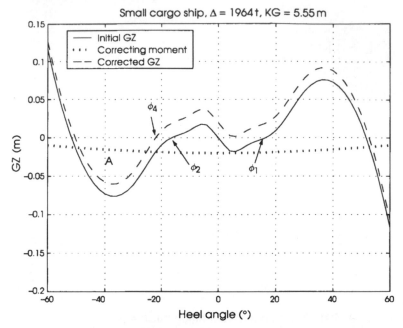

Figure 6.21 Correcting an angle of loll

6.12 The limitations of simple models

In Sections 6.3 and 6.4, we assumed that the water reaction to the heeling force acts at half-draught. This assumption is obviously arbitrary, but practice has proven it acceptable. A better evaluation would require an amount of calculations unacceptable in practical calculations. To find the exact location of the centre of pressure, it is necessary to take into account the exact hull-surface form. Moreover, the position of the centre of pressure can change with heel. In practice, stability calculations must be carried out for each change in load, in many cases by ship masters and mates. Under such circumstances computing resources are limited and one must be satisfied with an approximation of the centre of pressure consistent with other approximations assumed in the model. A documented discussion on the point of application of water reaction can be found in Wegner (1965). At this point it may be helpful to explain that the models developed in this chapter may be rough approximations of the reality, but they stand at the basis of national and international regulations that are compulsory. Stability regulations correspond to the notion of **codes of practice** as known in other engineering fields. All codes of practice accept simplifying assumptions that enable calculations with a reasonable amount of time and computing resources. Another situation occurs in research where more exact models must be assumed, powerful computer and experimenting resources are available, and more time is allowed.

Equation (6.2) developed in Section 6.3 yields a heeling arm equal to zero at the heel angle 90°. Such a result is obviously wrong as any vessel presents a sail area exposed to the wind even when lying on the side. Figure 1.103 in Henschke (1957) illustrates well this point. At small angles the results based on the curve proportional to $\cos^2 \phi$ differ little from those obtained with other approximations (see Chapters 8 and 10) and, therefore, they are acceptable for large vessels that do not heel much under wind, such as the capital ships of the US Navy. Smaller vessels tend to heel more under wind and then curves based on the $\cos^2 \phi$ assumption may become quite unrealistic.

The models developed in this chapter are based on further simplifications. In real life, water opposes the motions of a ship with forces that depend on the amplitude of motion, the speed of motion and the acceleration of motion. Assuming negligible roll velocity and acceleration, our models take into account only the moment that depends on the amplitude of heeling, that is the righting moment.

The moment that depends on the heeling speed, $\dot{\phi}$, is called **damping moment**. Damping causes **energy dissipation**. If a system that includes damping is displaced from its equilibrium position and then it is left to oscillate freely, the amplitude of oscillations will decrease with time and eventually will die out. The damping of the roll motion is mainly due to the generation of waves, but viscous effects may increase it and become important for certain bilge forms or if the vessel is fitted with bilge keels or a large keel.

The moments proportional to heel acceleration belong to a category of forces and moments called *added masses* because they can be collected together with the mass moment of inertia of the ship.

The evaluation of damping and added masses requires special computer programmes or model experiments. Neglecting damping and added masses leads to overestimation of dynamic heeling angles and this is on the **safe side**. Therefore, no stability regulation takes explicitly into account the effects of damping or added masses, but some regulations consider indirectly their influence by using different parameters for ships fitted with sharp bilges, bilge keels or deep keels.

Cardo *et al.* (1978), for example, discuss stability considering non-linear roll equations that include damping and added masses. Using advanced mathematical criteria, the authors reach the same qualitative results as those obtained in Section 6.7. An outline of the linear theory of ship motions is given in Chapter 12.

Last, but not least, we neglected until now the influence of waves, and we leave the discussion of this subject for Chapters 9–11.

6.13 Other modes of capsizing

Capsizing can be defined as the sudden transition of a floating body from a position of equilibrium to another position of equilibrium. Depending on the ship forms and loading, her second position can be on the side or with the keel up. If in the new position water can enter in large quantities, the ship will eventually sink. Often the process is so fast that many lives are lost. Sometimes no survivor remains to tell the story.

In Chapter 2, we saw that a floating body can capsize if the metacentric height is negative. In this chapter, we learnt that a vessel can capsize if the righting arm is too small in comparison with the heeling arm, or if the area under the righting-arm curve is too small in comparison with the area under the heeling-arm curve. In Chapter 9, we shall see that a ship can capsize because of the variation of the metacentric height and of the righting arm in waves that travel in the same direction as the ship (head or following seas) or at some angle with her. That dangerous phenomenon is called **parametric resonance** or **Mathieu effect**. What happens if the waves are parallel to the ship? Arndt (1960a) explains that a ship cannot capsize in regular, parallel waves. Adds Arndt, 'From practice we know cases in which captains put the ship parallel to the wave crests in order to reduce the effect of storms, neither in experiments could anyone cause until now a model to capsize in lateral, regular waves'. Otherwise seems to happen with freak, or breaking waves of great steepness whose impact on the ship side can be high enough to overturn the ship. Thus, for example, Morrall (1980) investigates the loss of the large stern trawler *Gaul*, and Dahle and Kjærland (1980) study the capsizing of the Norwegian research vessel *Helland-Hansen*. These studies support the hypothesis that the discussed disasters were due to high breaking waves.

It seems that the process of capsizing because of freak or breaking waves is not yet well understood and the methods proposed for its prediction are probabilistic (see Dahle and Myrhaug, 1996; Myrhaug and Dahle, 1994). Kat (1990) studied numerical models for the simulation of capsizing and Grochowalski (1989) describes a research on ship models. Probabilistic and simulation studies are beyond the scope of this book.

Another mode of capsizing is *broaching-to*; it is a dynamic phenomenon due to the loss of control in severe following or quartering seas. The ship enters into a forced turning that cannot be corrected by the rudder, heels and capsizes. Broaching-to is studied by Nicholson (1975), Spyrou (1995, 1996a,b).

It has been claimed that capsizing results from a combination of several factors. An example can be found in Hua (1996) who studied the capsize of the ferry *Herald of Free Enterprise* as a result of the interaction between heeling and turning motion, while great quantities of water were present on one deck.

6.14 Summary

The statical stability of ships is checked by comparing the righting-arm curve with the curves of heeling arms. A heeling arm is calculated by dividing a heeling moment by the ship displacement force. In general, a heeling-arm curve intersects the righting-arm curve at two points that correspond to positions of statical equilibrium. The equilibrium is stable only at the first position; there the slope of the righting-arm curve is larger than that of the heeling arm.

Heeling moments are caused by wind, by the centrifugal force developed in turning, by the crowding of passengers on one side of the ship, by towing or by the tension in a cable that links two vessels during a replenishment-at-sea operation.

The wind heeling arm is proportional to the square of the wind velocity and depends on the area of the lateral projection of the above-sea ship surface. We call that area 'sail area'. Assuming that the wind velocity is constant over the whole sail area, the wind heeling arm is proportional to the sail area. This assumption is acceptable for quick calculations. In reality, the wind speed increases with height above the sea level and this 'wind gradient' is taken into account in more exact calculations.

The heeling arm in turning is proportional to the square of the ship speed in turning, and inversely proportional to the radius of the turning circle. When the heeling moment appears or increases suddenly we must check the dynamical stability of the vessel. This situation can be caused by a gust of wind or by losing a mass on one side of the ship. The area under the righting arm up to the maximum angle reached momentarily by the ship is equal to the area under the heeling arm up to that point. The process depends on the angle of roll at which the sudden moment is applied. For a gust of wind, for example, the situation is worse if the ship is heeled to the windward side, than if the ship is caught by the gust with the lee side down. If the area available under the righting arm is smaller than the area under the heeling arm, the ship is lost.

The period of unresisted roll is proportional to the square root of the metacentric height. This imposes an upper limit on the $\overline{GM}$ value. If the roll period is too short, the roll motion is stiff; it is unpleasant for passengers and crew, and may be dangerous for equipment and cargo. If the motion is too tender, it may indicate a dangerously low metacentric height. A load displaced transversely reduces the stability when heeling to the same side as the load. Moving loads too decrease the stability. Thus, a load suspended so that it can move freely produces a virtual reduction of the metacentric height as if the load were moved to the point of suspension. A very common type of moving loads are liquids whose surfaces are free to move inside tanks or on the deck. The reduction of stability is proportional to the moment of inertia of the free surface about a barycentric axis parallel to the axis of ship inclination. The effect does not depend on the mass of the liquid (as long as the liquid surface does not reach the tank top or bottom) or the position of the tank. The usual way of reducing the free-surface effect is to subdivide the tanks by longitudinal bulkheads. Two other methods are to empty the tank or to fill it. In the latter case the effect of the thermal expansion of the liquid should be considered. Granular materials constitute another category of moving loads. Such loads stay in place until the heel angle exceeds a value characteristic for the material. This value is called angle of repose. The variation of stability reduction due to sliding loads follows a hysteresis loop. The effect of moving loads is a case of positive feedback.

If a ship is grounded in a region where the water level is descending, at a certain draught it can lose stability. The same happens with a ship on dock. The calculation of the critical draught is rather simple.

A ship with negative metacentric height can find a position of stable equilibrium, without capsizing, if the first part of the righting-arm curve lies above the tangent in the origin. This fixed angle of heel is called angle of loll. There are two angles of loll and they are symmetric about the origin. Under moderate perturbations, the ship can heel suddenly from one angle of loll to the other. This motion is different from a continuous roll and is characteristic for negative metacentric height. The angle of loll cannot be corrected by moving masses transversely; such an action can endanger the ship. Angles of loll should be corrected only by lowering the centre of gravity.

6.15 Examples

Example 6.1 – Wind pressure
Let us calculate the pressure corresponding to a wind speed of 70 knots. This is the value specified by the German Navy for evaluating the intact stability of vessels operating in open seas that are not exposed to tropical storms. Assuming an aerodynamic resistance coefficient equal to 1.2 and an air density equal to 1.27 kg m^{-1}, we obtain

$$p_W = c_W \frac{\rho}{2} V^2 = 1.2 \times \frac{1.27}{2} (0.514\,4 \times 70)^2 \ \frac{\text{kg}}{\text{m}^3} \left(\frac{\text{ms}^{-1}}{\text{knot}} \text{knot} \right)^2$$

$$= 987.99 \text{ kg m s}^{-2}$$

Rounding off yields 1 kN m^{-2}, or, using the SI term, 1 kPa. The conversion factor, 0.514 4, results from the definition of the **knot** as **nautical mile per hour**. Substituting SI units we divide 1852 m by 3600 s and obtain $1852/3600 = 0.514\,4 \text{ m s}^{-1}/\text{knot}$.

Example 6.2 – Calculating a wind heeling arm
Figure 6.22 is a simplified sketch of the sail area of the Ship *Lido 9* with the waterline corresponding to a draught of 1.85 m. To simplify calculations, the area is subdivided into five simple geometrical forms, namely rectangles and triangles. The calculations are carried out in the spreadsheet shown in Table 6.1.

If the stability of the vessel must be checked for a wind speed of 90 knots, we use the wind-pressure value of 1.5 MPa, as prescribed by the German Navy stability regulations for ships that can encounter tropical storms.

Example 6.3 – The statical stability curves of HMS Captain and HMS Monarch
In the night between 6 and 7 September 1870, a British fleet was sailing-off Cape Finisterre. The fleet was hit by a strong gale and one of the ships, HMS *Captain* capsized, but all other ships survived. The righting arms of HMS *Captain* are given in Anonymous (1872) and Attwood and Pengelly (1960), while

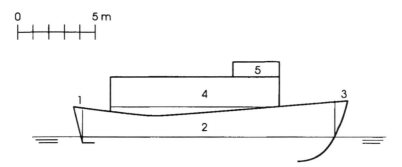

Figure 6.22 Calculation of sail area and its centroid

Table 6.1 Ship *Lido 9* – sail area for T = 1.85 m

Area component	Dimensions (m)	Area (m)	Centroid (m^2)	Moment (m^3)
1	0.6 × 2/2	0.60	1.33	0.80
2	2 × 16.4	32.80	1.00	32.80
3	0.8 × 2.4/2	0.96	1.60	1.54
4	2 × 11	22.00	3.00	66.00
5	1 × 3	3.00	4.50	13.50
Total		59.36	1.93	114.63

the latter book contains also the righting arms of HMS *Monarch*, a ship that was part of the same fleet and survived. The statical stability curves of the two ships are compared in Figure 6.23. The slopes in the origin of the curves show that both ships had practically the same initial metacentric height. The angle of vanishing stability of HMS *Monarch* was much larger than that of HMS *Captain*. The same was true for the areas under the righting-arm curves. The difference between those qualities was due mainly to a substantial difference in the freeboards.

Visual inspection of Figure 6.23 explains why HMS *Monarch* could survive the gust of wind that led to the capsizing of HMS *Captain*.

6.16 Exercises

Exercise 6.1 – stability in turning
Table 6.2 shows part of the cross-curves values of the small cargo ship exemplified in Section 6.11. The other data needed in this problem are the displacement volume, $\nabla = 2549 \, \text{m}^3$, the height of the centre of gravity above BL, $\overline{KG} = 5$ m, the ship length, $L_{\text{pp}} = 75.5$ m and the ship speed, $V = 16$ knots. Using the formulae given in Section 6.4, calculate the heeling arm in turning. Plot the

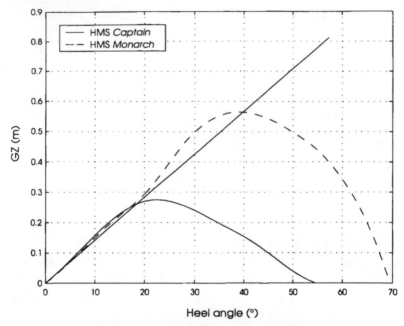

Figure 6.23 The statical stability curves of HMS *Captain* and HMS *Monarch*

heeling-arm curve over the righting arm and find the heel angle in turning. Next, consider a free-surface correction equal to $l_f = 0.04$ m, draw the corrected righting-arm curve, $\overline{GZ_{\text{eff}}}$, and see if the angle of heel is affected.

Hint: Use the tangent in origin when drawing the righting-arm curve.

Exercise 6.2 – Dynamical stability
The organizers of a boat race must throw a buoy from the starboard of a boat. The boat is rolling. Would you advise the organizers to throw the buoy while the starboard is down, or when the port side is down?

Table 6.2 Small cargo ship – partial cross-curves values

Heel angle ($^\circ$)	l_k (m)
10	0.918
20	1.833
30	2.717
45	3.847
60	4.653
75	5.007
90	4.994

Table 6.3 Small cargo ship – partial hydrostatic data

Draught, T (m)	∇ (m^3)	$\overline{KM}$ (m)	Draught, T (m)	∇ (m^3)	$\overline{KM}$ (m)
2.00	993	6.75	4.32	2549	5.16
2.20	1118	6.39	4.40	2609	5.16
2.40	1243	6.09	4.60	2757	5.16
2.60	1377	5.83	4.80	2901	5.17
2.80	1504	5.63	5.00	3057	5.18
3.00	1640	5.48	5.20	3210	5.20
3.20	1776	5.37	5.40	3352	5.23
3.40	1907	5.28	5.60	3507	5.27
3.60	2045	5.24	5.80	3653	5.31
3.80	2189	5.20	5.96	3786	5.34
4.00	2322	5.18	6.00	3811	5.36
4.20	2471	5.17	6.20	3972	5.42

Exercise 6.3 – Critical draught of grounded ship

Table 6.3 contains part of the hydrostatic data of the small cargo ship exemplified in the analysis of the angle of loll (Section 6.11).

1. The docking condition of the ship is characterized by the displacement volume, $\nabla = 1562.8\,m^3$ and $\overline{KG} = 5.34$ m. Find the critical draught at which props must be in place.
2. The data of the ship carrying a cargo of oranges and close to her destination (fuel tanks at minimum filling) are the displacement volume, $\nabla = 2979.4\,m^3$ and $\overline{KG} = 4.92$ m. Find the critical draught if the ship is grounded on the whole length of the keel.

Exercise 6.4 – Negative metacentric height

Using the data in Table 5.1, show that the vessel *Lido 9* cannot find an angle of loll if the metacentric height is negative.

5
Statical stability at large angles of heel

5.1 Introduction

Chapter 4 dealt with hull properties calculated as functions of draught, at constant trim and heel. We reminded then that the maritime terminologies of Romance languages have a concise term for the set of submerged hulls characterized as above. Thus, for example, the term in French is **carènes isoclines**. The first part of the term, 'iso', derives from the Greek 'isos' and means 'equal'. The meaning of the term 'isocline' is 'equal inclination' (see Figure 4.6 in Chapter 4). In this chapter, we are going to discuss the properties of submerged hulls as functions of heel, at constant displacement volume. Again, Romance languages have a concise term for the set of submerged hulls of a given vessel, having the same displacement volume. For example, the French term is **isocarènes**, while the Italian term is **isocarene**. The assumption of constant displacement volume recognizes the fact that while a ship heels and rolls, her weight remains constant. By virtue of Archimedes' principle, constant weight implies constant displacement volume.

The central notion in this chapter is the **righting arm**. We shall show how to calculate and represent the righting arm in a set of curves known as **cross-curves of stability**. Another topic is the plot of the righting arm as function of the heel angle, for a given displacement volume and a given height of the centre of gravity. This plot is called **curve of statical stability** and it is used to assess the ship stability.

5.2 The righting arm

In Figure 5.1, we consider a ship whose waterline in upright condition is $W_0 L_0$. The corresponding centre of buoyancy is B_0 and the centre of gravity G. Let us assume that the ship heels to starboard by an angle ϕ. The new waterline is $W_\phi L_\phi$ and the centre of buoyancy moves towards the submerged side, to the new position B_ϕ. The weight force, equal to Δ, passes through G and is vertical, that is perpendicular to $W_\phi L_\phi$. The buoyancy force, also equal to Δ, passes through

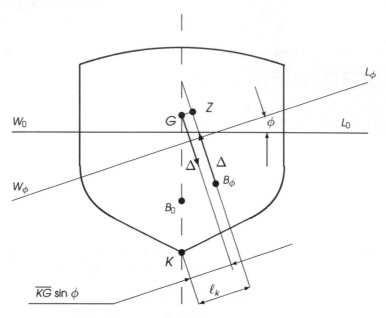

Figure 5.1 Definition of righting arm

B_ϕ and is also perpendicular to $W_\phi L_\phi$. The perpendicular from G to the line of action of the buoyancy force intersects the latter line in Z. The forces of weight and buoyancy produce a **righting moment** whose value is

$$M_R = \Delta \overline{GZ} \tag{5.1}$$

As Δ is a constant for all angles of heel, we can say that the righting moment is characterized by the **righting arm**, $\overline{GZ}$. From Figure 5.1, we write

$$\overline{GZ} = \ell_k - \overline{KG} \sin \phi \tag{5.2}$$

For reasons to be explained later, the distance ℓ_k is called **value of stability cross-curves**. This quantity results from hydrostatic calculations based on the ship lines. Such calculations are left today to the computer. The term $\overline{KG} \sin \phi$ depends on $\overline{KG}$, a quantity obtained from **weight calculations** as explained in Chapter 7. In European literature, the term ℓ_k is often described as 'lever arm of stability of form', while the term $\overline{KG} \sin \phi$ is called 'lever arm of stability of weight'.

It is important to note that ℓ_k is measured here from K, a point preferably chosen as the lowest keel point, or the projection of the lowest keel point on the midship section. The resulting ℓ_k value is thus always positive. This convention is practically standard in some European countries and, for its advantages, we follow it throughout this book. In American projects and computer programmes

ℓ_k is often measured from one of the positions of the centre of gravity, G. For example, the reference point can be the centre of gravity for the full-load condition (see, for example, Lewis, 1988, pp. 78–79). When proceeding so, the designer must define in the clearest way the position of the reference point.

The relationship between the value of the stability cross-curves, ℓ_k, and the angle of heel, ϕ, is not linear and, in general, cannot be defined explicitly. For small angles of heel a linear expression for the righting arm, $\overline{GZ}$, can be derived from Figure 5.2:

$$\overline{GZ} = \overline{GM} \sin \phi \qquad (5.3)$$

But, what do we mean by 'small angle'? The answer is given by the same Figure 5.2. Equation (5.3) holds true as long as the metacentre, M, does not move visibly from its initial position. Thus, for many ships an angle equal to $5°$ is small, while for a few others even $15°$ may be a small angle. The value depends on both ship forms and loading condition. More insight on this point can be gained by looking at the metacentric evolutes shown in Chapter 2. A further criterion for the 'smallness' of the heel angle will be given in the next section.

A useful way of plotting the ℓ_k values is shown in Figure 5.3. There, the ℓ_k curves are plotted as functions of the displacement volume, ∇, for a set of constant heel-angle values. Thus, we have a curve for $\phi = 10°$, one for $\phi = 20°$, and so on. To use Eq. (5.2) for a given displacement volume, say ∇_0, it is

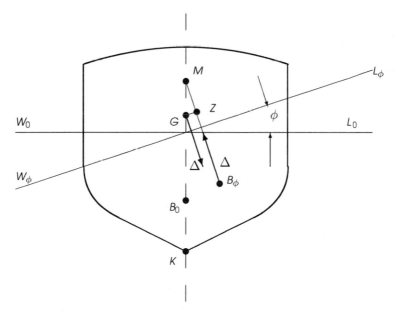

Figure 5.2 Righting arm, $\overline{GZ}$, at small angles of heel

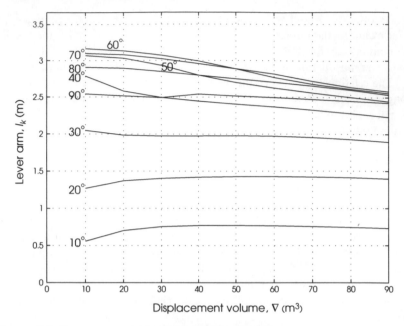

Figure 5.3 Cross-curves of stability of Ship *Lido 9*

necessary to draw the vertical line $\nabla = \nabla_0$ and read the values where this line 'crosses' the curves. Hence the term **cross-curves of stability**.

5.3 The curve of statical stability

The plot of the righting arm, $\overline{GZ}$, calculated from Eq. (5.2), as function of the heel angle, ϕ, at constant ∇ and $\overline{KG}$ values is called **curve of statical stability**. Such diagrams are used to evaluate the stability of the ship in a given loading condition. For a full appreciation, it is necessary to compare the righting arm with the various **heeling arms** that can endanger stability. We discuss several models of heeling arms in Chapter 6. An example of statical-stability curve is shown in Figure 5.4; it is based on Table 5.1. The table can be calculated in an electronic spreadsheet, or in MATLAB as shown in Biran and Breiner (2002, Example 2.9).

Let us identify some properties of the righting-arm curves. One important value is the maximum $\overline{GZ}$ value and the heel angle where this value occurs. For example, in Figure 5.4 the maximum righting arm value is 1.009 m and the corresponding heel angle is 50°. Another important point is that in which the $\overline{GZ}$ curve crosses zero. The corresponding ϕ value is called **angle of vanishing stability**. In our example, the righting-arm curve crosses zero at an angle greater

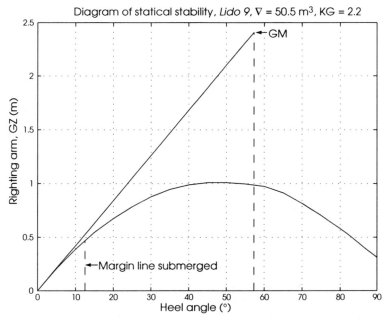

Figure 5.4 Statical-stability curve

Table 5.1 Ship *Lido 9* – Righting arm, $\overline{GZ}$, for $\nabla = 50.5\,\mathrm{m^3}$, $\overline{KG} = 2\,\mathrm{m}$

Heel angle	ℓ_p	$\overline{KG}\sin\phi$	$\overline{GZ}$	Heel angle	ℓ_p	$\overline{KG}\sin\phi$	$\overline{GZ}$
(°)	(m)	(m)	(m)	(°)	(m)	(m)	(m)
0	0.000	0.000	0.000	50	2.694	1.685	1.009
5	0.396	0.192	0.204	55	2.799	1.802	0.997
10	0.770	0.382	0.388	60	2.879	1.905	0.974
15	1.115	0.569	0.546	65	2.908	1.994	0.914
20	1.427	0.752	0.675	70	2.883	2.067	0.816
25	1.713	0.930	0.783	75	2.828	2.125	0.703
30	1.977	1.100	0.877	80	2.747	2.167	0.580
35	2.208	1.262	0.946	85	2.641	2.192	0.449
40	2.402	1.414	0.988	90	2.513	2.200	0.313
45	2.564	1.556	1.008				

$\overline{KM} = 4.608\,\mathrm{m};\ \overline{KG} = 2.200\,\mathrm{m};\ \overline{GM} = 2.408\,\mathrm{m}$

than 90°, in a region outside the plot frame. The angle of vanishing stability can often occur at less than 90°, as shown, for example, in Figure 6.23.

A very useful property refers to the tangent in the origin of the righting-arm curve. The slope of this tangent is given by

$$
\begin{aligned}
|\tan \alpha|_{\phi=0} \;&\approx\; \left| \frac{\mathrm{d}(\overline{GM})}{\mathrm{d}\phi} \right|_{\phi=0} \\[2mm]
&=\; \frac{\mathrm{d}\overline{GM}}{\mathrm{d}\phi}\sin 0 \;+\; \overline{GM}_0 \cos 0 = \overline{GM}_0
\end{aligned}
\tag{5.4}
$$

Equation (5.4) yields a simple rule for drawing the tangent:

> In the curve of statical stability, at the heel angle 1 rad (approximately 57.3°) draw a vertical and measure on it a length equal to that of $\overline{GM}$. Draw a line from the origin of coordinates to the end of the measured segment. This line is tangent to the $\overline{GZ}$ curve.

From the triangle formed by the heel-angle axis, the vertical at 1 rad, and the tangent in origin, we find the slope of the line defined as above; it is equal to $\overline{GM}/1$, that is the same as yielded by Eq. (5.4). The tangent in the origin of the righting-arm curve should always appear in the curve of statical stability; it gives an immediate, visual indication of the $\overline{GM}$ magnitude, and it is a check of the correctness of the curve. We strongly recommend **not** to try the inverse operation, that is to 'fit' a tangent to the curve and measure the resulting $\overline{GM}$ value. This would amount to graphic differentiation, a procedure that is neither accurate nor stable.

Figure 5.4 lets us give another appreciation of what small angle means: we can consider as small those heel angles for which the curve of the righting arm can be confounded with the tangent in its origin. In our example, this holds true for angles up to 7–8°.

For any angle of heel, ϕ, we can rewrite Eq. (5.4) as

$$
\frac{\mathrm{d}\overline{GZ}}{\mathrm{d}\phi} = \overline{ZM_\phi}
\tag{5.5}
$$

where Z is as previously the foot of the perpendicular from G to the line of action of the buoyancy force and M_ϕ is the metacentre corresponding to the heel angle ϕ. The geometric construction of this tangent is similar to that of the tangent in origin. For a proof of this result see, e.g. Bîrbănescu-Biran (1979).

5.4 The influence of trim and waves

Once it was usual to calculate the cross-curves of stability at constant trim, i.e. for the ship on even keel. This approach was justified before the appearance of computers and Naval Architectural software. However, Eq. (2.28), developed in Chapter 2, shows that the longitudinal position of the centre of buoyancy changes if the heel angle is large. It happens so because at large heel angles the waterplane area ceases to be symmetric about the centreline. If the centre of buoyancy moves along the ship, while the position of the centre of gravity is constant, the trim changes too. Therefore, cross-curves calculated at constant trim may not represent actual stability condition. Jakić (1980) has shown that trim can greatly influence the values of cross-curves and, therefore, that influence should be taken into account. The stability regulations, BV 1033, of the German Navy require, indeed, the calculation of the cross-curves at the trim induced by heel. Modern computer programmes for Naval Architecture include this option.

As we shall show in Chapter 9, waves perpendicular or oblique to the ship velocity influence the values of cross-curves and can cause a very dangerous effect called parametric resonance. This effect too must be taken into account and modern computer programmes can calculate cross-curves on waves. The stability regulations of the German Navy take into account the variation of the righting arm in waves (see Arndt, 1965; Arndt, Brandl, and Vogt, 1982).

5.5 Summary

In this chapter, we dealt with the righting moment at large angles of heel, $M_{\rm R} = \Delta\overline{GZ}$. The quantity $\overline{GZ}$, called righting arm, is the length of the perpendicular drawn from the centre of gravity, G, to the line of action of the buoyancy force. We assume that the ship heels at constant displacement. This is the desired situation in which the ship neither loses loads nor takes water aboard. Then, the factor Δ is constant and the variation of the righting moment with heel is described by the variation of the righting arm $\overline{GZ}$. The value of the righting arm is calculated from

$$\overline{GZ} = \ell_k - \overline{KG}\sin\phi$$

where ℓ_k, called value of stability cross-curve, is the distance from the reference point K to the line of action of the buoyancy force, $\overline{KG}$, the distance of the centre of gravity from the same point K, and ϕ, the heel angle. It is recommended to take the point K as the lowest hull point. The values of the stability cross-curves, ℓ_k, are usually represented as functions of the displacement volume, with the heel angle as parameter.

One can read in this plot the values corresponding to a given displacement volume, calculate with them the righting arm and plot its values against the heel angle. This plot is called curve of statical stability and it is used to appreciate the stability of the ship, at a given displacement and height of the centre of gravity. To check the correctness of the righting-arm curve, it is recommended to draw the tangent in the origin. To do this, one should draw a vertical line at the angle of 1 rad and measure on the vertical a length equal to the metacentric height, $\overline{GM}$. The tangent is the line that connects the origin of coordinates to the point found as in the previous sentence.

The trim changes as the ship heels. That effect should be taken into account when calculating cross-curves of stability. Another influence to be taken into account is that of waves.

Table 5.2 summarizes the main terms related to stability at large angles of heel. As in Chapter 1, we note by 'Fr' the French term, by 'G' the German term, and by 'I' the Italian term. Old symbols used once in those languages are given between parentheses.

Table 5.2 Terms related to stability at large angles of heel

English term	Symbol	Computer notation	Translations (old European symbol)
Centre of buoyancy	B		Fr centre de carène (C), G Verdrängungsschwepunkt (F), I centro di carena
Centre of gravity	G		Fr centre de gravité, G Massenschwerpunkt, I centro di gravità
Curve of statical stability			Fr courbe de stabilité, G Stabilitätskurve, I curva di stabilità
Heel angle (positive starboard down)	ϕ	HELANG	Fr angle de bande, angle de gîte, G Krängungswinkel, I angolo di inclinazione traversale, sbandamento
Keel point – reference point on BL	K		F point le plus bas de la carène, G Kielpunkt, I intersezione della linea base con la sezione maestra
Projected centre of gravity	Z		G Projizierte Massenschwerpunkt
Righting lever	$\overline{GZ}$	GZ	F bras de levier (GK), G Aufrichtenden Hebelarm, I braccio radrizzante
Value of stability cross-curve	ℓ_k	LK	Fr pantocarènes, bras de levier du couple de redressement, G Pantocarenenwert bezogen auf K
z-coordinate of centre of gravity	$\overline{KG}$	ZKG	Fr distance du centre de gravité à la ligne d'eau zéro, G z-Koordinate des Massenschwerpunktes, I distanza verticale del centro di gravità

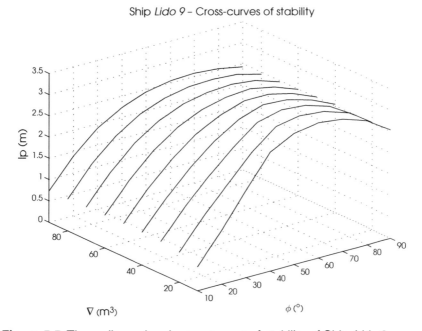

Figure 5.5 Three-dimensional cross-curves of stability of Ship *Lido 9*

5.6 Example

Figure 5.5 is a three-dimensional representation of the cross-curves of the Ship *Lido 9*.

5.7 Exercises

Exercise 5.1
Plot in one figure the righting-arm curves and the tangents in origin of the Ship *Lido 9*, for $\nabla = 50.5\,\mathrm{m}^3$ and $\overline{KG}$-values 1.8, 2.0, 2.4 and 2.6 m. Comment the influence of the centre-of-gravity height.

Exercise 5.2
Draw the curve of statical stability of the Ship *Lido 9* for a displacement in sea water $\Delta = 35.3$ t and a height of the centre of gravity $\overline{KG} = 2.1$ m. Use data in Tables 4.2 and 5.1.

7
Weight and trim calculations

7.1 Introduction

All models of stability require the knowledge of the displacement mass, Δ, and of the height of the centre of gravity (vertical centre of gravity), $\overline{KG}$. Stevin's law (see Subsection 2.3.2) shows that the ship trim is determined by the longitudinal position of the centre of gravity, LCG. The three quantities, Δ, $\overline{KG}$ and LCG are calculated by summing up the masses of all ship components and their moments about a horizontal and a transverse plane. The centre of gravity of a ship in upright condition is situated in the plane of port-to-starboard symmetry of the ship (centreline plane); therefore, the coordinate of the centre of gravity about this plane is zero. However, individual mass ship components may not be symmetrical about the centreline plane and it is necessary to calculate their moments about that plane and ensure that the transverse coordinate (y-coordinate) of the ship's centre of gravity is zero. It is usual to call the latter coordinate **transverse centre of gravity** and note it by **TCG**. Thus, we have a consistent notation for the triple of coordinates LCG, VCG, TCG. Systematic calculations of displacements and centres of gravity are known as **weight calculations** and they are the subject of the first part of this chapter. Recent literature and standards deal with masses rather than weights. We follow this trend in our book, but use the term *weight calculations* because it is rooted in tradition.

Another subject of this chapter is the calculation of the trim and of the forward and aft draughts. As mentioned in Chapter 6, the trim affects the ship stability. Also, a ship trimmed at a large angle can look unpleasant to the eye. Above all, the trim determines the forward and aft draughts and thus affects certain ship functions. For example, the aft draught must be large enough to ensure sufficient propeller submergence and avoid cavitation.

Frequently weight calculations are based on approximate or insufficient data. The sources of uncertainty are explained in this chapter when introducing the notions of **reserve** and **margin** of displacement and of $\overline{KG}$. Because of these uncertainties, statutory regulations require an experimental validation of the coordinates of the centre of gravity, and of the corresponding metacentric height, $\overline{GM}$, for all new buildings or for vessels that underwent alterations that can influence their stability. This validation is carried out in the **inclining experiment**, also known in some shipyards as **stability test**.

7.2 Weight calculations

7.2.1 Weight groups

A vessel is composed of hundreds, sometimes thousands of mass items. To systematize calculations it is necessary to organize them into **weight groups**. The first subdivision is into two main sets: **lightship** and **deadweight**. The lightship (less frequently known as **lightweight**) is the mass of the empty ship; it is composed of the hull, the outfit, and the machinery masses, including the liquids in the machinery and various systems, but not those in tanks or storage spaces. The deadweight is the sum of the masses of crew, cargo and passengers, fuel, lubricating oil, provisions, water, stores and spare parts. The usual abbreviation for deadweight is DWT. In simpler terms, the deadweight is the weight that the ship 'carries'.

One should make a distinction between the term *lightship* used as above, and its homophone that designs a ship provided with a strong light and used to mark a position.

In the first stages of ship design, known as **preliminary design**, the lightship masses and their centres of gravity are estimated by empirical equations, based on statistics of similar ships, or are derived from the masses of a given **parent ship**. This subject is treated in books on **ship design** such as Kiss (1980), Schneekluth (1980), Schneekluth and Bertram (1998) and Watson (1998). For merchant ships, the lightship groups are the hull, the outfit and the machinery. The classification of warship weight groups may be somewhat different. Thus, the classification system of the US Navy, SWBS, distinguishes the following main weight groups: hull structure, propulsion plant, electric plant, command and surveillance, auxiliary systems, outfit and furnishings, armament.

As the design progresses by successive iterations, the weight estimations are refined by subdividing the weight groups into subgroups, the subgroups into lower-lever subgroups, and so on. Thus, the hull mass is subdivided into hull and superstructure, then the hull into bottom, sides, decks, bulkheads etc. The machinery components are first subdivided into main, or propulsion machinery, and auxiliary machinery. In the final stages it is possible to calculate the masses and centres of gravity of individual items from detailed drawings or from data provided by equipment suppliers.

The procedure described above requires a classification of the various weight groups, subgroups and so on that ensures that no item is forgotten and that no item belongs to two groups. Readers who like mathematics may say that the weight groups shall be **disjoint**. Such readers can also see that such a classification system can be described by a **tree graph** (see Bîrbănescu-Biran, 1988). Several authorities and organizations engaged in ship design and construction have developed their own classification systems. An example of classification system for merchant ships is shown in Kiss (1980). As mentioned above, the classification system adopted by the US Navy is known as SWBS, an acronym for **Ship Work Breakdown Structure**.

The main deadweight item is the **cargo**; it is pre-specified by the owner. The number of crew members depends on the functions to be carried aboard: frequently a minimum is prescribed by regulations. The masses of fuel, lubricating oil, and water result from the required ship speed and range, two characteristics specified by the owner.

To compensate for the uncertainties in weight estimation in the first design stages, Naval Architects introduce a weight item called **reserve**, or **weight margin**. Some regulations consider also a $\overline{KG}$ margin; that is the calculated height of the centre of gravity, $\overline{KG}$, is increased by a certain amount ensuring that stability calculations fall on the safe side. As the ship design progresses, the uncertainties are reduced and so must be the weight reserve and the $\overline{KG}$ margin.

When the detailed ship project is delivered for construction, all weight and centres of gravity are supposed to be exact; however, a 'building' weight reserve and a $\overline{KG}$ margin are still included in weight calculations. By doing so designers take into account acceptable tolerances in plate, profile and pipe thicknesses, tolerances in metal densities, and changes in the catalogues of suppliers.

Even when the ship is delivered to the owner, weight calculations still include 'commissioning' margins that take into account future equipment additions, trapping of water in places from where it cannot be pumped out, and weight increase due to rust and paint. Certain codes of practice, such as the stability regulations of the US Navy and those of the German Federal Navy, impose well-defined margin values.

7.2.2 Weight calculations

Once the ship is built and in service, the lightship displacement and its centre of gravity are taken in calculations as constants. For each possible **loading case**, that is for each combination of cargo and other deadweight items, the masses of those items and their moments are added to those of the lightship. The calculations yield the displacement and the coordinates of the centre of gravity of the loading case under consideration. To give an example, we return to the data of the small cargo ship considered in Chapter 6. Figure 7.1 shows the calculations corresponding to the load case **Homogeneous cargo, departure**. By **departure condition** we mean the ship leaving the port, with all the fuel, lubricating oil and provisions.

Table 7.1 was calculated in MS Excel. Alternatively, the calculations can be performed in MATLAB. Then, the weight data can be stored in a matrix, for example in the format

$$m_i \ kg_i \ lcg_i$$

where m_i is the mass of the ith weight item, kg_i, its vertical centre of gravity, and lcg_i, its longitudinal centre of gravity. An example of calculations for the loading case considered in Figure 7.1 is

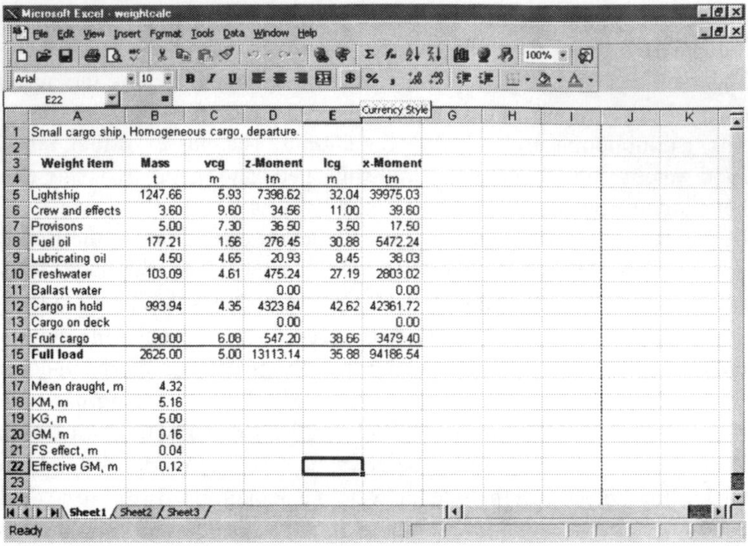

Small cargo ship, Homogeneous cargo, departure.

Weight item	Mass	vcg	z-Moment	lcg	x-Moment
	t	m	tm	m	tm
Lightship	1247.66	5.93	7398.62	32.04	39975.03
Crew and effects	3.60	9.60	34.56	11.00	39.60
Provisons	5.00	7.30	36.50	3.50	17.50
Fuel oil	177.21	1.56	276.45	30.88	5472.24
Lubricating oil	4.50	4.65	20.93	8.45	38.03
Freshwater	103.09	4.61	475.24	27.19	2803.02
Ballast water			0.00		0.00
Cargo in hold	993.94	4.35	4323.64	42.62	42361.72
Cargo on deck			0.00		0.00
Fruit cargo	90.00	6.08	547.20	38.66	3479.40
Full load	2625.00	5.00	13113.14	35.88	94186.54
Mean draught, m	4.32				
KM, m	5.16				
KG, m	5.00				
GM, m	0.16				
FS effect, m	0.04				
Effective GM, m	0.12				

Figure 7.1 A spreadsheet of weight calculations

Table 7.1 Results of inclining experiment

Inclining moment (tm)	Heel angle tangent	Inclining moment (tm)	Heel angle tangent
1156.9	0.0187	−1136.5	−0.0179
1156.9	0.0185	−1136.5	−0.0180
1156.9	0.0179	−1136.5	−0.0185
771.5	0.0126	−757.5	−0.0119
771.5	0.0126	−757.5	−0.0120
771.5	0.0121	−757.5	−0.0124
386.3	0.0065	−379.4	−0.0057
386.3	0.0064	−379.4	−0.0060
386.3	0.0062	−379.4	−0.0065
1.1	0.0004	−0.2	0
1.1	0.0005	−0.2	0
1.1	0	−0.2	−0.0006

```
Wdata = [
 1247.66  5.93  32.04
    3.60  9.60  11.00
    5.00  7.30   3.50
  177.21  1.56  30.88
    4.50  4.65   8.45
  103.09  4.61  27.19
  993.94  4.35  42.62
   90.00  6.08  38.66 ];
```

```
format bank, format compact
Displ = sum(Wdata(:, 1))
Displ = 2625.00
KG    = Wdata(:, 1)'*Wdata(:, 2)/Displ
KG    = 5.00
LCG   = Wdata(:, 1)'*Wdata(:, 3)/Displ
LCG   = 35.88
```

Unless all calculations are carried out by a computer programme, the results of weight calculations are used as described below:

1. The mean draught, T_m, corresponding to the calculated displacement, is read in the hydrostatic curves.
2. The trimming moment is calculated as

$$M_{\text{trim}} = \Delta(LCG - LCB) \qquad (7.1)$$

where the LCG value corresponding to T_m is found in the hydrostatic curves. The moment to change trim, MCT, corresponding to T_m, is read from the hydrostatic curves and the trim is calculated as shown in Section 7.3. If the trim is small one can go to the next step, otherwise it is advisable to continue the calculations using the Bonjean curves or to resort to a computer programme.
3. The height of the metacentre above BL, $\overline{KM}$, corresponding to T_m, is read in the hydrostatic curves.
4. The metacentric height is calculated as

$$\overline{GM} = \overline{KM} - \overline{KG}$$

5. The free-surface effects of the tanks filled with liquids are added up and their sum is subtracted from the metacentric height to find the effective metacentric height, $\overline{GM}_{\text{eff}}$.
6. The righting levers, $\overline{GZ}$, are calculated, and the effective righting levers are obtained by subtracting the free-surface effect

$$
\begin{aligned}
\overline{GZ} &= l_k - \overline{KG}\sin\phi \\
\overline{GZ}_{\text{eff}} &= \overline{GZ} - l_F\sin\phi
\end{aligned} \qquad (7.2)
$$

7. The data are used to plot the statical stability curve.

With older computer programmes, such as ARCHIMEDES, the displacement and the coordinates of the centre of gravity can be used as input to obtain the mean draught and the trim of the ship. The accuracy is good even for large trim values. In recent computer programmes the user has to input the degree of filling of cargo holds and of the various tanks and the computer carries on all weight and hydrostatic calculations. This subject is discussed in Chapter 13.

7.3 Trim

7.3.1 Finding the trim and the draughts at perpendiculars

In Figure 7.2 we consider a ship initially on even keel; the corresponding water-line is $W_0 L_0$. Let us assume that the ship trims reaching a new waterline, $W_\theta L_\theta$. If the trim angle, θ, is small (for normal loading conditions it is always small), the intersection line of the two waterlines, $W_0 L_0$ and $W_\theta L_\theta$, passes through the centre of flotation, F, of the initial waterplane. The midship draught of the ship on even keel, T_m, can be read in the hydrostatic curves at the intersection of the displacement curve and the vertical corresponding to the given displacement. For that draught we read the moment to change trim, MCT. We calculate the trim, in m, as

$$\text{trim} = T_F - T_A = \frac{\Delta(LCG - LCB)}{MCT} \tag{7.3}$$

The trim angle is given by

$$\tan \theta = \frac{T_F - T_A}{L_{\text{pp}}} \tag{7.4}$$

From Figure 7.2, we see that

$$T_A = T_m - LCF \tan \theta = T_m - LCF \frac{\text{trim}}{L_{\text{pp}}} \tag{7.5}$$

and

$$T_F = \text{trim} + T_A = T_m + \text{trim} \left(1 - \frac{LCF}{L_{\text{pp}}}\right) \tag{7.6}$$

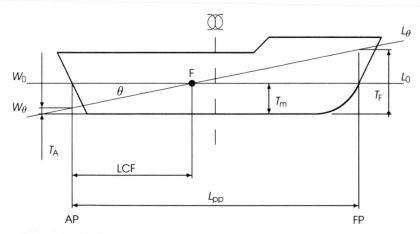

Figure 7.2 Finding the forward and aft draughts

To give an example we consider again the loading case of the small cargo ship analyzed in Subsection 7.2.2. In Tables 6.2 and 7.3 (See Exercise 7.1) we find $T_m = 4.32$ m, $LCB = 0.291$ m, $LCF = -0.384$ m, and $MCT = 3223$ mt m^{-1}. We know that the length between perpendiculars is $L_{pp} = 75.40$ m. In the table LCB is measured from midship, positive forwards. As LCG is measured from AP, we calculate

$$\frac{75.40}{2} + 0.291 = 37.99 \text{ m}$$

and the trim

$$\frac{\Delta(LCG - LCB)}{MCT} = \frac{2625(35.88 - 37.99)}{3223} = -1.72 \text{ m}$$

The ship is trimmed by the stern. In Table 7.3, LCF is measured from the midship, positive forward; the value measured from AP is

$$\frac{75.40}{2} - 0.384 = 37.32 \text{ m}$$

and we calculate

$$
\begin{aligned}
T_A &= 4.32 - 37.32\frac{-1.72}{75.4} = 5.17 \\
T_F &= -1.72 + 5.17 = 3.45
\end{aligned}
$$

where the results are in m.

7.3.2 Equilibrium at large angles of trim

For small angles of trim, Stevin's law yields $LCB = LCG$ where both lengths are measured from the same origin. As Figure 7.3 shows, when the trim is large, things are not so simple and the heights of the centres of buoyancy and gravity must be taken into account. In Figure 7.3 we assume again that both LCB and LCG are measured in the same system and from AP and write

$$LCG + (\overline{KG} - \overline{KB})\tan\theta = LCB \tag{7.7}$$

The longitudinal centre of gravity, LCG, is always measured in a system fixed in the ship. Some computer programmes may measure LCB in a system fixed in space. Therefore, when using the output of a computer programme it is necessary to read carefully the definitions used by it.

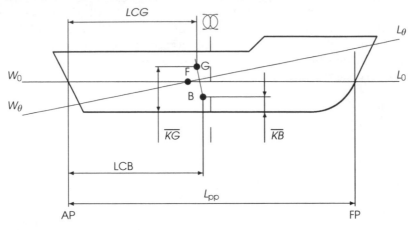

Figure 7.3 Equilibrium at large trim

7.4 The inclining experiment

Because of the importance of this subject we give here the term in three foreign languages:

French   Expérience de stabilité
German   Krängungsversuch
Italian   Prova di stabilità

It is usual to carry out the inclining experiment a short time before the completion of the ship. The vessel must float in calm water and the work must be done while no wind is blowing. The number of persons aboard should be limited to that strictly necessary for the experiment; their masses and positions should be exactly recorded. Tank fillings and free surfaces in tanks should be well known. Free surfaces should not reach tank bottoms or ceilings for the expected heel and trim angles. All draught marks should be read, i.e. forward, at midship, at stern, both on starboard and on port side. Good practice requires to put a glass pipe before the draught mark and to read the draught value corresponding to the water level in the pipe. This procedure minimizes errors due to small waves. The water density should be read at several positions around the ship.

Figure 7.4 shows a common set-up for the inclining experiment. A plumb line with a bob B is hung at A. The bob is immersed in a water tank that serves as an oscillation damper. A mass p is displaced transversely a distance d. The resulting heel angle, assumed small, is given by

$$\tan \theta = \frac{pd}{\Delta \overline{GM}} \tag{7.8}$$

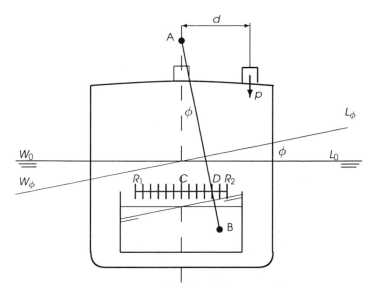

Figure 7.4 Set-up for the inclining experiment

The deflection of the plumb line is measured on a graduated batten $R_1 R_2$ and is used to calculate

$$\tan \theta = \frac{\overline{CD}}{\overline{AC}} \tag{7.9}$$

A recommended practice is to displace the mass once to starboard and measure $\tan \theta_S$, then to port and measure $\tan \theta_P$. The value to be substituted into Eq. (7.8) is

$$\tan \theta = \frac{\tan \theta_S + \tan \theta_P}{2}$$

It is recommended to repeat the set-up in Figure 7.4 at three stations along the ship. The masses used for inclining the vessel should be chosen so that the heel angles fall within that range in which Eq. (7.8) is applicable. Moore (1967) recommends angles of $1°$ for very large vessels, $1.5°$ for ships of $120\,\mathrm{m}$ length, and 2–$3°$ for small vessels. Kastner (1989) cites German regulations that require heel angles ranging between 1 and $3.5°$. Equation (7.8) can be used for the estimation of suitable masses.

According to Hansen (1985) the length of the plumb line should be chosen so that the length measured on the batten should be maximum 150–$200\,\mathrm{mm}$. Writes Hansen, 'In general, long pendulums used on stiff ships and short pendulums used on tender ships result in about the same accuracy in measuring the ship list.' Kastner (1989) studies the dynamics of a compound pendulum consisting of the ship and the plumb line. A long plumb line ensures a good resolution in reading the graduation on the batten. On the other hand, a long plumb line can yield a large dynamic response to small-amplitude ship motions and increase reading errors. Kastner concludes that a length of 1.146 m is sufficient.

Today, the set-up shown in Figure 7.4 can be replaced by electronic instruments that measure the heel angle (inclinometres, gyroscopic platforms) whose output can be fed directly to an on-board computer. A common way of checking the accuracy of the results consists in plotting the tangents of heel angles, arc tan θ, against the heeling moments, pd. Equation (7.8) shows that the ideal plot should be a straight line. Years ago Naval Architects fitted by eye a straight line passing through the plotted points. Nowadays computers and many hand calculators yield easily a **least-squares fit**. Example 1 shows how to do it.

When analyzing the results of the inclining experiment, Eq. (7.8) is rewritten as

$$\Delta \overline{GM} = \frac{pd}{\tan \theta}$$

The interpretation of the results of inclining experiments requires the knowledge of the displacement, Δ, and of the height of the metacentre above the baseline, $\overline{KM}$. If the trim is small, one can read the desired values in the hydrostatic curves, entering them with the measured mean draught, T_m, as input. Hansen (1985) quotes the limits imposed on the trim by the US Navy and the US Coast Guard. The recommended value for naval vessels is 0.67%, and for commercial ships 1% of the ship length. If the trim is not small one can use the Bonjean curves or a computer programme for hydrostatic calculations. When drawing the waterline on the Bonjean curves we must not forget that, in general, the forward and the aft draught marks are not placed in the transverse planes of the forward and aft perpendiculars. Therefore, the values read on the marks must be adjusted and extrapolated to the FP and AP positions.

A computer programme for hydrostatic calculations can be used if the offsets of the ship are stored in the required input format. Then, it is sufficient to run the programme for the mean draught and the trim read during the inclining experiment.

The ship hull behaves like a beam that can deflect under bending moments. Bending moments arise from differences between the longitudinal distribution of masses and that of hydrostatic pressures. Deflections of the hull beam also can be caused by differences between thermal expansions of the deck and of the bottom. The deflection can be calculated as the difference between the average of forward and aft draught and the draught T_m measured at midship

$$d = T_M - \frac{T_F - T_A}{2} \tag{7.10}$$

Various authorities and authors publish formulae for calculating an **equivalent draught** that allows the calculation of the displacement of a deflected hull. For example, Hansen (1985) uses a rather complicated formula recommended by NAVSEA, a design authority of the US Navy. Ziha (2002) analyzes the displacement change due to hull deflection and proposes ways of taking it into account. Hervieu (1985) simplifies the problem by assuming a parabolic elastic line (the deflected shape of the beam). Then, for a rectangular waterplane and vertical sides in the region of the actual waterline (wall-sided hull) the added or lost volume equals

$$\delta\Delta = \frac{2}{3}A_W d$$

where A_W is the waterplane area, and d is the deflection. In most cases the waterplane area is not rectangular, but still in a first approximation we can use as equivalent draught

$$T_{eq} = T_M + \frac{2}{3}d$$

The sign of d results from Eq. (7.10). The equivalent draught is used as input to hydrostatic curves.

We think that with present-day computers, and even hand calculators, it is possible to obtain with little effort and in a reasonable time more exact hydrostatic data. Moreover, assuming that the equivalent draught yields a good approximation of the displacement, what about the height of the metacentre above the baseline?

It is easy to calculate the hydrostatic data of a deflected hull by using the Bonjean curves. To do so one must simply draw a waterline passing through the three measured draughts, that is the forward, the midship and the aft draughts. The exact shape of the waterline is not known, but for small hull deflections that line cannot differ much from the shape taken by a drafting spline. Once the waterline is drawn, the Naval Architect has to read the Bonjean curves and use the readings as explained in Section 4.4. If a computer programme is available, and the ship offsets are stored in the required input format, one has to run the programme option for hydrostatic calculations in waves. The input wave length is equal to twice the waterline length. The input wave height (trough-to-crest) to be considered is equal to twice the hull deflection. If $T_m > (T_F + T_A)/2$, a bending situation known as **sagging**, the wave crest shall be placed in the midship section. This case is exemplified in Figure 7.5. The upper figure (a), shows what happens in reality. The lower figure (b), shows the corresponding computer input. If $T_M < (T_F + T_A)/2$, a bending situation known as **hogging**, the wave trough shall be placed in the midship section. The midship draught and the trim measured during the experiment shall be those supplied as input.

Example 1 shows an analytic treatment of the results of the inclining experiment; it yields the product $\Delta\overline{GM}$ corresponding to the ship loading during the test. As described above, the displacement, Δ, is read in the hydrostatic curves, or is calculated from Bonjean curves or by a computer programme. Thus, one obtains the metacentric height, $\overline{GM}$, of the same ship loading. The height of

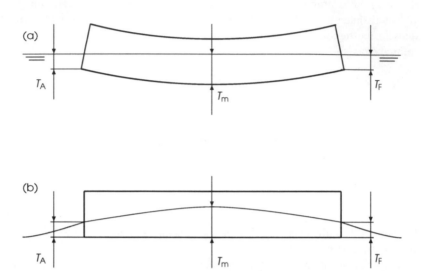

Figure 7.5 Deflected hull – sagging: (a) actual condition (b) computer input

the metacentre above the baseline, $\overline{KM}$, is obtained in the same way as the displacement, that is from the hydrostatic curves, by integrating values read in the Bonjean curves, or by running a computer programme. The height of the centre of gravity above baseline is calculated as

$$\overline{KG} = \overline{KM} - \overline{GM}$$

For small trim angles we can assume that the x-coordinates of the centre of gravity and of the centre of buoyancy are equal, that is $LCG = LCB$; otherwise see Subsection 7.3.2. The longitudinal centre of buoyancy is obtained in the same way as the displacement. At this point the Naval Architect knows the displacement and the centre of gravity of the ship loaded as during the inclining experiment. To calculate the data of the lightship one must first subtract the masses and the moments of the items that do not belong to the lightship, but were aboard during the test. Such items are, for example, the masses used to incline the ship. Next, one has to add the masses and the moments of the items that belong to the lightship, but were not yet assembled at the time of the inclining experiment. Sometimes the authorities that must approve the ship have their own inclining experiment regulations. Alternatively, the designer may be asked to abide by certain codes of practice that include provisions for the inclining experiment. Then, it is imperious to read those regulations before carrying on the work. An example of such regulations is the standard F1321-92 developed by ASTM (2001).

7.5 Summary

Stability and trim calculations require the knowledge of the displacement and of the position of the centre of gravity. To calculate these quantities it is necessary to organize the ship masses into weight groups. The sum of the weight groups that do not change during operation is called lightship displacement; for merchant vessels it is the sum of hull, outfit and machinery masses. The sum of the masses that are carried in operation according to the different loading cases is called deadweight; it includes the crew and its equipment, the cargo and passengers, the fuel, the lubricating oil, the fresh water, and the stores.

To find the displacement of a given loading case it is necessary to add the masses of the lightship and the deadweight items carried on board in that case. To find the coordinates of the centre of gravity, LCG, and VCG ($\overline{KG}$), it is necessary to sum up the moments of the above masses with respect to a transverse plane for the first, a horizontal plane for the second. The calculations can be conveniently carried out in an electronic spreadsheet or by software such as MATLAB.

Once the displacement, Δ, is known, one can find the corresponding mean draught, T_m, by reading the hydrostatic curves. These curves also yield the values of the longitudinal centre of buoyancy, LCB, the longitudinal centre of flotation, LCF, and the moment to change trim by 1 m, MCT. If the trim is small it can be found from

$$T_F - T_A = \frac{\Delta(LCG - LCB)}{MCT}$$

For normal loading situations the trim is always small. Then, the trimmed waterline, $W_\theta L_\theta$, intersects the waterlines of the ship on even keel, $W_0 L_0$, along a line passing through the centre of flotation, F, of $W_0 L_0$. To obtain the forward draught, T_F, and the aft draught, T_A, it is necessary to add to, or subtract from the mean draught a part of the trim proportional to the distance of the respective perpendicular from the centre of flotation

$$T_A = T_m - \text{trim}\, \frac{LCF}{L_{pp}} \ (\text{m})$$

$$T_F = T_m + \text{trim}\left(1 - \frac{LCF}{L_{pp}}\right)(\text{m})$$

If the trim is large, the heights of the centres of buoyancy and flotation must be taken into account.

Because of uncertainties in the calculation of masses and centres of gravity, it is necessary to validate them experimentally. This is done in the inclining experiment, an operation to be carried out for new buildings and for ships that underwent substantial changes. The ship is brought in sheltered waters and when no wind is blowing. A known mass, p, is displaced transversely a known distance, d, and the tangent of the resulting heel angle, $\tan\theta$, is measured. The statistical analysis of several inclining tests yields the product

$$\Delta\overline{GM} = \frac{pd}{\tan\theta}$$

The displacement, Δ, is found as a function of the draughts measured during the experiment. If a hull deflection is measured it must be taken into account. The vertical centre of gravity is calculated as

$$\overline{KG} = \overline{KM} - \overline{GM}$$

If the trim is large the hydrostatic curves cannot be used. The Bonjean curves are helpful here, as is a computer programme. Both Bonjean curves and computer programmes can be used to calculate the effect of hull deflection.

7.6 Examples

Example 7.1 – Least-squares fit of the results of an inclining experiment
The results of the inclining experiment presented here are taken from an example in Hansen (1985), but are converted into SI units. The data are plotted as points in Figure 7.6. At a first glance it seems reasonable to fit a straight line whose slope equals the mean of $pd/\tan\theta$ values. In this example, some trials performed with

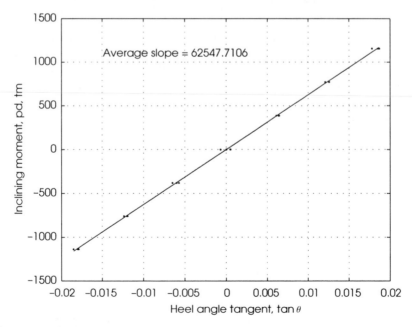

Figure 7.6 A plot of the results of an inclining experiment

very small pd values produced zero heel-angle tangents. Those cases must be discarded when averaging because they yield $pd/\tan\theta = \infty$. After eliminating the pairs corresponding to zero heel-angle tangents, we calculate the mean slope and obtain 53 679.638. The reader can easily verify that the line having this slope is far from being satisfactory. Available programmes for linear least-squares interpolation cannot be used because, in general, they fit a line having an equation of the form

$$y = c_1 x + c_2$$

Obviously, in our case the line must pass through the origin, that is $c_2 = 0$. Therefore, let us derive by ourselves a suitable procedure.

To simplify notations let x_i be the tangents of the measured heeling angles, and y_i the corresponding inclining moments. As said, we want to fit to the measured data a straight line passing through the origin

$$y = Mx \tag{7.11}$$

The error of the fitted point to the ith measured point is

$$y_i - Mx_i \tag{7.12}$$

We want to minimize the sum of the squares of errors

$$e = \sum (y_i - Mx_i)^2 \tag{7.13}$$

To do this we differentiate e with respect to M and equal the derivative to zero

$$\sum x_i(y_i - Mx_i) = 0 \tag{7.14}$$

The solution is

$$M = \frac{\sum x_i y_i}{\sum x_i^2} \tag{7.15}$$

An example of a MATLAB script file that plots the data, calculates the slope, M, and plots the fitted line is

```
%INCLINING   Analysis of Inclining Experiment
% Format of data is [ moment tangent ],
% initial units [ ft-tons - ]

incldata = [
   . . .
   . . . ];
% separate data
```

```
moment   = incldata(:, 1); tangent = incldata(:, 2);
plot(tangent, moment, 'k.'), grid
ylabel('Inclining moment, pd, tm')
xlabel('Heel angle tangent, tan\theta')
hold on
tmin = min(tangent); tmax = max(tangent);
M    = sum(tangent.*moment)/sum(tangent.^2);
Mmin = M*tmin; Mmax = M*tmax;
plot( [ tmin tmax ], [ Mmin Mmax ], 'k-')
text(-0.015, 1100, ['Average slope = ' num2str(M)])
hold off
```

Above, the user has to write the data of the inclining experiment in the matrix incldata. The MATLAB programme shown here can be easily transformed so that the user can input the name of a separate file that stores the incldata matrix.

7.7 Exercises

Exercise 7.1 – Small cargo ship homogeneous load, arrival
Using the data in Table 7.2 calculate the loading case **homogeneous cargo, arrival**, of the small cargo ship earlier encountered in this book. By **arrival** we mean the situation of the ship entering the port of destination with the fuel, the lubricating oil and the provisions consumed in great part. Using data in Tables 6.2 and 7.3 calculate the trim, the mean draught and the draughts at perpendiculars.

Table 7.2 Small cargo ship – homogeneous cargo, arrival

Weight item	Mass (t)	VCG (m)	LCG (m)
Lightship	1247.66	5.93	32.04
Crew and effects	3.60	9.60	11.00
Provisions	1.00	7.00	3.50
Fuel oil	27.74	2.17	23.15
Lubricating oil	3.49	0.62	17.08
Fresh water	8.70	1.61	9.75
Ballast water	248.87	0.55	39.62
Cargo in hold	993.94	4.35	42.62
Fruit cargo	90.00	6.08	38.66

Table 7.3 Small cargo ship – partial hydrostatic data, 2

Draught, T (m)	MCT (m)	LCB from midship (m)	LCF from midship (m)	Draught, T (m)	MCT (m)	LCB from midship (m)	LCF from midship (m)
2.00	2206	0.607	0.518	4.32	3223	0.291	−0.384
2.20	2296	0.600	0.460	4.40	3260	0.272	−0.430
2.40	2382	0.590	0.398	4.60	3336	0.225	−0.560
2.60	2470	0.575	0.330	4.80	3413	0.180	−0.698
2.80	2563	0.557	0.260	5.00	3485	0.131	−0.839
3.00	2645	0.537	0.190	5.20	3567	0.083	−0.960
3.20	2732	0.510	0.119	5.40	3639	0.033	−1.066
3.40	2824	0.480	0.041	5.60	3716	−0.018	−1.158
3.60	2906	0.442	−0.035	5.80	3793	−0.067	−1.231
3.80	2293	0.406	−0.017	5.96	3863	−0.108	−1.281
4.00	3085	0.360	−0.210	6.00	3880	−0.118	−1.293
4.20	3167	0.319	−0.314	6.20	3951	−0.167	−1.348

Exercise 7.2

Check that substituting in $T_F - T_A$ the expressions given by Eqs. (7.5) and (7.6) we obtain, indeed, the trim.

8

Intact stability regulations I

8.1 Introduction

In the preceding chapters, we presented the laws that govern the behaviour of floating bodies. We learnt how to find the parameters of a floating condition and how to check whether or not that condition is stable. The models we developed allow us to check the stability of a vessel under the influence of various heeling moments. At this point we may ask what is satisfactory stability, or, in simpler terms, how much stable a ship must be. Analyzing the data of vessels that behaved well, and especially the data of vessels that did not survive storms or other adverse conditions, various researchers and regulatory bodies prescribed criteria for deciding if the stability is satisfactory. In this chapter, we present examples of such criteria. To use picturesque language, we may say that in Chapters 2–7 we described *laws of nature*, while in this chapter we present *man-made laws*. Laws of nature act independently of man's will and they always govern the phenomena to which they apply. Man-made laws, in our case stability regulations, have another meaning. Stability regulations prescribe criteria for approving ship designs, accepting new buildings, or allowing ships to sail out of harbour. If a certain ship fulfils the requirements of given regulations, it does not mean that the ship can survive all challenges, but her chances of survival are good because stability regulations are based on considerable experience and reasonable theoretic models. Conversely, if a certain ship does not fulfill certain regulations, she must not necessarily capsize, only the risks are higher and the owner has the right to reject the design or the authority in charge has the right to prevent the ship from sailing out of harbour. Stability regulations are, in fact, **codes of practice** that provide reasonable safety margins. The codes are compulsory not only for designers and builders, but also for ship masters who must check if their vessels meet the requirements in a proposed loading condition.

The codes of stability presented in this chapter take into consideration only phenomena discussed in the preceding chapters. The stability regulations of the German Federal Navy are based on the analysis of a phenomenon discussed in Chapter 9; therefore, we defer their presentation until Chapter 10. For obvious reasons, it is not possible to include in this book all existing stability regulations; we only choose a few representative examples. Neither is it possible to present all the provisions of any single regulation. We only want to draw the attention of the reader to the existence of such codes of practice, to show how the models

developed in the previous chapters are applied, and to help the reader in understanding and using the regulations. Technological developments, experience
accumulation, and, especially major marine disasters can impose revisions of
existing stability regulations. For all the reasons mentioned above, before checking the stability of a vessel according to given regulations, the Naval Architect
must read in detail their newest, official version.

All stability regulations specify a number of loading conditions for which
calculations must be carried out. Some regulations add a sentence like 'and any
other condition that may be more dangerous'. It is the duty of the Naval Architect
in charge of the project to identify such situations, if they exist, and check if the
stability criteria are met for them.

8.2 The IMO code on intact stability

The Inter-Governmental Maritime Consultative Organization was established in
1948 and was known as IMCO. That name was changed in 1982 to **IMO – International Maritime Organization**. The purpose of IMO is the inter-governmental
cooperation in the development of regulations regarding shipping, **maritime
safety**, navigation, and the prevention of marine pollution from ships. IMO is an
agency of the United Nations and has 161 members. The regulations described
in this section were issued by IMO in 1995, and are valid 'for all types of ships
covered by IMO instruments' (see IMO, 1995). The intact stability criteria of
the code apply to 'ships and other marine vehicles of 24 m in length and above'.
Countries that adopted these regulations enforce them by issuing corresponding
national ordinances. Also, the Council of the European Community published
the Council Directive 98/18/EC on 17 March 1998.

8.2.1 Passenger and cargo ships

The code uses frequently the terms **angle of flooding, angle of downflooding**;
they refer to the smallest angle of heel at which an opening that cannot be closed
weathertight submerges. Passenger and cargo ships covered by the code shall
meet the following general criteria:

1. The area under the righting-arm curve should not be less than 0.055 m rad up
   to 30°, and not less than 0.09 m rad up to 40° or up to the angle of flooding
   if this angle is smaller than 40°.
2. The area under the righting-arm curve between 30° and 40°, or between 30°
   and the angle of flooding, if this angle is less than 40°, should not be less than
   0.03 m rad.
3. The maximum righting arm should occur at an angle of heel preferably
   exceeding 30°, but not less than 25°.
4. The initial metacentric height, $\overline{GM_0}$, should not be less than 0.15 m.

These requirements are inspired by Rahola's work cited in Section 6.1. Example 8.1 illustrates their application. Passenger ships should meet two further requirements. First, the angle of heel caused by the crowding of passengers to one side should not exceed 10°. The mass of a passenger is assumed equal to 75 kg. The centre of gravity of a standing passenger is assumed to lie 1 m above the deck, while that of a seated passenger is taken as 0.30 m above the seat. The second additional requirement for passenger ships refers to the angle of heel caused by the centrifugal force developed in turning. The heeling moment due to that force is calculated with the formula

$$M_{\mathrm{T}} = 0.02 \frac{V_0^2}{L_{\mathrm{WL}}} \Delta \left(\overline{KG} - \frac{T_{\mathrm{m}}}{2} \right) \tag{8.1}$$

where V_0 is the service speed in m s^{-1}. Again, the resulting angle shall not exceed 10°. The reason for limiting the angle of heel is that at larger values passengers may panic. The application of this criterion is exemplified in Figure 8.1 and Example 8.3.

In addition to the general criteria described above, ships covered by the code should meet a **weather criterion** that considers the effect of a beam wind applied when the vessel is heeled windwards. We explain this criterion with the help of Figure 8.2.

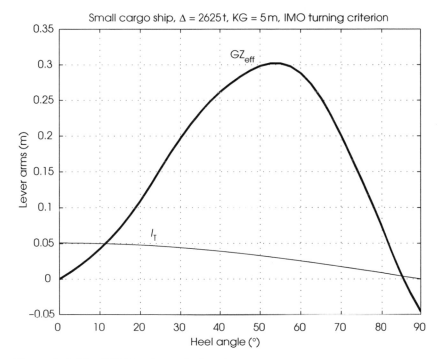

Figure 8.1 The IMO turning criterion

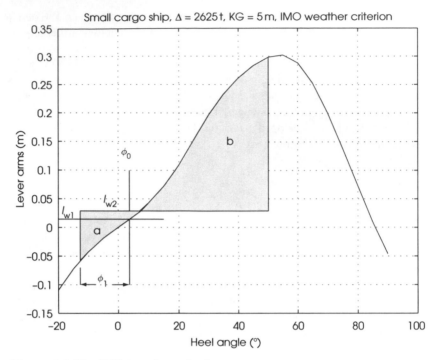

Figure 8.2 The IMO weather criterion

The code assumes that the ship is subjected to a constant wind heeling arm calculated as

$$\ell_{w1} = \frac{PAZ}{1000g\Delta} \qquad (8.2)$$

where $P = 504\,\mathrm{N\,m^{-2}}$, A is the projected lateral area of the ship and deck cargo above the waterline, in $\mathrm{m^2}$, Z is the vertical distance from the centroid of A to the centre of the underwater lateral area, or approximately to half-draught, in m, Δ is the displacement mass, in t, and $g = 9.81\,\mathrm{m\,s^{-2}}$. Unlike the model developed in Section 6.3 (model used by the US Navy), IMO accepts the more severe assumption that the wind heeling arm does not decrease as the heel angle increases. The code uses the notation θ for heel angles; we shall follow our convention and write ϕ. The static angle caused by the wind arm ℓ_{w1} is ϕ_0. Further, the code assumes that a wind gust appears while the ship is heeled to an angle ϕ_1 windward from the static angle, ϕ_0. The angle of roll is given by

$$\phi_1 = 109kX_1X_2\sqrt{rs} \qquad (8.3)$$

where ϕ_1 is measured in degrees, X_1 is a factor given in Table 3.2.2.3-1 of the code, X_2 is a factor given in Table 3.2.2.3-2 of the code, and k is a factor defined as follows:

- $k = 1.0$ for round-bilge ships;
- $k = 0.7$ for a ship with sharp bilges;
- k as given by Table 3.2.2.3-3 of the code for a ship having bilge keels, a bar keel or both.

As commented in Section 6.12, by using the factor k, the IMO code considers indirectly the effect of damping on stability. More specifically, it acknowledges that sharp bilges, bilge keels and bar keels reduce the roll amplitude. By assuming that the ship is subjected to the wind gust while heeled windward from the static angle, the dynamical effect appears more severe, as explained in Section 6.6 and the lower plot of Figure 6.5.

The factor r is calculated from

$$r = 0.73 + 0.6 \, \frac{\overline{OG}}{T_{\mathrm{m}}} \tag{8.4}$$

where $\overline{OG}$ is the distance between the waterline and the centre of gravity, positive upwards. The factor s is given in Table 3.2.2.3-4 of the code, as a function of the roll period, T. The code prescribes the following formula for calculating the roll period, in seconds,

$$T = \frac{2CB}{\sqrt{GM}_{\mathrm{eff}}} \tag{8.5}$$

where

$$C = 0.373 + 0.023 \left(\frac{B}{T_{\mathrm{m}}} \right) - 0.043 \left(\frac{L_{\mathrm{WL}}}{100} \right) \tag{8.6}$$

The code assumes that the lever arm of the wind gust is

$$\ell_{\mathrm{w2}} = 1.5 \, \ell_{\mathrm{w1}} \tag{8.7}$$

Plotting the curve of the arm ℓ_{w2} we distinguish the areas a and b. The area b is limited to the right at $50°$ or at the angle of flooding, whichever is smaller. The area b should be equal to or greater than the area a. This provision refers to dynamical stability, as explained in Section 6.6. When applying the criteria described above, the Naval Architect must use values corrected for the free-surface effect, that is $\overline{GM}_{\mathrm{eff}}$ and $\overline{GZ}_{\mathrm{eff}}$. The free-surface effect is calculated for the tanks that develop the greatest moment, at a heel of $30°$, while half full. The code prescribes the following equation for calculating the free-surface moment

$$M_{\mathrm{F}} = vb\gamma k\sqrt{\delta} \tag{8.8}$$

where v is the tank capacity in m^3, b is the maximum breadth of the tank in m, γ is the density of the liquid in $t\,m^{-3}$, δ is equal to the block coefficient of the tank, $v/b\ell h$, with h, the maximum height and ℓ, the maximum length, and k, a coefficient given in Table 3.3.3 of the code as function of b/h and heel angle. The contribution of small tanks can be ignored if $M_F/\Delta_{\min} < 0.01$ m at $30°$. We would like to remind the reader that present computer programmes for hydrostatic calculations yield values of the free-surface lever arms for any tank form described in the input, and for any heel angle. It is our opinon that, when available, such values should be preferred to those obtained with Eq. (8.8).

The code specifies the loading cases for which stability calculations must be performed. For example, for cargo ships the criteria shall be checked for the following four conditions:

1. Full-load departure, with cargo homogeneously distributed throughout all cargo spaces.
2. Full-load arrival, with 10% stores and fuel.
3. Ballast departure, without cargo.
4. Ballast arrival, with 10% stores and fuel.

8.2.2 Cargo ships carrying timber deck cargoes

Section 4.1 of the code applies to cargo ships that carry on their deck timber cargo extending longitudinally between superstructures and transversally on the full deck breadth, excepting a reasonable gunwale. Where there is no limiting superstructure at the aft, the cargo should extend at least to the after end of the aftermost hatch. For such ships the area under the righting-arm curve should not be less than 0.08 m rad up to $40°$ or up to the angle of flooding, whichever is smaller. The effective metacentric height should be positive in all stages of loading, voyage and unloading. The calculations should take into account the absorption of water by the deck cargo, and the water trapped within the cargo.

8.2.3 Fishing vessels

Section 4.2 of the code applies to decked seagoing vessels; they should fulfill the first three general requirements described in Subsection 8.2.1, while the metacentric height should not be less than 0.35 m for single-deck ships. If the vessel has a complete superstructure, or the ship length is equal to or larger than 70 m, the metacentric height can be reduced with the agreement of the government under whose flag the ship sails, but it should not be less than 0.15 m. The weather criterion applies in full to ships of 45 m length and longer. For fishing vessels whose length ranges between 24 and 45 m the code prescribes a wind gradient such that the pressure ranges between 316 and 504 Nm^{-2} for heights of 1–6 m above sea level. Decked vessels shorter than 30 m must have a minimum metacentric height calculated with a formula given in paragraph 4.2.6.1 of the code.

8.2.4 Mobile offshore drilling units

Section 4.6 of the code applies to mobile drilling units whose keels were laid after 1 March 1991. The wind force is calculated by considering the shape factors of structural members exposed to the wind, and a height coefficient ranging between 1.0 and 1.8 for heights above the waterline varying from 0 to 256 m. The area under the righting-arm curve up to the second static angle, or the downflooding angle, whichever is smaller, should exceed by at least 40% the area under the wind arm. The code also describes an alternative intact-stability criterion for two-pontoon, column-stabilized semi-submersible units.

8.2.5 Dynamically supported craft

A vessel is a **dynamically supported craft (DSC)** in one of the following cases:

1. If, in one mode of operation, a significant part of the weight is supported by other than buoyancy forces.
2. If the craft is able to operate at Froude numbers, $F_n = V/\sqrt{gL}$, equal or greater than 0.9.

The first category includes air-cushion vehicles and hydrofoil boats. Hydrofoil boats float, or sail, in the **hull-borne** or **displacement mode** if their weight is supported only by the buoyancy force predicted by Archimedes' principle. At higher speeds hydrodynamic forces develop on the foils and they balance an important part of the boat weight. Then, we say that the craft operates in the **foil-borne** mode.

Section 4.8 of the code applies to DSC operating between two ports situated in different countries. The requirements for hydrofoil boats are described in Subsection 4.8.7 of the code. The heeling moment in turning, in the displacement mode, is calculated as

$$M_R = \frac{0.196 V_0^2 \Delta \overline{KG}}{L}$$

where V_0 is the speed in turning, in $\mathrm{m\,s^{-1}}$, and M_R results in kN m. The formula is valid if the radius of the turning circle lies between $2L$ and $4L$. The resulting angles of inclination should not exceed $8°$.

The wind heeling moment, in the displacement mode, in kN m, should be calculated as

$$M_V = 0.001 P_V A_V Z$$

and is considered constant within the whole heeling range. The area subjected to wind pressure, A_V, is called here **windage area**. The wind pressure, P_V, corresponds to force 7 on the Beaufort scale. For boats that sail 100 nautical miles from the land, Table 4.8.7.1.1.4 of the code gives P_V values ranging between 46 and 64 Pa, for heights varying from 1 to 5 m above the water-line. The **windage area lever**, Z, is the distance between the waterline and the centroid of the windage area. A minimum capsizing moment, M_C, is calculated as shown in paragraph 4.8.7.1.1.5.1 of the code and as illustrated in Figure 8.3. The curve of the righting arm is extended to the left to a roll angle ϕ_z averaged from model or sea tests. In the absence of such data, the angle is assumed equal to $15°$. Then, a horizontal line is drawn so that the two grey areas shown in the figure are equal. The ordinate of this line defines the value M_C. According to the theory developed in Section 6.6 the ship capsizes if this moment is applied dynamically. The stability is considered sufficient if $M_C/M_V \geq 1$.

The code also prescribes criteria for the transient and foil-borne modes. Such criteria consider the forces developed on the foils, a subject that is not discussed in this book.

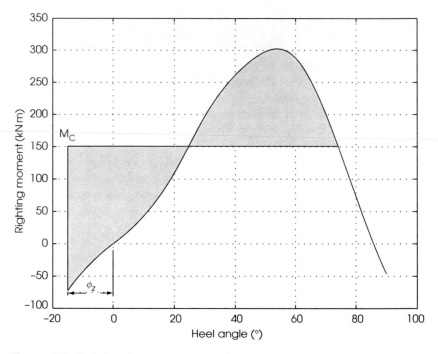

Figure 8.3 Defining the minimum capsizing moment of a dynamically supported craft (DSC)

8.2.6 Container ships greater than 100 m

Section 4.9 of the code defines a form factor C depending on the main dimensions of the ship and the configuration of hatches (Figure 4.9-1 in the code). The minimum values of areas under the righting-arm curve are prescribed in the form a/C, where a is specified for several heel intervals.

8.2.7 Icing

Chapter 5 of the code bears the title 'Ice considerations'. The following values, prescribed for fishing vessels, illustrate the severity of the problem. Stability calculations should be carried out assuming ice accretion (this is the term used in the code) with the surface densities:

* $30 \, \mathrm{kg \, m^{-2}}$ on exposed weather decks and gangways;
* $7.5 \, \mathrm{kg \, m^{-2}}$ for projected lateral areas on each side, above the waterplane.

The code specifies the geographical areas in which ice accretion can occur.

8.2.8 Inclining and rolling tests

Chapter 7 of the code contains the instructions for carrying on inclining experiments for all ships covered by the regulations, and roll-period tests for ships up to 70 m in length. The relationship between the metacentric height, $\overline{GM}_0$, and the roll period, T, is given as

$$\overline{GM}_0 = \left(\frac{fB}{T}\right)^2$$

where B is the ship breadth.

An interesting part of the Annex refers to the plot of heel-angle tangents against heeling moments; it explains the causes of deviations from a straight line, such as free surfaces of liquids, restrictions of movements, steady wind or wind gust.

8.3 The regulations of the US Navy

In 1944, an American fleet was caught by a tropical storm in the Pacific Ocean. In a short time three destroyers capsized, a fourth one escaped because a funnel broke down under the force of the wind. This disaster influenced the development of stability regulations for the US Navy. They were first published by Sarchin and Goldberg in 1962. These regulations were subsequently adopted by other navies.

The intact stability is checked under a wind whose speed depends on the service conditions. Thus, all vessels that must withstand tropical storms should be checked for winds of 100 knots. Ocean-going ships that can avoid the centre of tropical storms should be checked under a wind of 80 knots, while coastal vessels that can avoid the same dangers should be checked for winds of 60 knots. Coastal vessels that can be called to anchorage when expecting winds above Force 8, and all harbour vessels should be checked under the assumption of 60-knots winds.

We explain the weather criterion in Figure 8.4. The righting arm, $\overline{GZ}$, is actually the effective righting arm, $\overline{GZ}_{\text{eff}}$, calculated by taking into account the free-surface effect. The wind arm is obtained from the formula

$$l_V = \frac{0.017V_w^2 A\ell \cos^2 \phi}{1000\Delta} \tag{8.9}$$

where V_w is the wind velocity in knots, A, the sail area in m^2, ℓ, the distance between half-draught and the centroid of the sail area in m, and Δ, the displacement in t. The first angle of static equilibrium is ϕ_{st1}. The criterion for static stability requires that the righting arm at this angle be not larger than 0.6 of the maximum righting arm. To check dynamical stability the regulations assume that the ship is subjected to a gust of wind while heeled 25° to the windward of ϕ_{st1}. We distinguish then the area a between the wind heeling arm and the

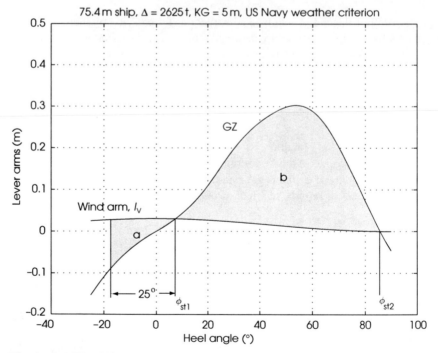

Figure 8.4 The US Navy weather criterion

righting-arm curves up to ϕ_{st1}, and the area b between the two curves, from the first static angle, ϕ_{st1}, up to the second static angle, ϕ_{st2} (see Figure 8.4), or up to the angle of downflooding, whichever is less (see Figure 8.5). The ratio of the area b to the area a should be at least 1.4. A numerical example of the application of the above criteria is shown in Example 8.4.

The designer can take into account the wind gradient, that is the variation of the wind speed with height above the waterline. Then, the 'nominal' wind speed defined by the service area is that measured at 10 m (30 ft) above the waterline. Performing a regression about new data presented by Watson (1998) we found the relationship

$$\frac{V_W}{V_0} = 0.73318h^{0.13149} \tag{8.10}$$

where V_W is the wind speed at height h, V_0 is the nominal wind velocity, and h is the height above sea level, in m. In Figure 8.6, the points indicated by Watson (1998) appear as asterisks, while the values predicted by Eq. (8.10) are represented by the continuous line. An equation found in literature has the form $V_W/V_0 = (h/10)^b$. Regression over the data given by Watson yielded $b = 0.73318$, but the resulting curve fitted less well than the curve corresponding to Eq. (8.10).

To apply the wind gradient one has to divide the sail area into horizontal strips and apply in each strip the wind ratio yielded by Eq. (8.10). Let R_i be that ratio

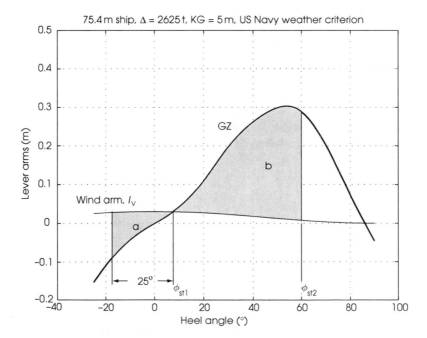

Figure 8.5 The US Navy weather criterion, downflooding angle 60°

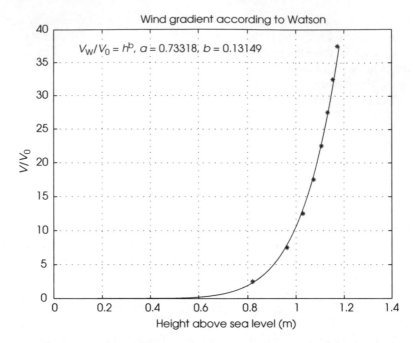

Figure 8.6 Wind gradient

for the ith strip. The results for the individual strips should be integrated by one of the rules for numerical integration. The coefficient in Eq. (8.9) should be modified to 0.0195 and then, the wind arm is given by

$$\ell_{\mathrm{V}} = \frac{0.0195 V_0^2}{1000\Delta} h \left(\sum \alpha_i R_i^2 A_i \ell_i \right) \cos^2 \phi \tag{8.11}$$

where V_0 is the nominal wind speed, h is the common height of the horizontal strips, α_i is the trapezoidal multiplier, A_i is the area of the ith strip, and ℓ_i, the vertical distance from half-draught to the centroid of the ith strip. It can be easily shown that

$$\ell_i = \frac{2i - 1}{2} h + \frac{T}{2} \tag{8.12}$$

To explain the criterion for stability in turning we use Figure 8.7. The heeling arm due to the centrifugal force is calculated from

$$\ell_{\mathrm{TC}} = \frac{V^2(\overline{KG} - T/2)}{gR} \cos \phi \tag{8.13}$$

where V is the ship speed in $\mathrm{m\,s^{-1}}$ and R is the turning radius in m.

Ideally, R should be taken as one half of the **tactical diameter** measured from model or sea tests at full scale. Where this quantity is not known, an estimation

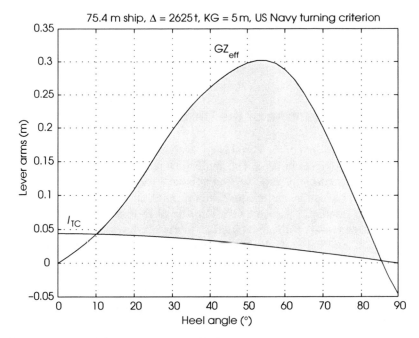

Figure 8.7 The US Navy turning criterion

must be made. In Section 6.4, we described an empirical formula developed for this aim, in Section 8.4, about the UK Navy, we give another approximate relationship. The stability is considered satisfactory if

1. the angle of heel does not exceed $15°$;
2. the heeling arm at the angle of static equilibrium is not larger than 0.6, the maximum righting-arm value;
3. the grey area in the figure, called **reserve of dynamical stability** is not less than 0.4 of the whole area under the positive righting-arm curve.

If the downflooding angle is smaller than the second static angle, the area representing the reserve of stability should be limited to the former value. An application of the above criteria is given in Example 8.5.

Another hazard considered in the regulations of the US Navy is the lifting of heavy weights over the side. The corresponding heeling arm is yielded by

$$l_W = \frac{wa}{\Delta} \cos \phi \qquad (8.14)$$

where w is the lifted mass, a is the transverse distance from the centreline to the boom end, and Δ is the displacement mass including w. The criteria of stability are the same as those required for stability in turning.

The crowding of personnel to one side causes an effect similar to that of a heavy weight lifted transversely to one side. The heeling arm is yielded by Eq. (8.14), assuming that the personnel moved to one side as far as possible when five men crowd in one square metre. Again, the stability is considered sufficient if the requirements given for stability in turning are met.

8.4 The regulations of the UK Navy

The stability standard of the Royal Navy evolved from the criteria published by Sarchin and Goldberg in 1962. The first British publication appeared in 1980 as NES 109. The currently valid version is Issue 4 (see MoD, 1999a). The document should be read in conjunction with the publication SSP 42 (MoD, 1999b). The British standard is issued by the Ministry of Defence, shortly MoD, and is applicable to vessels with a military role, to vessels designed to MoD standards but without a military role, and to auxiliary vessels. Vessels with a military role are exposed to enemy action or to similar dangers during peacetime exercises. We shall discuss here only the provisions related to such vessels. The standard NES 109 has two parts, the first dealing with conventional ships, the second with unconventional vessels. The second category includes:

1. monohull vessels of rigid construction having a speed in knots larger than $4\sqrt{L_{\mathrm{WL}}}$, where the waterline length is measured in m;
2. multi-hull vessels;
3. dynamically supported vessels.

In this book, we briefly discuss only the provisions for conventional vessels. According to NES 109 the displacement and $\overline{KG}$ values used in stability calculations should include growth margins. For warships the weight growth margin should be 0.65% of the lightship displacement, for each year of service. The $\overline{KG}$ margin should be 0.45% of the lightship $\overline{KG}$, for each year of service.

The shape of the righting-arm curve should be such that:

- the area under the curve, up to $30°$, is not less than 0.08 m rad;
- the area up to $40°$ is not less than 0.133 m rad;
- the area between $30°$ and $40°$ is not less than 0.048 m rad;
- the maximum $\overline{GZ}$ is not less than 0.3 m and should occur at an angle not smaller than $30°$.

One can immediately see that all these requirements are considerably more severe than those prescribed by IMO 95 for merchant ships.

The stability under beam winds should be checked for the following wind speeds:

- 90 knots for ocean-going vessels;
- 70 knots for ocean-going or coastal vessels that can avoid extreme conditions;

- 50 knots for coastal vessels that can be called to anchorage to avoid winds over Force 8, and for harbour vessels.

These values are lower than those required by the US Navy and partially coincide with those specified by the German Navy. The angle of heel caused by the wind should not exceed 30°. The criterion for statical stability is the same as that of the US Navy, that is, the righting arm at the first static angle should not be greater than 0.6, the maximum righting arm. As in the American regulations, it is assumed that the ship rolls 25° windwards from the first static angle, and it is required that the reserve of stability should not be less than 1.4 times the area representing the wind heeling energy. Figure 1.3 in the UK regulations shows that the area representing the reserve of stability is limited at the right by the downflooding angle. When checking stability in turning the corresponding ship speed should be 0.65 times the speed on a straight-line course. If no better data are available, it should be assumed that the radius of turning equals 2.5 times the length between perpendiculars. The angle of heel in turning should be less than 20°, a requirement less severe than that of the US Navy. The static criterion, regarding the value of the righting arm at the first static angle, and the dynamic criterion, regarding the reserve of stability, are the same as those of the US Navy.

To check stability when lifting a heavy mass over the side, the heeling arm should be calculated from

$$l_W = \frac{w(a \cos \phi + d \sin \phi)}{\Delta} \tag{8.15}$$

where a is the horizontal distance of the tip of the boom from the centreline, and d is the height of the point of suspension above the deck. Stability is considered sufficient if the following criteria are met:

1. The angle of heel is less than 15°.
2. The righting arm at the first static angle is less than half the maximum righting arm.
3. The reserve of stability is larger than half the total area under the righting-arm curve. The area representing the reserve of stability is limited at the right by the angle of downflooding.

It can be easily seen that criteria 2 and 3 are more stringent than those of the US Navy.

The NES 109 standard also specifies criteria for checking stability under icing. A thickness of 150 mm should be assumed for all horizontal decks, with an ice density equal to $950 \, \text{kg m}^{-3}$. Only the effect on displacement and $\overline{KG}$ should be considered, and not the effect on the sail area.

8.5 A criterion for sail vessels

The revival of the interest for large sailing vessels and several accidents justified new researches and the development of codes of stability for this category of ships. Thus, the UK Department of Transport sponsored a research carried out at the Wolfson Unit for Marine Transportation and Industrial Aerodynamics of the University of Southampton (Deakin, 1991). The result of the research is the code of stability described in this section. A more recent research is presented by Cleary, Daidola and Reyling (1996). The authors compare the stability criteria for sailing ships adopted by the US Coast Guard, the Wolfson Unit, the Germanischer Lloyd, the Bureau Veritas, the Ateliers & Chantiers du Havre, and Dr Ing Alimento of the University of Genoa. These criteria are illustrated by applying them to one ship, the US Coast Guard training barque *Eagle*, formerly *Horst Wessel* built in 1936 in Germany.

In this section, we describe the intact stability criteria of 'The code of practice for safety of large commercial sailing & motor vessels' issued by the UK Maritime and Coastguard Agency (Maritime, 2001). The code 'applies to vessels in commercial use for sport or pleasure ... that are 24 metres in load line length and over ... and that do not carry cargo and do not carry more than 12 passengers.' For shorter sailing vessels, the UK Marine Safety Agency published another code, namely 'The safety of small commercial sailing vessels.'

The research carried out at the Wolfson Unit yielded a number of interesting results:

1. Form coefficients of sail rigs vary considerably and are difficult to predict. We mean here the coefficient c in

$$p = \frac{1}{2}c\rho v^2$$

   where p is the pressure, ρ, the air density, and V, the speed of the wind component perpendicular to the sail.
2. The wind-arm curve behaves like $\cos^{1.3} \phi$.
3. Wind gusts do not build up instantly, as conservatively assumed (see Section 6.6). The wind speed of gusts due to atmospheric turbulence are unlikely to exceed 1.4 times the hourly mean, have rise times of 10 to 20 s and durations of less than a minute. Other gusts, due to other atmospheric phenomena, are known as **squalls** and they can be much more dangerous. Because the rise-up times of significant gusts are usually larger than the natural roll periods of sailing vessels, ships do not respond as described in Section 6.6, but have time to find equilibrium positions close to the intersection of the gust-arm curve and the righting-arm curve.
4. Sails considerably increase the damping of the roll motion, limiting the response to a wind gust and enhancing the effect described above. Thus, the

heel angle caused by a wind gust is smaller than that predicted by the balance of areas representing wind energy and righting-arm work (Section 6.6).

Based on the above conclusions, the criterion of intact stability adopted by the UK Maritime and Coastguard Agency does not consider the sail rig and the wind moment developed on it. The code simply provides the skipper with a means for appreciating the maximum allowable heel angle under a steady wind, if wind gusts are expected. Sailing at the recommended angle will avoid the submergence under gusts of openings that could lead to ship loss.

The code defines the downflooding angle as the angle at which openings having an 'aggregate area' whose value in metres is greater than $\Delta/1500$, submerge. The displacement, Δ, is measured in t. Deakin (1991) explains that under his assumptions the mass of water flowing through the above openings during 5 minutes equals the ship displacement. No ship is expected to float after a flooding of this extent, and five minutes are considered a maximum reasonable time of survival. For those who wish to understand Deakim's reasoning we remind that the flow through an orifice is proportional to the orifice area multiplied by the fluid speed

$$Q = a\, c_V \sqrt{2gh}$$

where a is the orifice area, c_V, a discharge coefficient always smaller than 1, g, the acceleration of gravity, and h, the level of water above the orifice. The authors of the code assume $c_V = 1$ and $h = 1$ m. We calculate

$$Q = \frac{\Delta}{1500} \times 1 \times \sqrt{2 \times 9.81 \times 1} = 0.003\Delta\,\mathrm{m}^3\,\mathrm{s}^{-1}$$

It follows that in sea water 5.5 minutes are required for a mass of water equal to the displacement mass.

We use Figure 8.8 to describe the criterion for intact stability. The righting-arm curve is marked GZ; it is based on the data of an actual training yacht. At the downflooding angle we measure the value of the righting arm, $\overline{GZ}_\mathrm{f}$. We assume here the downflooding angle $\phi_\mathrm{f} = 60°$. We calculate a gust-wind lever in upright condition

$$WLO = \frac{\overline{GZ}_\mathrm{f}}{\cos^{1.3}\phi_\mathrm{f}}$$

The dashed line curve represents the gust arm. Under the assumptions that the gust speed is 1.4 times the speed of the steady wind, the pressure due to steady wind is one half that of the gust, and so is the corresponding heeling arm. Therefore, we draw the 'derived curve' as the dash-dot line beginning at $WLO/2$

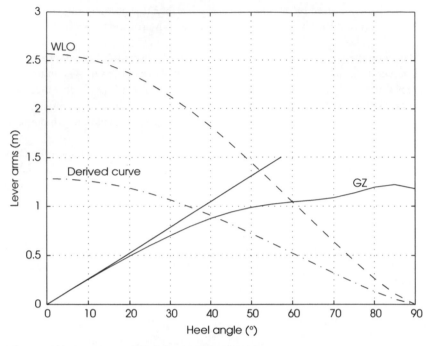

Figure 8.8 Intact stability criterion for sail ships

and proportional to $\cos^{1.3} \phi$. This curve intercepts the GZ curve at the angle of
steady heel, here a bit larger than 40°.
The code requires that:

1. The $\overline{GZ}$ curve should have a positive range not shorter than 90°.
2. If the downflooding angle is larger than 60°, ϕ_f should be taken as 60°.
3. The angle of steady heel should not be less than 15°.

8.6 A code of practice for small workboats and pilot boats

The regulations presented in this section (see Maritime, 1998) apply to small
UK commercial sea vessels of up to 24 m load line length and that carry cargo
and/or not more than 12 passengers. The regulations also apply to service or
pilot vessels of the same size. By 'load line length' the code means either 96%
of the total waterline length on a waterline at 85% depth, or the length from the
fore side of the stern to the axis of the rudder stock on the above waterline.

The lightship displacement to be used in calculations should include a margin
for growth equal to 5% of the lightship displacement. The x-coordinate of the

centre of gravity of this margin shall equal LCG, and the z-coordinate shall equal either the height of the centre of the weather deck amidships or the lightship $\overline{KG}$, whichever is the higher. Curves of statical stability shall be calculated for the following loading cases:

- loaded departure, 100% consumables;
- loaded arrival, 10% consumables;
- other anticipated service conditions, including possible lifting appliances.

The stability is considered sufficient if the following two criteria are met in addition to criteria 1–4 in Subsection 8.2.1.

1. The maximum of the righting-arm curve should occur at an angle of heel not smaller than 25°.
2. The effective, initial metacentric height, $\overline{GM}_{\text{eff}}$, should not be less than 0.35 m.

If a multihull vessel does not meet the above stability criteria, the vessel shall meet the following alternative criteria:

1. If the maximum of the righting-arm curve occurs at 15°, the area under the curve shall not be less than 0.085 m rad. If the maximum occurs at 30°, the area shall not be less than 0.055 m rad.
2. If the maximum of the righting-arm curve occurs at an angle ϕ_{GZmax} situated between 15° and 30°, the area under the curve shall not be less than

$$A = 0.055 + 0.002(30° - \phi_{\text{GZmax}}) \tag{8.16}$$

where A is measured in m rad.
3. The area under the righting-arm curve between 30° and 40°, or between 30° and the angle of downflooding, if this angle is less than 40°, shall not be less than 0.03 m rad.
4. The righting arm shall not be less than 0.2 m at 30°.
5. The maximum righting arm shall occur at an angle not smaller than 15°.
6. The initial metacentric height shall not be less than 0.35 m.

The intact stability of new vessels of less than 15 m length that carry a combined load of passengers and cargo of less than 1000 kg is checked in an inclining experiment. The passengers, the crew without the skipper, and the cargo are transferred to one side of the ship, while the skipper may be assumed to stay at the steering position. Under these conditions the angle of heel shall not exceed 7°.

For vessels with a watertight weather deck the freeboard shall be not less than 75 mm at any point. For open boats the freeboard to the top of the gunwale shall not be less than 250 mm at any point.

8.7 Regulations for internal-water vessels

8.7.1 EC regulations

The European prescriptions for internal-navigation ships are contained in directive 82/714/CEE of October 1982. In September 1999, a proposal for modifications was submitted to the European parliament. The proposal details the internal waterways of Europe for which it is valid.

Intact stability is considered sufficient if:

- the heel angle due to the crowding of passengers on one side does not exceed $10°$;
- the angle of heel due to the combined effect of crowding, wind pressure and centrifugal force does not exceed $12°$.

In calculations it should be assumed that fuel and water tanks are half full. The considered wind pressure is $0.1 \, \mathrm{kN \, m^{-2}}$. At the angles of heel detailed above, the minimum freeboard should not be less than 0.2 m. If lateral windows can be opened, a minimum safety distance of 0.1 m should exist.

8.7.2 Swiss regulations

The Swiss regulations for internal navigation are contained in an ordinance of 8 October 1978. Some modifications are contained in an ordinance of 9 March 2001 of the Swiss Parliament (Der Schweizerische Bundesrat). According to them cargo ships should be tested under a wind pressure of $0.25 \, \mathrm{kN \, m^{-2}}$. The heeling moment in turning, in kN m, should be calculated as

$$M_{\mathrm{TC}} = \frac{cV^2\Delta}{L_{\mathrm{WL}}} \left(\overline{KG} - \frac{T}{2} \right)$$

where $c \geq 0.4$ is a coefficient to be supplied by the builder or the operator. Stability is considered sufficient if under the above assumptions the heeling angle does not exceed $5°$ and the deck side does not submerge. The metacentric height should not be less than 1 m. The required wind pressure is definitely lower than that required for sea-going ships. On the other hand, the other requirements are more stringent.

8.8 Summary

The IMO Code on Intact Stability applies to ships and other marine vehicles of 24 m length and above. The metacentric height of passenger and cargo ships should be at least 0.15 m, and the areas under the righting-arm curve, between certain heel angles, should not be less than the values indicated in the document. Passenger vessels should not heel in turning more than 10°. In addition, passenger and cargo ships should meet a weather criterion in which it is assumed that the vessel is subjected to a wind arm that is constant throughout the heeling range. The heeling arm of wind gusts is assumed equal to 1.5 times the heeling arm of the steady wind. If a wind gust appears while the ship is heeled windwards by an angle prescribed by the code, the area representing the reserve of buoyancy should not be less than the area representing the heel energy. The former area is limited to the right by the angle of downflooding or by 50°, whichever is less.

The IMO code contains special requirements for ships carrying timber on deck, for fishing vessels, for mobile offshore drilling units, for dynamically supported craft, and for containerships larger than 100 m. The code also contains recommendations for inclining and for rolling tests.

The stability regulations of the US Navy prescribe criteria for statical and dynamical stability under wind, in turning, under passenger crowding on one side, and when lifting heavy weights over the side. The static criterion requires that the righting arm at the first static angle should not exceed 60% of the maximum righting arm. When checking dynamical stability under wind, it is assumed that the ship rolled 25° windwards from the first static angle. Then, the area representing the reserve of stability should be at least 1.4 times the area representing the heeling energy. When checking stability in turning, or under crowding or when lifting heavy weights, the angle of heel should not exceed 15° and the reserve of stability should not be less than 40% of the total area under the righting-arm curve.

The stability regulations of the UK Navy are derived from those of the US Navy. In addition to static and dynamic criteria such as those mentioned above, the UK standard includes requirements concerning the areas under the righting-arm curve. The minimum values are higher than those prescribed by IMO for merchant ships. While the wind speeds specified by the UK standard are lower than those in the US regulations, the stability criteria are more severe.

A quite different criterion is prescribed in the code for large sailing vessels issued by the UK Ministry of Transport. As research proved that wind-pressure coefficients of sail rigs cannot be predicted, the code does not take into account the sail configuration and the heeling moments developed on it. The document presents a simple method for finding a heel angle under steady wind, such that the heel angle caused by a gust of wind would be smaller than the angle leading to downflooding and ship loss. The steady heel angle should not exceed 15°, and the range of positive heeling arms should not be less than 90°.

Additional regulations mentioned in this chapter are a code for small work-boats issued in the UK, and codes for internal-navigation vessels issued by the European Parliament and by the Swiss Parliament.

8.9 Examples

Example 8.1 – Application of the IMO general requirements for cargo and passenger ships

Let us check if the small cargo ship used in Subsection 7.2.2 meets the IMO general requirements. We assume the same loading condition as in that section. The vessel was built four decades before the publication of the IMO code for intact stability; therefore, it is not surprising if several criteria are not met. Table 8.1 contains the calculation of righting-arm levers and areas under the righting-arm curve. Figure 8.2 shows the corresponding statical stability curve. The areas under the righting-arm curve are obtained by means of the algorithm described in Section 3.4. The analysis of the results leads to the following conclusions:

1. The area under the $\overline{GZ}_{\text{eff}}$ curve, up to 30°, is 0.043 m rad, less than the required 0.055. The area up to 40° equals 0.084 m rad, less than the required 0.09 m rad. The area between 30° and 40° equals 0.04 m rad, more than the required 0.03 m rad.

Table 8.1 Small cargo ship – the IMO general requirements

Heel angle (°)	ℓ_{p} (m)	$(\overline{KG} + \ell_{\text{F}})\sin\phi$ (m)	$\overline{GZ}_{\text{eff}}$ (m)	Area under righting arm (m^2)
0.0	0.000	0.000	0.000	0.000
5.0	0.459	0.439	0.019	0.001
10.0	0.918	0.875	0.043	0.004
15.0	1.377	1.304	0.072	0.009
20.0	1.833	1.724	0.109	0.017
25.0	2.283	2.130	0.153	0.028
30.0	2.717	2.520	0.197	0.043
35.0	3.124	2.891	0.233	0.062
40.0	3.501	3.240	0.262	0.084
45.0	3.847	3.564	0.283	0.107
50.0	4.159	3.861	0.298	0.133
55.0	4.431	4.129	0.302	0.159
60.0	4.653	4.365	0.288	0.185
65.0	4.821	4.568	0.253	0.208
70.0	4.937	4.736	0.201	0.228
75.0	5.007	4.868	0.139	0.243
80.0	5.036	4.963	0.073	0.252
85.0	5.030	5.021	0.009	0.256
90.0	4.994	5.040	−0.046	0.254

2. The righting arm lever equals 0.2 m at 30°; it meets the requirement at limit.
3. The maximum righting arm occurs at an angle exceeding the required 30°.
4. The initial effective metacentric height is 0.12 m, less than the required 0.15 m.

Example 8.2 – Application of the IMO weather criterion for cargo and passenger ships

We continue the preceding example and illustrate the application of the weather criterion to the same ship, in the same loading condition. The main dimensions are $L = 75.4$, $B = 11.9$, $T_m = 4.32$, and the height of the centre of gravity is $\overline{KG} = 5$, all measured in metres. The sail area is $A = 175\,\text{m}^2$, the height of its centroid above half-draught $Z = 4.19\,\text{m}$, and the wind pressure $P = 504\,\text{N m}^{-2}$. The calculations presented here are performed in MATLAB keeping the full precision of the software, but we display the results rounded off to the first two or three digits. To keep the precision we define at the beginning the constants, for example $L = 75.4$, and then call them by name, for example L.

The wind heeling arm is calculated as

$$l_{w1} = \frac{PAZ}{1000g\Delta} = 0.014\,\text{m}$$

The lever of the wind gust is

$$l_{w2} = 1.5 l_{w1} = 0.022\,\text{m}$$

We assume that the bilge keels are 15 m long and 0.4 m deep; their total area is

$$A_k = 2 \times 15 \times 0.4 = 12\,\text{m}^2$$

To enter Table 3.2.2.3-3 of the code we calculate

$$\frac{A_k \times 100}{L \times B} = 1.337$$

Interpolating over the table we obtain $k = 0.963$. To find X_1 we calculate $B/T_m = 2.755$ and interpolating over Table 3.2.2.3-1 we obtain $X_1 = 0.94$. To enter Table 3.2.2.3-2 we calculate the block coefficient

$$C_B = \frac{2635}{(1.03 \times L \times B \times T_m)} = 0.66$$

Interpolation yields $X_2 = 0.975$. The height of centre of gravity above water-line is

$$\overline{OG} = KG - T_m = 0.68$$

In continuation we calculate

$$r = 0.73 + \frac{0.6 \times \overline{OG}}{T_m} = 0.824$$

To find the roll period we first calculate the coefficient

$$C = 0.373 + 0.023 \times \left(\frac{B}{T_m}\right) - 0.043 \times \left(\frac{L}{100}\right) = 0.404$$

With $\overline{GM}_{\text{eff}} = 0.12$ m the formula prescribed by the code yields the roll period

$$T = \frac{2CB}{\sqrt{\overline{GM}_{\text{eff}}}} = 27.752\,\text{s}$$

With this value we enter Table 3.2.2.3-4 and retrieve $s = 0.035$. Then, the angle of roll windwards from the angle of statical stability, under the wind arm l_{w1}, is

$$\phi_1 = 109kX_1X_2\sqrt{rs} = 16.34°$$

Visual inspection of Figure 8.2 shows that the weather criterion is met. This fact is explained by the low sail area of the ship.

Example 8.3 – The IMO turning criterion
To illustrate the IMO criterion for stability in turning we use the data of the same small cargo ship that appeared above. Cargo ships are not required to meet this criterion, but we can assume, for our purposes, that the ship carries more than 12 passengers.

The ship length is $L = 75.4$ m, the mean draught $T_m = 4.32$ m, the ship speed $V_0 = 16$ knots, and the vertical centre of gravity $\overline{KG} = 5.0$ m. The speed in m s^{-1} is

$$V_0 = 16 \times 0.5144 = 8.23 \text{ m s}^{-1}$$

and the heel arm due to the centrifugal force is

$$l_T = 0.02\frac{V_0^2}{L}\left(\overline{KG} - \frac{T_m}{2}\right) = 0.051 \text{ m}$$

Figure 8.1 shows the resulting statical stability curve. We see that the heel angle is slightly larger than 11°.

Example 8.4 – The weather criterion of the US Navy
To allow comparisons between various codes of stability we use again the data of the small cargo ship that appeared in the previous examples. We initiate the calculations by defining the wind speed, $V_W = 80$ knots, the sail area, $A = 175 \text{ m}^2$, the height of its centroid above half-draught, $\ell = 4.19$ m, and the displacement, $\Delta = 2625$ t. The corresponding stability curve is shown in Figure 8.4. The wind heeling arm is given by

$$l_V = \frac{0.017V_w^2 A\ell \cos^2\phi}{1000\Delta} = 0.03\cos^2\phi \text{ (m)}$$

At the intersection of the righting-arm and the wind-arm curves we find the first static angle, $\phi_{st1} \approx 7.5°$, and the righting arm at that angle equals 0.03 m.

Rolling 25° windwards from the first static angle the ship reaches $-17.5°$. The second static angle is $\phi_{st2} = 85.7°$. The ratio of the $\overline{GZ}$ value at the first static angle to the maximum $\overline{GZ}$ is 0.03/0.302, that is close to 0.1 and smaller than the maximum admissible 0.6. The area b equals 0.235 m rad, and the area a equals 0.024 m rad. The ratio of the area b to the area a is nearly 10, much larger than the minimum admissible 1.4. We conclude that the vessel meets the criteria of the US Navy.

Example 8.5 – The turning criterion of the US Navy
We continue the calculations using the data of the same ship as above. We assume the speed of 16 knots, and the vertical centre of gravity, $\overline{KG} = 5$ m, as in Example 8.3. In the absence of other recommendations we consider, as in NES 109, that the speed in turning is 0.65 times the speed on a straight-line course, that is

$$V_0 = 0.65 \times 16 \times 0.5144 = 5.35 \, \mathrm{m\,s}^{-1}$$

Also, we assume that the radius of the turning circle equals 2.5 times the waterline length

$$R = \frac{2.5 \times 75}{4} = 188.5 \, \mathrm{m}$$

Then, the heeling arm in turning is given by

$$l_{TC} = \frac{V_0^2(\overline{KG} - T/2)}{gR} \cos \phi = 0.044 \cos \phi \, \mathrm{m}$$

Drawing the curves as in Figure 8.7 we find that the first static angle is $\phi_{st1} = 10.3°$, and the corresponding righting arm equals 0.044 m. The ratio of this arm to the maximum righting arm is 0.044/0.302 = 0.15, less than the maximum admissible 0.6. The reserve of dynamical stability, that is the grey area in Figure 8.4, equals 0.205 m rad, while the total area under the positive righting-arm curve is 0.256 m rad. The ratio of the two areas equals 0.8, the double of the minimum admissible 0.4. We conclude that the ship meets the criteria of the US Navy.

8.10 Exercises

Exercise 8.1 – IMO general requirements
Let us refer to Example 8.1. Find the $\overline{KG}$ value for which the general requirement 4 is fulfilled. Check if with this value the first general requirement is also met.

Exercise 8.2 – The IMO turning criterion
Return to the example in Section 8.9 and find the limit speed for which the turning criterion is met.

Exercise 8.3 – The IMO turning criterion

Return to the example in Exercise 8.2 and check if with the vertical centre of gravity, $\overline{KG}$, found in Section 8.10 the turning criterion is met.

Exercise 8.4 – The US-Navy turning criterion

Return to Example 8.4 and redo the calculations assuming a wind speed of 100 knots.

Exercise 8.5 – The code for small vessels

Check that for $\phi_{\mathrm{GZmax}} = 15°$ and $30°$, Eq. (8.16) yields the values specified in criterion 1 for multihull vessels (Section 8.6).

9
Parametric resonance

9.1 Introduction

Up to this chapter we assumed that the sea surface is plane. Actually, such a situation never occurs in nature, not even in the sheltered waters of a harbour. Waves always exist, even if very small. Can waves influence ship stability? And if yes, how? Arndt and Roden (1958) and Wendel (1965) cite French engineers that discussed this question at the end of the nineteenth century (J. Pollard and A. Dudebout, 1892, *Théorie du Navire*, Vol. III, Paris). In the 1920s, Doyère explained how waves influence stability and proposed a method to calculate that influence. After 1950 the study of this subject was prompted by the sinking of a few ships that were considered stable.

At a first glance *beam seas* – that is waves whose crests are parallel to the ship – seem to be the most dangerous. In fact, parallel waves cause large angles of heel; loads can get loose and endanger stability. However, it can be shown that the resultant of the weight force and of the centrifugal force developed in waves is perpendicular to the wave surface. Therefore, a correctly loaded vessel will never capsize in parallel waves, unless hit by large breaking waves. Ships **can capsize** in *head seas* – that is waves travelling against the ship – and especially in *following seas* – that is waves travelling in the same direction with the ship. This is the lesson learnt after the sinking of the ship *Irene Oldendorff* in the night between 30 and 31 December 1951. Kurt Wendel analyzed the case and reached the conclusion that the disaster was due to the variation of the righting arm in waves. Divers that checked the wreck found it intact, an observation that confirmed Wendel's hypothesis. Another disaster was that of *Pamir*. Again, the calculation of the righting arm in waves surprised the researchers (Arndt, 1962).

Kerwin (1955) analyzed a simple model of the variation of $\overline{GM}$ in waves and its influence on ship stability. His investigations included experiments carried out at Delft and he reports difficulties that we attribute to the equipment available at that time.

To confirm the results of their calculations, researchers from Hamburg carried out model tests in a towing tank (Arndt and Roden, 1958) and with self-propelled models on a nearby lake (Wendel, 1965). *Post-mortem* analysis of other marine disasters showed that the righting arm was severely reduced when the ship was on the wave crest. Sometimes it was even negative.

Paulling (1961) discussed 'The transverse stability of a ship in a longitudinal seaway'.

Storch (1978) analyzed the sinking of thirteen king-crab boats. In one case he discovered that the righting arm on wave crest must have been negative, and in two others, greatly reduced.

Lindemann and Skomedal (1983) report a ship disaster they attribute to the reduction of the stability in waves. On 1 October 1980 the RO/RO (*roll-on/roll-off*) ship *Finneagle* was close to the Orkney Islands and sailing in *following seas*, that is with waves travelling in the same direction as the ship. All of a sudden three large roll cycles caused the ship to heel up to 40°. It is assumed that this large angle caused a container to break loose. Trimethylphosphate leaked from the container and reacted with the acid of a car battery. Because of the resulting fire the ship had to be abandoned.

Chantrel (1984) studied the large-amplitude motions of an offshore supply buoy and attributed them to the variation of properties in waves leading to the phenomenon of parametric resonance explained in this chapter. Interesting experimental and theoretical studies into the phenomenon of parametric resonance of trimaran models were performed at the University College of London, within the framework of Master's courses supervised by D.C. Fellows (Zucker, 2000).

The influence of waves on ship stability can be modelled by a linear differential equation with periodic coefficients known as the **Mathieu equation**. Under certain conditions, known as parametric resonance, the response of a system governed by a Mathieu equation can be unstable, that is, grow beyond any limits. For a ship, unstable response means capsizing. This is a new mode of ship capsizing; the first we learnt are due to **insufficient metacentric height** and to **insufficient area under the righting-arm curve**. This chapter contains a practical introduction to the subjects of parametric excitation and resonance known also as **Mathieu effect**.

9.2 The influence of waves on ship stability

In this section we explain why the metacentric height varies when a wave travels along the ship. We illustrate the discussion with data calculated for a 29-m fast patrol boat (further denoted as FPB) whose offsets are described by Talib and Poddar (1980). For hulls like the one chosen here the influence of waves is particularly visible. Figure 9.1 shows an outline of the boat and the location of three stations numbered 36, 9, and 18. This is the original numbering in the cited reference. The shapes of those sections are shown in Figure 9.2. We calculated the hydrostatic data of the vessel for the draught 2.5 m, by means of the same ARCHIMEDES programme that Talib and Podder used. The waterline corresponding to the above draught appears as a solid line in Figures 9.1 and 9.2. Let us see what happens in waves. Calculations and experiments show that the maximum influence of longitudinal waves on ship stability occurs when the

Figure 9.1 Wave profiles on a fast patrol boat outline – S = still water, T = wave trough, C = wave crest

wave length is approximately equal to that of the ship waterline. Consequently, we choose a wave length

$$\lambda = L_{pp} = 27.3\,\text{m}$$

The wave height prescribed by the German Navy is

$$H = \frac{\lambda}{10 + 0.05\lambda} = \frac{27.3}{10 + 0.05 \times 27.3} = 2.402\,\text{m}$$

The dot-dot lines in Figures 9.1 and 9.2 represent the waterline corresponding to the situation in which the **wave crest** is in the midship-section plane. We say that the ship is on **wave crest**. In Figure 9.2 we see that in the midship section the waterline lies above the still-water line. The breadth of the waterline almost does not change in that section. In sections 36 and 18 the waterline descends below the still-water position. In section 18 the breadth decreases. This effect occurs in a large part of the forebody. In the calculation of the metacentric radius, $\overline{BM}$, breadths enter at the third power (at constant displacement!). Therefore, the overall result is a decrease of the metacentric radius.

The dash-dash lines in Figures 9.1 and 9.2 represent the situation in which the position of the wave relative to the ship changed by half a wave length. The **trough** of the wave reached now the midship section and we say that the ship is in a **wave trough**. In Figure 9.2 we see that the breadth of the waterline increased significantly in the plane of station 18, decreased insignificantly in the midship section, and increased slightly in the plane of station 36. The overall effect is an increase of the metacentric radius.

A quantitative illustration of the effect of waves on stability appears in Figure 9.3. For some time the common belief was that the minimum metacentric radius occurs when the ship is on a wave crest. It appeared, however, that for

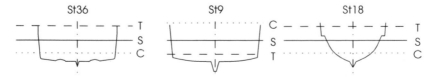

Figure 9.2 Wave profiles on FPB transverse sections – S = still water, T = wave trough, C = wave crest

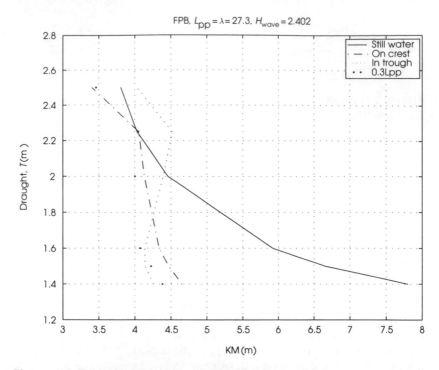

FPB, $L_{pp} = \lambda = 27.3$, $H_{wave} = 2.402$

Figure 9.3 The influence of waves on KM

forms like those of the FPB the minimum occurs when the wave crest is approximately $0.3L_{pp}$ astern of the midship section. Calculations carried out by us
for various ship forms showed that the relationships can change. Figure 9.3
shows, indeed, that for draughts under 1.6 m, $\overline{KM}$ is larger on wave crest than
in wave trough. Similar conclusions can be reached for the righting-arm curves
in waves. For example, the righting arm in wave trough can be the largest in a
certain heeling-angle range, and ceases to be so outside that range. The reader is
invited to use the data in Exercise 1 and check the effect of waves on the righting
arm of another vessel, named *Ship No. 83074* by Poulsen (Poulsen, 1980).

More explanations of the effect of waves on righting arms can be found in
Wendel (1958), Arndt (1962) and Abicht (1971). Detailed stability calculations
in waves, for a training ship, are described by Arndt, Kastner and Roden (1960),
and results for a cargo vessel with $C_B = 0.63$, are presented by Arndt (1964).
A few results of calculations and model tests for ro–ro ships can be found in
Sjöholm and Kjellberg (1985).

To develop a simple model of the influence of waves we assume that the wave
is a periodic function of time with period T. Then, also $\overline{GM}$ is a periodic function
with period T. We write

$$\overline{GM(t)} = \overline{GM_0} + \delta\overline{GM(t)}$$

where

$$\overline{\delta GM(t)} = \overline{\delta GM(t + T)}$$

for any t. In Section 6.7 we developed a simple model of the free rolling motion. To include the variation of the metacentric height in waves we can rewrite the roll equation as

$$\ddot{\phi} + \frac{g}{i^2}(\overline{GM_0} + \overline{\delta GM})\phi = 0$$

Going one step further we assume that the wave is harmonic (*regular wave*) so that the free rolling motion can be modelled by

$$\ddot{\phi} + \frac{g}{i^2}(\overline{GM_0} + \overline{\delta GM}\ \cos \omega_e t)\phi = 0 \qquad (9.1)$$

This is a Mathieu equation; those of its properties that interest us are described in the following section.

9.3 The Mathieu effect – parametric resonance

9.3.1 The Mathieu equation – stability

A general form of a differential equation with periodic coefficients is *Hill's equation*:

$$\ddot{x} + h(t)x = 0,$$

where $h(t) = h(t + T)$. In the particular case in which the periodic function is a cosine we have the *Mathieu equation*; it is frequently written as

$$\ddot{\phi} + (\delta + \epsilon \cos 2t)\phi = 0 \qquad (9.2)$$

This equation was studied by Mathieu (Émile-Léonard, French, 1835–1900) in 1868 when he investigated the vibrational modes of a membrane with an elliptical boundary. Floquet (Gaston, French, 1847–1920) developed in 1883 an interesting theory of linear differential equations with periodic coefficients. Since then many other researchers approached the subject; a historical summary of their work can be found in McLachlan (1947).

A rigorous discussion of the Mathieu equation is beyond the scope of this book; for more details the reader is referred to specialized books, such as Arscott (1964), Cartmell (1990), Grimshaw (1990) or McLachlan (1947). A comprehensive bibliography on 'parametrically excited systems' and a good theoretic treatment are given by Nayfeh and Mook (1995). For our purposes it is sufficient to explain the conditions under which the equation has stable solutions. By 'stable' we understand that the response, ϕ, is *bounded*. Correspondingly, 'unstable' means that the response grows beyond any boundaries. For a ship whose rolling motion is governed by the Mathieu equation, unstable response simply means that the ship capsizes. The reader may be familiar with the condition of stability of an ordinary, linear differential equation with constant coefficients: A system is stable if all the poles of the transfer function have negative real parts (Dorf and Bishop, 2001). This is not the condition of stability of the Mathieu equation; the behaviour of its solutions depends on the parameters ϵ and δ. This behaviour can be explained with the aid of Figure 9.4. In this figure, sometimes known as *Strutt diagram*, but attributed by McLachlan (1947) to Ince, the horizontal axis represents the parameter δ, and the vertical axis, the parameter ϵ. The $\delta - \epsilon$ plane is divided into two kinds of regions. For δ, ϵ combinations that fall in the grey areas, the solutions of the Mathieu equation are stable. The δ, ϵ points in white regions and on the boundary curves correspond to unstable solutions. The diagram is symmetric about the δ axis; for our purposes it is sufficient to show only half of it.

The theory reveals the following properties of the Strutt-Ince diagram.

- The lines separating stable from unstable regions intercept the δ axis in points for which

$$\delta = \frac{n^2}{4}, \qquad n = 0, 1, 2, 3, \ldots$$

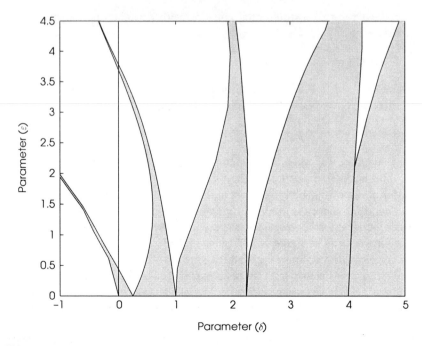

Figure 9.4 Strutt-Ince diagram, $\delta - \epsilon$ plane

- As δ grows larger, so do the stable regions.
- As ϵ grows, the stable regions become smaller. Remember, ϵ is the 'disturbance'.

Cesari (1971) considers the equation

$$\ddot{x} + (\sigma^2 + \epsilon \cos \omega t)x = 0 \qquad (9.3)$$

The natural frequency of the 'undisturbed' equation – that is for $\epsilon = 0$ – is $\sigma/2\pi$, while the frequency of the periodic disturbance is $\omega/2\pi$. With the transformation

$$\omega t = 2t_1 \qquad (9.4)$$

we calculate

$$
\begin{aligned}
\dot{x} &= \frac{dx}{dt_1}\frac{dt_1}{dt} = \frac{\omega}{2}\frac{dx}{dt_1} \\
\ddot{x} &= \frac{d\dot{x}}{dt_1}\frac{dt_1}{dt} = \frac{\omega^2}{4}\frac{d^2x}{dt_1^2}
\end{aligned}
\qquad (9.5)
$$

Substituting Eqs. (9.4) and (9.5) into Eq. (9.3) yields an equation in the standard form

$$\ddot{x} + (\delta_1 + \epsilon_1 \cos 2t_1)x = 0$$

where

$$\delta_1 = \frac{4\sigma^2}{\omega^2}, \qquad \epsilon_1 = \frac{4\epsilon}{\omega^2}$$

The general aspect of the $\delta_1 - \epsilon_1$ plane is shown in Figure 9.5. Visual inspection shows us that for small ϵ values the danger of falling into an unstable region is greater in the neighbourhoods of $\delta_1 = 1^2,\ 2^2,\ 3^2,\ \ldots$. This means that for small ϵ parametric resonance occurs at circular frequencies $\omega = 2\sigma/n^2$, where $n = 1,\ 2,\ 3,\ \ldots$. The first dangerous situation is met when $\omega = 2\sigma$. We reach the important conclusion that the danger of parametric resonance is greatest when the frequency of the perturbation equals twice the natural frequency of the undisturbed system.

This statement is rephrased in terms of ship-stability parameters in Example 9.1 where σ becomes the natural roll circular frequency, ω_0, of the ship, and ω becomes ω_E, the frequency of encounter, that is the frequency with which the ship encounters the waves. This theoretical conclusion was confirmed by basin tests.

Surprising as it may seem, the phenomenon of parametric excitation is well known. The main character in Molière's *Le bourgeois gentilhomme* has been

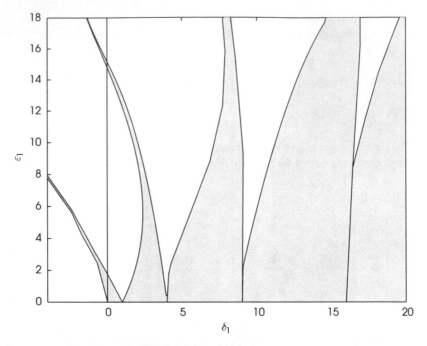

Figure 9.5 Strutt-Ince diagram, $\delta_1 - \epsilon_1$ plane

writing prose for many years without being aware of it. Similarly, readers are certainly familiar with parametric excitation since their childhood. Here are, indeed, three well-known examples.

The motion of a pendulum is stable. However, if the point from which the pendulum hangs is moved up and down periodically, with a suitable amplitude and frequency, the pendulum can be caused to overturn.

Try to 'invert' a pendulum so that its mass is concentrated above the centre of oscillation. The pendulum will fall. Still, at circus we see clowns that keep a long rod clasped in their hand, as shown in Figure 9.6(a). The rod can be *stabilized* by moving the hand up and down with a suitable amplitude and frequency.

A third, familiar example of parametric excitation is that of a swing. To increase the amplitude of motion the person on the swing kneels close to the extreme positions and stands up in the middle position (Figure 9.6(b)). Thus, the distance between the hanging point and the centre of gravity of the person varies periodically. The swing behaves like a *pendulum with varying length*.

More examples of parametrically excited systems can be found in Den Hartog (1956). That author also studies a case in which the periodic function is a *rectangular ripple* whose analytic treatment is relatively simple and allows the derivation of an explicit condition of stability.

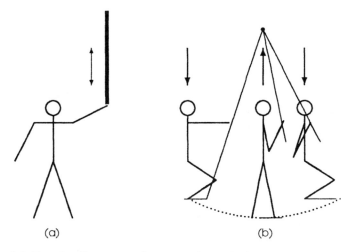

(a)   (b)

Figure 9.6 Two familiar uses of parametric excitation

9.3.2 The Mathieu equation – simulations

In this section we show how to simulate the behaviour of the Mathieu equation and give four examples that illustrate the conclusions reached in the preceding section. To solve numerically the Mathieu equation we define

$$\phi_1 = \phi, \qquad \phi_2 = \dot{\phi}_1$$

and replace Eq. (9.2) by the first-order system

$$\dot{\phi}_1 = \phi_2, \qquad \dot{\phi}_2 = -(\delta + \epsilon \cos 2t)\phi_1 \qquad\qquad (9.6)$$

The following MATLAB function, written on a file mathieu.m, calculates the derivatives in Eq. (9.6):

```
%MATHIEU          Derivatives of Mathieu equation.
      function     dphi = mathieu(t1, phi, d1, e1)


dphi(1)  =   [ phi(2); -(d1 + e1*cos(2*t1))*phi(1);
```

We write a second MATLAB function, mathisim.m, that calls the function mathieu:

```
function ms = mathisim(omega_0, epsilon,
                  omega_e, tf)

%MATHISIM Simulates the Mathieu equation
```

```
clf                                % clean window

d1 = 4*omega_0^2/omega_e^2;
e1 = 4*epsilon/omega_e^2;
w0 = [ 0.1; 0.0 ];                 % initial conditions;
ts = [ 0; tf ];                    % time span
hmathieu = @mathieu;
[ t1, phi ] = ode45(hmathieu, ts, w0, [], d1, e1);

t = 2*t1/omega_e;
subplot(2, 2, 1), plot(t, phi(:, 1)), grid
ns = num2str(omega_0);
nd = num2str(d1);
ne = num2str(e1);
no = num2str(omega_e);
title('Time domain'), ylabel('\phi')
subplot(2, 2, 3), plot(t, phi(:, 2)), grid
xlabel('t'), ylabel('\phi''')
subplot(2, 2, 2), plot(phi(:, 1), phi(:, 2));
                                   % phase plot
grid
title('Phase plan'), xlabel('\phi'),
                                ylabel('\phi''')
text(phi(1, 1), phi(1, 2), 'start')
subplot(2, 2, 4), axis off
text(0.1, 0.66, [ '\omega_0 = ' ns ' ,
                  \delta_1 = ' nd ])
text(0.1, 0.33, [ '\epsilon_1 = ' ne ' ,
                  \omega_e = ' no ])
```

Figures 9.7–9.10 show results of simulations carried out by means of the function mathisim. Figure 9.7 corresponds to the parameters

$$\sigma = 4, \qquad \epsilon_1 = 0, \qquad \omega = \pi/4$$

In this case we deal with the well-known equation

$$\ddot{\phi} + \delta\phi = 0$$

whose solution is a sinusoid with circular frequency $\sqrt{\delta}$:

$$\phi = C_1 \sin(\sqrt{\delta}t + C_2)$$

The constants C_1, C_2 can be found from the initial conditions of the problem. The first derivative, $\dot{\phi}$, shown in the second subplot, is also a sinusoid:

$$\dot{\phi} = C_1\sqrt{\delta} \cos(\sqrt{\delta}t + C_2)$$

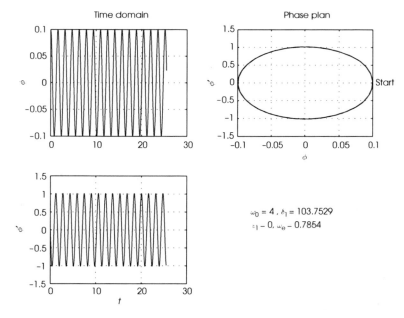

Figure 9.7 Simulation of Mathieu equation; sinusoidal response

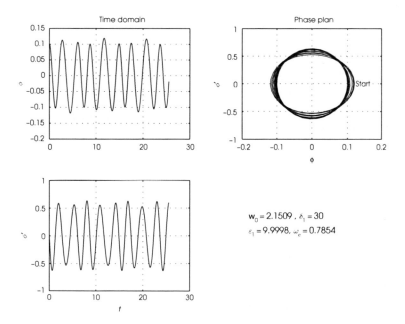

Figure 9.8 Simulation of Mathieu equation; stable response

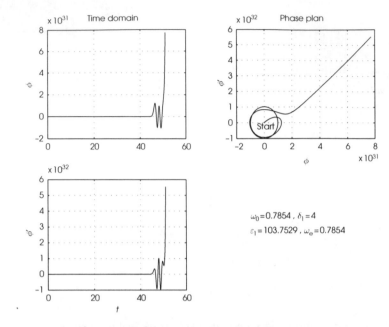

Figure 9.9 Simulation of Mathieu equation; unstable response

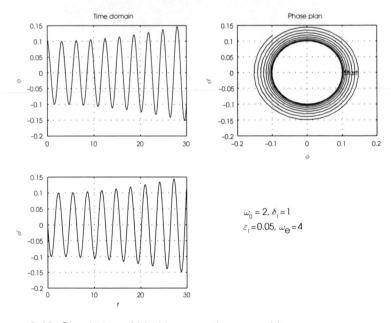

Figure 9.10 Simulation of Mathieu equation; unstable response

The third subplot is the *phase plane* of the motion. The curve is an ellipse. Indeed, simple calculations show us that

$$\frac{\phi^2}{C_1^2} + \frac{\dot{\phi}^2}{(\sqrt{\delta}C_2)^2} = 1$$

The run parameters that generate Figure 9.8 are

$$\sigma = 2.1509, \qquad \epsilon = 1.5421, \qquad \omega = \pi/4$$

These values define in Figures 9.4 and 9.5 a point in a stable region. As the simulation shows, the solution is bounded, periodic, but not sinusoidal.

The run parameters that generate Figure 9.9 are

$$\sigma = \pi/4, \qquad \epsilon = 16, \qquad \omega = \pi/4$$

These values define in Figures 9.4 and 9.5 a point in an unstable region. As the simulation shows, the solution is unbounded. This can be best seen in the phase plane where the start of the curve is marked by the word 'start'.

The run parameters for Figure 9.10 are

$$\sigma = 2, \qquad \epsilon = 0.2, \qquad \omega = 4$$

These values define in the Strutt diagram a point in an unstable region, very close to a boundary curve. As the simulation shows, the solution is periodic and steadily growing. This can be best seen in the phase plane where the start of the curve is marked by the word 'start'. The case shown in this figure corresponds to the most dangerous condition of parametric resonance, $\omega = 2\sigma$.

9.3.3 Frequency of encounter

When judging ship stability, the frequency to be used in the Mathieu equation is the number of waves 'seen' by the ship in one time unit. This is the **frequency of encounter**, ω_E; to calculate it we use Figure 9.11. Let v be the ship speed, c, the **wave celerity**, that is the speed of the wave, λ, the wave length, ω_w, the wave circular frequency, and α, the angle between ship speed and wave celerity. By convention, $\alpha = 180°$ in head seas and $0°$ in following seas. The relative speed between ship and wave is

$$c - v \cos \alpha$$

The ship encounters wave crests (or wave troughs) at time intervals equal to

$$T_E = \frac{\lambda}{c - v \cos \alpha}$$

This is the **period of encounter**. By definition, the wave circular frequency is

$$\omega_w = \frac{2\pi}{T_w}$$

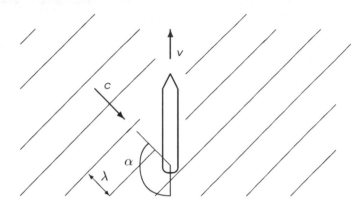

Figure 9.11 Calculating the frequency of encounter

where T_w is the *wave period*. Similarly, the *circular frequency of encounter* is defined by

$$\omega_E = \frac{2\pi}{T_E}$$

In wave theory (see, for example, Faltinsen, 1993; Bonnefille, 1992) it is shown that the relationship between wave length and wave circular frequency, in water of infinite depth, is

$$\lambda = \frac{2\pi g}{\omega_w^2}$$

Putting all together we obtain

$$\omega_E = \omega_w - \frac{\omega_w^2}{g} v \cos \alpha \qquad (9.7)$$

9.4 Summary

Longitudinal and quartering waves influence the stability of ships and other floating bodies. The moment of inertia of the waterline surface in waves differs from that of the waterplane in still water and, consequently, so do the metacentric height and the righting arms. The way in which those quantities vary depends on the ship form; however, it can be said that in many cases the righting moment in wave trough is larger than in still water, while on wave crest it is smaller. If the wave is periodic, also the variation of the righting arm is periodic. Then, into the equation of rolling developed in Chapter 6 we must add to the coefficient of the roll angle a term that is a periodic function of time:

$$\ddot{\phi} + \frac{g}{i^2}(\overline{GM_0} + \delta\overline{GM} \cos \omega_e)\phi = 0$$

For small heel angles the above equation can be reduced to the canonical form of the Mathieu equation

$$\ddot{\phi} + (\delta + \epsilon \cos 2t)\phi = 0$$

The condition of stability is not the same as for a linear differential equation with constant coefficients. In other words, the condition of positive metacentric height, $\overline{GM} > 0$, is no longer sufficient. The theory of differential equations with periodic coefficients shows that the plane of the parameters δ and ϵ can be subdivided into regions so that if in one of them the solution of the Mathieu equation is stable, in the adjacent regions it is not. This means that for certain pairs $[\delta, \epsilon]$ the solution is unstable and we say that parametric resonance occurs. The partition of the $\delta - \epsilon$ plane into stable and unstable regions can be best visualized in the Strutt-Ince diagram. Thus, it can be easily discovered that even for small ϵ values a particularly dangerous situation arises when the frequency of the periodic coefficient is twice the natural frequency of the system without periodic excitation.

Parametric excitation occurs in several systems we are familiar with. Thus, the amplitude of oscillation of a swing can be increased by periodically changing the position of the centre of gravity of the person on the swing. As another example, a conventional pendulum is usually stable, but it can be forced to overturn if the point of hanging is moved up and down with appropriate frequency and amplitude. Conversely, an inverted pendulum, although inherently unstable, can be stabilized by applying a suitable periodic motion to its centre of oscillation.

Ships have capsized although they fulfilled the criteria of stability commonly accepted at the time of the disaster. Post-mortem analysis of some cases pinpointed the Mathieu effect as the cause of capsizing. The surprising discovery was that the righting arm could be negative on wave crest.

The analysis of the Mathieu effect confirms a fact well known to experienced seafarers: following seas are more dangerous than head seas. In fact, when the direction of the waves is the same as that of the ship, the relative velocity is small and the time interval in which the stability is reduced is longer. Then, there is more time to develop large heeling angles. Still worse, in following seas the effect of reduced stability can be enhanced by waves flowing over the deck. The latter effect will increase the height of the centre of gravity because it means an extra weight loaded high up on the ship. It also adds a free-surface effect.

9.5 Examples

Example 9.1 – Parametric resonance in ship stability
In this example we are going to explain the significance of the parameters δ and ϵ for ship stability. In Chapter 6 we developed the equation of free roll

$$\ddot{\phi} + \frac{g\overline{GM}}{i^2} = 0 \tag{9.8}$$

The natural, circular roll frequency is

$$\omega_0 = \frac{[g\overline{GM}]^{1/2}}{i} \qquad (9.9)$$

Let us assume that the wave produces a periodic variation of the metacentric height equal to

$$\delta\overline{GM} \cos \omega_E$$

Where ω_E is the circular frequency of encounter. With this assumption and with the notation introduced by Eq. (9.9) we rewrite Eq. (9.8) as

$$\ddot{\phi} + \left(\omega_0^2 + \frac{g\delta\overline{GM}}{i^2} \cos \omega_E\, t \right) \phi = 0 \qquad (9.10)$$

Following Cesari (1971) we use the substitution $\omega_E t = 2t_2$ and proceeding like in Subsection 9.3.1 we transform Eq. (9.10) to

$$\frac{\omega_E^2}{4} \cdot \frac{d^2\phi}{dt_1^2} + \left(\omega_0^2 + \frac{g\delta\overline{GM}}{i^2} \cos 2t_1 \right) \phi = 0 \qquad (9.11)$$

Substituting Eq. (9.9) we obtain

$$\frac{d^2\phi}{dt_1^2} + \left\{ 4 \left(\frac{\omega_0}{\omega_E} \right)^2 + 4 \frac{g\delta\overline{GM}}{\overline{GM}} \left(\frac{\omega_0}{\omega_E} \right)^2 \cos 2t_1 \right\} \phi = 0 \qquad (9.12)$$

Equation (9.12) can be brought to the standard Mathieu form with

$$\delta 1 = 4 \left(\frac{\omega_0}{\omega_E} \right)^2, \qquad \epsilon_1 = 4 \frac{g\delta\overline{GM}}{\overline{GM}} \left(\frac{\omega_0}{\omega_E} \right)^2 \qquad (9.13)$$

We know that the most dangerous situation occurs at $\delta_1 = 1$, that is for $\omega_E = 2\omega_0$.

Example 9.2 – Sail ship in longitudinal waves

The righting-arm curve in still-water shown Figure 9.12 was calculated for an actual training yacht. We assume that the righting-arm curves on wave crest and in wave trough, and the wind heeling arm are as shown in the figure. It is obvious that while advancing in waves the yacht will roll between points A and B. Thus, the Mathieu effect induces roll in head or following seas, a behaviour that is not predicted by the conventional roll equation. Readers involved in yachting may have experienced the phenomenon.

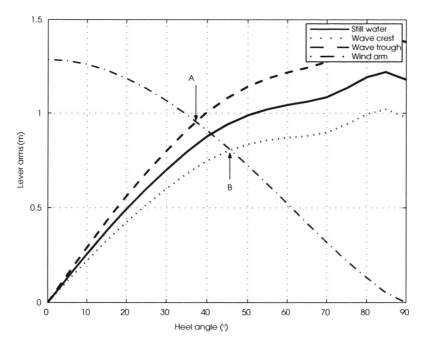

Figure 9.12 Sail ship in longitudinal waves

9.6 Exercise

Exercise 9.1 – Ship 83074, levers of stability in seaway
Table 9.1 shows the cross-curves of stability of the *Ship No. 83074* for a displacement volume equal to 20000 m^3. Plot in the same graph the curves for still water, in wave trough and on wave crest.

Table 9.1 Levers of stability of Ship 83074, 20000 m^3

Heel angle (°)	Wave trough (m)	Still water (m)	Wave crest (m)
0	0.000	0.000	0.000
10	2.617	2.312	2.309
20	4.985	4.606	4.635
30	6.912	6.759	6.892
45	9.095	9.361	9.235
60	9.734	10.447	10.073
75	10.783	10.425	9.917

10
Intact stability regulations II

10.1 Introduction

We give in this section a simplified overview of the BV 1033 regulations of the German Federal Navy, as an example of philosophy different from those illustrated in Chapter 8. These are the only effectively applied regulations that consider the Mathieu effect. There were proposals to consider parametric excitation in other codes of ship stability; to our best knowledge they remained proposals. Our description follows the 1977 edition of the regulations and includes updatings received as personal communications in the 1980s. As we suggested for other regulations, for checks of stability that must be submitted for approval, it is highly recommended to inquire about the latest, complete edition of BV 1033 and consult it for updatings and specific details.

10.2 The regulations of the German Navy

Kurt Wendel wrote in 1961 the first draft of stability regulations for the German Federal Navy. Wendel issued in 1964 a new edition known as BV 103. An early detailed explanation of the regulations and their background is due to Arndt (1965). His paper was soon translated into English by the British Ship Research Association and appeared as BSRA Translation No. 5052. An updated version of the regulations was published in 1969 and since then they are known as BV 1033. As pointed out by Brandl (1981), the German regulations were adopted by the Dutch Royal Navy (see, for example, Harpen, 1971) and they also served in the design of some ships built in Germany for several foreign navies.

In Chapter 9 we mentioned experiments performed by German researchers before the publication of the regulations. The authors continued to experiment after the implementation of BV 1033 and thus confirmed the validity of the requirements and showed that the German regulations and the regulations of the U.S. Navy confer to a large extent equivalent safety against capsizing. For details we refer the reader to Brandl (1981) and Arndt, Brandl and Vogt (1982).

Righting arms are denoted in BV 1033 by the letter h; heeling arms, by k. Thus, k_W is the wind heeling arm, k_D the heeling arm in turning, and so on.

10.2.1 Categories of service

Stability requirements vary according to the intended use of the ships. The regulations of the German Federal Navy classify vessels into five categories, as explained below:

Group A There are no limitations to the area of operation of ships belonging to this category. Calculations for Group A should be carried out for a wind velocity equal to 90 knots.

Group B This category includes ships that can avoid winds whose velocity exceeds 70 knots. Examples of corresponding areas of operation are the North Atlantic, the North sea, the Baltic sea, and the Mediterranean sea. The wind velocity to be considered for this group is 70 knots.

Group C The category consists of coastal vessels that can reach a harbour if a storm warning is received. Stability calculations shall be based on a wind velocity of 50 knots.

Group D It consists of ships and decked boats that operate as harbour and estuarial craft. The wind velocity to be considered is 40 knots.

Group E In this category enter open boats intended for coastal and harbour operation, within well-defined geographical limits. Stability calculations shall assume a wind velocity equal to 20 knots.

10.2.2 Loading conditions

The BV 1033 regulations require the verification of stability in a number of loading conditions. We shall exemplify here only three of them. The detailed description of the loading cases involves the term *empty ship ready for operation*. By this the regulations mean the ship with fuel, feed water and lubricating oil in machines, piping, weapons and other systems, if necessary also with fixed ballast.

Loading case 0 – Empty ship
The weight groups to be included are

- Empty ship ready for operation.
- Crew and personal effects.

Loading case 1 – Limit displacement with ballast water
The weights to be included are

- Empty ship ready for operation.
- Crew and personal effects.
- 10% consumables and provisions.

- 10% fresh water or 50% if a fresh water generator with a capacity of minimum 25 l per head and day is on board.
- 10% feed water or 50% if a fresh water generator is available.
- 10% fuel.
- 50% lubricating oil.
- 33% munitions, where launching tubes and weapons are charged, and the rest of the ammunition is in the corresponding storing places.
- Aircraft.
- 33% or 100% deliverable or transported loads, whichever is worst for stability.
- Ballast water, if necessary for stability.

Loading case 1A – Limit displacement for ships to be checked with 90 or 70 knot wind
Same as case 1, except:

- Fuel and lubricating oil as necessary for stability, but not less than 10% fuel and 50% lubricating oil.
- No ballast water.

10.2.3 Trochoidal waves

According to BV 1033, stability on waves should be checked in **trochoidal waves**. This wave form has been used also for other naval-architectural calculations, mainly those of *longitudinal bending*.

The trochoidal wave theory is the oldest among wave theories; it was elaborated in 1804 by Gerstner (Franz Joseph von, lived in Bohemia, 1756–1832). Rankine (William John Macquam, Scot, 1820–1872) gave an independent formulation in 1863. This theory assumes that each water particle moves along a circular path. For example, in Figure 10.1 the water particles shown as black circles move along circles with centres lying on the x-axis and having the radius r. Let the x coordinate of the first shown circle be 0, and consider a particle on a circle whose centre has the x coordinate equal to x_0. This particle rotated with an

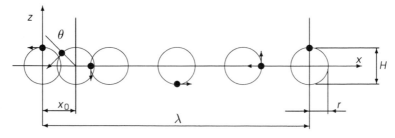

Figure 10.1 The generation of the trochoidal wave

angle θ relative to the particle on the first circle. The phase angle θ is proportional to the distance x_0, that is

$$\theta = 2\pi \frac{x_0}{\lambda} \tag{10.1}$$

The resulting wave form is the trochoid; it looks very much like the actual surface of *swells*. The assumption that water particles move on circular paths also corresponds to a simple observation. A floating object, such as a piece of wood thrown in a swell, describes a circular motion in a vertical plane. The mean position of the floating object does not change. In trochoidal wave theory, the wave particles do not travel, it is only the *wave form* that travels. Figure 10.2 shows two phases of a trochoidal wave. The dimensions are those prescribed by the BV 1033 regulations for a 114-m ship, such as that described in Example 10.2. From Figure 10.1 and Eq. (10.1) we deduce that for two wave particles separated by a distance $x = \lambda$ the phase angles θ differ by 2π. In other words, the z coordinates of two points separated by a distance λ are equal. The quantity λ is the **wave length**. In the same figure we see that the **trough-to-crest wave height** equals $H = 2r$.

To draw a trochoidal wave we need the following information:

- the equation of the trochoid;
- the position of the axis of orbit centres with respect to the still-water line.

To obtain the above-mentioned information we are going to use another definition of the trochoid:

> The trochoid is the curve generated by a circle that rolls, without sliding, on the underside of a straight line.

The equations of the trochoidal wave are

$$
\begin{aligned}
x &= R\theta - r\sin\theta = \frac{\lambda}{2\pi}\theta - \frac{H}{2}\sin\theta \\
z &= r\cos\theta = \frac{H}{2}\cos\theta
\end{aligned}
\tag{10.2}
$$

The trochoidal wave has a sharp form near the crest and is flatter in the trough. Therefore, the still-water line must lie below the line of orbit centres by some

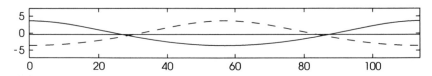

Figure 10.2 The trochoidal wave suiting the Maestral example

distance a. As the volume of water above the still-water line must equal that below the same line, we can write

$$\int_0^{2\pi} (z - a)\mathrm{d}x = 0 \tag{10.3}$$

We can separate the above integral into two integrals that we calculate separately. The first integral is

$$\begin{aligned} \int_0^{2\pi} z\mathrm{d}x &= \int_0^{2\pi} r\cos\theta(R - r\cos\theta)\mathrm{d}\theta \\ &= -\pi r^2 \end{aligned} \tag{10.4}$$

The second integral is

$$\int_0^{2\pi} a\mathrm{d}x = ax\big|_0^{2\pi} = 2\pi aR \tag{10.5}$$

Equating Eqs. (10.4) and (10.5) we obtain

$$a = -\frac{r^2}{2R} \tag{10.6}$$

We mention here, without proving, two interesting hydrodynamical properties of the trochoidal wave.

1. *Motion decay with depth*

   The radius of orbits decays exponentially with depth. For a given depth h, the amplitude of the orbital motion is

   $$r_{\mathrm{h}} = re^{-h/\mathrm{R}} \tag{10.7}$$

   The amplitude on the sea bottom should be zero. In our model this only happens at an infinite depth; therefore, the trochoidal wave model is correct only in infinite depth seas. However, let us calculate the radius of the orbit at a depth equal to half a wave length:

   $$r_{-\lambda/2} = r\exp\left(\frac{-2\pi R}{2R}\right) \approx 0.0043r$$

   that is practically zero.

2. *Virtual gravity*

   A water particle moving along a circular orbit is subjected to two forces:

   - its weight, mg;
   - a centrifugal force, $mr\omega^2$, where ω is the angular velocity of the particle. It can be shown that $\omega^2 = g/R$.

In the trough the two forces add up to

$$mg \left(1 + \frac{r}{R} \right)$$

while on a wave crest the result is

$$mg \left(1 - \frac{r}{R} \right)$$

Thus, a floating body experiences the action of a **virtual gravity** acceleration whose value varies between $g(1 - r/R)$ and $g(1 + g/R)$. One wave height-to-length ratio frequently employed in Naval Architecture is 1/20. With this value the apparent gravity varies between $0.843g$ and $1.157g$.

The variation of apparent gravity, and consequently of buoyancy, in waves is known as the *Smith effect*, after the name of the researcher who described it first in 1883. The reduction of virtual gravity on wave crest was considered another cause of loss of stability in waves. To quote Attwood and Pengelly (1960):

> This is the explanation of the well-known phenomenon of the tenderness of sailing boats on the crest of a wave.

As the vessel seems to weigh less on the crest, so does the righting moment that is the product of displacement and righting arm. As the wind moment does not change, a boat '*of sufficient stiffness in smooth water, is liable to be blown over to a large angle and possibly capsize.*'

On the other hand, Devauchelle (1986) considers that in real seas, characterized by the irregularity of waves (see Chapter 12), the effect of virtual gravity variation can be neglected. Model tests described by Wendel (1965) revealed that the influence of the orbital motion can be neglected when compared with the effect of the variation of the waterline in waves. Calculations carried out when investigating the loss of a trawler showed that in that case the Smith effect was completely negligible for heel angles up to 20° (Morrall, 1980).

More details on the theory of trochoidal waves can be found in Attwood and Pengelly (1960), Bouteloup (1979), Susbielles and Bratu (1981), Bonnefille (1992) and Rawson and Tupper (1994). To conclude this section, we state the characteristics of the wave specified by the BV 1033 regulations:

wave form	trochoidal
wave length	equal to ship length, that is, $\lambda = L$
wave height	$H = \lambda/(10 + 0.05\lambda)$

The relationship between wave length and height is based on statistics and probabilistic considerations. We may mention here that a slightly different relationship was proposed for merchant ships by the maritime registers of the former German Democratic Republic and of Poland (Helas, 1982):

$$H = \frac{L_{pp}}{4.14 + 0.14L_{pp}}$$

A value frequently used by other researchers is $H = \lambda/20$.

We described here the trochoidal wave because the BV regulations require its use in stability calculations, while other codes of practice specify this wave for bending-moment calculations. Other wave theories are preferred in other branches of Naval Architecture. Thus, in Chapter 12, we introduce the sinusoidal waves. There is no great difference in shape between the trochoidal and the sine wave, but some other properties are significantly different.

10.2.4 Righting arms

The cross-curves of stability shall be calculated in still water and in waves. For the latter, ten *wave phases* shall be considered. More specifically, the calculations shall be performed with the wave crest at distances equal to $0.5L$, $0.4L$, ... $0L$, ... $- 0.4L$ from midship. The average of the cross-curves in waves shall be compared with the cross-curves in still water and the smaller values shall be used in the calculation of righting arms. The BV 1033 regulations denote by h_G the righting arm in still water, and by h_S the righting arm in waves. It is easy to remember the latter notation if we relate the subscript S to the word 'seaway', the translation of the German term 'Seegang'. The reason for considering the mean of the righting arms in waves, and not the smallest values, is that, in general, there is not enough time for the Mathieu effect to fully develop.

Most ships are not symmetric about a transverse plane (notable exceptions are Viking ships and some ferries). Therefore, during heeling the centre of buoyancy travels in the longitudinal direction causing trim changes. According to the German regulations this effect must be considered in the calculation of cross-curves. In the terminology of BV 1033 the calculations shall be performed with *trim compensation*. The data in Table 9.1 and in Example 10.2 are calculated with trim compensation.

10.2.5 Free liquid surfaces

The German regulations consider the influence of free liquid surfaces as a heeling arm, rather than a quantity to be deducted from metacentric height and righting arms. The first formula to be used is

$$k_F = \frac{\sum\limits_{j=1}^{n} \rho_j i_j}{\Delta} \sin \phi \qquad (10.8)$$

where, as shown in Chapter 5, n is the number of tanks or other spaces containing free liquid surfaces, ρ_j, the density of the liquid in the jth tank, and i_j, the moment of inertia of the free liquid surface, in the same tank, with respect to a baricentric axis parallel to the centreline. As convened, Δ is the mass displacement.

If k_F calculated with Formula 10.8 exceeds 0.03 m at $30°$, an exact calculation of the free surface effect is required. The formula to be used is

$$k_F = \frac{1}{\Delta} \sum_{j=1}^{n} p_j b_j \qquad (10.9)$$

where p_j is the mass of the liquid in the jth tank and b_j, the actual transverse displacement of the centre of gravity of the liquid at the heel angle considered. Obviously, calculations with Formula 10.9 should be repeated for enough heel angles to allow a satisfactory plot of the k_F curve.

10.2.6 Wind heeling arm

The wind heeling arm is calculated from the formula

$$k_w = \frac{A_w(z_A - 0.5T_m)}{g\Delta} p_w(0.25 + 0.75 \cos^3 \phi) \qquad (10.10)$$

where A_w is the sail area in m^2; z_A, the height coordinate of the sail area centroid, in m, measured from the same line as the mean draught; T_m, the mean draught, in m; p_w, the wind pressure, in kN/m^2; $g\Delta$, the ship displacement in kN. The wind pressure is taken from Table 10.1, which contains rounded off values.

The sail area, A_w, is the lateral projection of the ship outline above the sea surface. The BV 1033 regulations allow for the multiplication of area elements by aerodynamic coefficients that take into account their shape. For example, the area of circular elements should be multiplied by 0.6.

Arndt (1965) attributes Formula 10.10 to Kinoshita and Okada who published it in the proceedings of a symposium held at Wageningen in 1957. The above equation yields non-zero values at $90°$ of heel; therefore, as pointed out by Arndt, it gives realistic values in the heel range $60°$–$90°$.

Table 10.1 Wind pressures

knots	m/s	Beaufort	Pressure kN/m^2 (kPa)
90	46	14	1.5
70	36	12	1.0
50	26	10	0.5
40	21	8	0.3
20	10	5	0.1

10.2.7 The wind criterion

With reference to Figure 10.3, let us explain how to apply the wind criterion of the BV 1033 regulations.

1. Plot the heeling arm, k_F, due to free liquid surfaces.
2. Draw the curve of the wind arm, k_W, by measuring from the k_F curve upward.
3. Find the intersection of the $k_F + k_W$ curve with the curve of the righting arm, h; it yields the angle of static equilibrium, ϕ_{ST}.
4. Look at a reference angle, ϕ_{REF}, defined by

$$\phi_{REF} = \begin{cases} 35° & \text{if } \phi_{ST} \leq 15° \\ 5° + 2\phi_{ST} & \text{otherwise} \end{cases} \qquad (10.11)$$

5. At the reference angle, ϕ_{REF}, measure the difference between the righting arm, h, and the heeling arm, $k_F + k_W$. This difference, h_{RES}, called **residual arm**, shall not be less than the value yielded by

$$h_{RES} = \begin{cases} 0.1 & \text{if } \phi_{ST} \leq 15° \\ 0.01\phi_{ST} - 0.05 & \text{otherwise} \end{cases} \qquad (10.12)$$

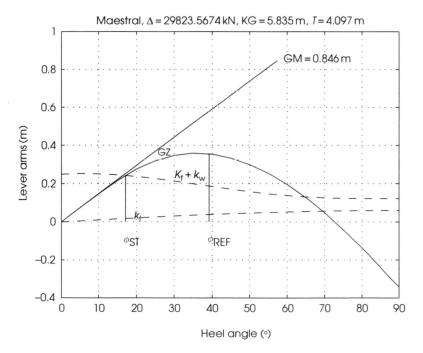

Figure 10.3 Statical stability curve of the example Maestral, according to BV1033

The explicit display of the free liquid surface effect as a heeling arm makes it possible to compare its influence to that of the wind and take correcting measures, if necessary. For example, a too large surface effect, compared to the wind arm, can mean that it is desirable to subdivide some tanks. The heel angle caused by winds up to Beaufort 10 shall not exceed 15°. The reader may have observed that the regulations assume a wind blowing perpendicularly on the centreline plane, while the waves run longitudinally. Arndt, Brandl and Vogt (1982) write:

> This combination is accounting for the fact that even strong winds may change their direction in short time only, whereas the waves are proceeding in the direction in which they were excited. Waves and winds from different directions can be observed especially near storm centres...

Figure 10.3 was plotted with the help of the function described in Example 10.1. Example 10.2 details the data used in the above-mentioned figure. Both examples can provide a better insight into the techniques of BV 1033.

10.2.8 Stability in turning

The heeling arm due to the centrifugal force developed in turning is calculated from

$$k_D = \frac{c_D v^2 (\overline{KG} - 0.5T_m)}{g L_{DWL}} \cos \phi \qquad (10.13)$$

where v is the speed of approach, in $m\,s^{-1}$, and L_{DWL}, the length of the design waterline, in m. The value of this speed should not exceed $0.5\sqrt{gL_{DWL}}$. The coefficient c_D can be used in the design stage when neither speed in turning, nor turning diameter are known. Recommended values are $c_D = 0.3$ for Froude numbers smaller than 1, and $c_D = 0.18$ for faster vessels. When basin or sea trials have been performed, their results shall be used to calculate the actual value of the coefficient. The meaning of the coefficient c_D can be explained as follows. Usually, in the first design stages neither the speed in turning, V_{TC}, nor the radius of the turning circle, R_{TC}, is known. The speed in turning is smaller than the speed in straight-line sailing; therefore, let us write

$$V_{TC} = c_V V, \qquad c_V < 1$$

The radius of the turning circle is usually a multiple of the ship length. Let us write

$$R_{TC} = c_R L_{DWL}, \qquad c_R > 1$$

The factor $V_{\mathrm{TC}}^2/R_{\mathrm{TC}}$ in the equation of the centrifugal force (see Section 6.4) can be written as

$$\frac{c_{\mathrm{V}}^2 V^2}{c_{\mathrm{R}} L_{\mathrm{DWL}}} = c_{\mathrm{D}} \frac{V^2}{L_{\mathrm{DWL}}}$$

with $c_{\mathrm{D}} = c_{\mathrm{V}}^2/c_{\mathrm{R}}$.

Stability in turning is considered satisfactory if the heel angle does not exceed 15°.

10.2.9 Other heeling arms

Other heeling arms can act on the ship, for instance, hanging loads or crowding of passengers on one side. The following data shall be considered in calculating the latter. The mass of a passenger, including 5 kg of equipment, shall be taken to be equal to 80 kg. The centre of gravity of a person shall be assumed as placed at 1 m above deck. Finally, a passenger density of 5 men per square metre shall be considered in general, and only 3 passengers per square metre for craft in Group E.

Replenishment at sea requires some connection between two vessels. A transverse pull develops; it can be translated into a heeling arm. A transverse pull also can appear during towing. The German regulations contain provisions for calculating these heeling arms. The heel angle caused by replenishment at sea or by crowding of passengers shall not exceed 15°.

10.3 Summary

In Chapter 9 we have shown that longitudinal and quartering waves affect stability by changing the instantaneous moment of inertia that enters into the calculation of the metacentric radius. This effect is taken into account in the stability regulations of the German Federal Navy and it has been proposed to consider it also for merchant ships (Helas, 1982). As shown in Chapter 9, German researchers were the first to investigate parametric resonance in ship stability. They also took into consideration this effect when they elaborated stability regulations for the German Federal Navy. These regulations, known as BV 1033, require that the righting arm be calculated both in still water and in waves. More specifically, cross-curves shall be calculated for ten *wave phases*, that is for ten positions of the wave crest relative to the midship section. The average of those cross-curves shall be compared with the cross-curves in still water and the smaller values shall be used in the stability diagram.

In the German regulations, the criterion of stability under wind regards the difference between the righting arm and the wind heeling arm. This difference, $h_{\mathrm{RES}} = \overline{GZ} - k_{\mathrm{w}}$, is called *residual arm*. If the angle of static equilibrium is ϕ_{ST}, stability shall be checked at a *reference angle*, ϕ_{REF}, defined by

$$\phi_{REF} = \begin{cases} 35° & \text{if } \phi_{ST} \leq 15° \\ 5° + 2\phi_{ST} & \text{otherwise} \end{cases}$$

At this reference angle, the residual arm shall be not smaller than the value given by

$$h_{RES} = \begin{cases} 0.1 & \text{if } \phi_{ST} \leq 15° \\ 0.01\phi_{ST} - 0.05 & \text{otherwise} \end{cases}$$

Finally, a few words about ship forms. Traditionally ship forms have been chosen as a compromise between contradictory requirements of reduced hydrodynamic resistance, good seakeeping qualities, convenient space arrangements and stability in still water. The study of the Mathieu effect has added another criterion: small variation of righting arms in waves. A formulation of this subject can be found in Burcher (1979). Pérez and Sanguinetti (1995) experimented with models of two small fishing vessels of similar size but different forms. They show that the model with round stern and round bilge displayed less metacentric height variation in wave than the model with transom stern.

10.4 Examples

Example 10.1 – Computer function for BV 1033
In this example we describe a function, written in MATLAB 6, that automatically checks the wind criterion of BV 1033. The input consists of four arguments: cond, w, sail, V. The argument cond is an array whose elements are:

1. the displacement, Δ, in kN;
2. the height of the centre of gravity above BL, $\overline{KG}$, in m;
3. the mean draft, T, in m;
4. the height of the metacentre above BL, $\overline{KM}$, in m;
5. the free-surface arm in upright condition, $k_F(0)$, in m.

The argument w is a two dimensional array whose first column contains heel angles, in degrees, and the second column, the lever arms w, in metres. For instance, the following lines are taken from Example 10.2:

```
Maestral = [
0          0
5          0.582
...        ...
90         5.493 ];
```

The argument sail is an array with two elements: the sail area, in m², and the height of the sail-area centroid above BL, in m. Finally, the argument V is the prescribed wind speed, in knots. Only wind speeds specified by BV 1033 are valid arguments.

After calling the function with the desired arguments, the user is prompted to enter the name of the ship under examination. This name will be printed within the title of the stability diagram and in the heading of an output file containing the results of the calculation. In continuation a first plot of the statical-stability curve is presented, together with a cross-hair. The user has to bring the cross-hair on the intersection of the righting-arm and heeling-arm curves. Then, the diagram is presented again, this time with the angle of equilibrium and the angle of reference marked on it. The output file, bv1033.out, is a report of the calculations; among others it contains a comparison of the actual residual arm with the required one.

```
function    [ phiST, hRES ] = bv1033(cond, w, sail, V)
%BV1033 Stability calculations acc. to BV 1033.

clc                           % clean window
Delta = cond(1);              % displacement, kN
KG    = cond(2);              % CG above BL, m
T     = cond(3);              % mean draft, m
KM    = cond(4)               % metacentre above BL, m
kf0   = cond(5);              % free-surface arm, m
heel  = w(:, 1)*pi/180;       % heel angle, deg
lever = w(:, 2);              % arm of form stability, m
A     = sail(1);              % sail area, sq m
z     = sail(2);              % its centroid above BL, m
GZ    = lever - KG*sin(heel); % righting arm
%         choose wind pressure acc. to wind speed
switch  V
          case 90
                  p = 1.5;
          case 70
                  p = 1.0;
          case 50
                  p = 0.5;
          case 40
                  p = 0.3;
          case 20
                  p = 0.1;
          otherwise
                  error('Incorrect wind speed')
end
kf      = kf0*sin(heel);            % free-surface arm, m
%         calculate wind arm in upright condition
kw0     = A*(z - 0.5*T)*p/Delta;
%         calculate wind arm at given heel angles
kw      = kw0*(0.25 + 0.75*cos(heel).^3);
%%%%%%%%%%%%%%%%% Initialize output file  %%%%%%%%%%%%%%%%
sname   = input('Enter ship name ', 's')
fid     = fopen('BV1033.out', 'w');
fprintf(fid, 'Stability of ship %s acc. to BV 1033\n', sname);
fprintf(fid, 'Displacement ................. %9.3f kN\n', Delta);
```

```
fprintf(fid, 'KG ......................... %9.3f m\n', KG);
GM     = KM - KG;              % metacentric height, m
fprintf(fid, 'Metacentric height, GM ....... %9.3f m\n', GM);
fprintf(fid, 'Mean draft, T ................ %9.3f m\n', T);
fprintf(fid, 'Free-surface arm ............. %9.3f m\n', kf0);
fprintf(fid, 'Sail area .................... %9.3f sq m\n', A);
fprintf(fid, 'Sail area centroid above BL .. %9.3f m\n', z);
fprintf(fid, 'Wind pressure ................ %9.3f MPa\n', p);
phi    = w(:, 1);             % heel angle, deg
fprintf(fid, '  Heel    Righting  Heeling \n');
fprintf(fid, '  angle      arm       arm    \n');
fprintf(fid, '   deg        m         m      \n');
harm   = kf + kw;            % heeling arm, m
report = [ phi'; GZ'; harm' ];    % matrix to be printed
fprintf(fid, '%6.1f %11.3f %11.3f \n', report);
plot(phi, GZ, phi, kf, phi, harm, [ 0 180/pi ], [ 0 GM ])
hold on
t1     = [sname ', \Delta = ' num2str(Delta) ' kN, KG = ', ];
t1     = [ t1 num2str(KG) 1 ' m, T = ' num2str(T) ' m' ];
title(t1)
xlabel('Heel angle, degrees')
ylabel('Lever arms, m')
text(phi(5), 1.1*kf(5), 'k_f')
text(phi(7), 1.1*(kf(7)+kw(7)), 'K_f + k_w')
text(phi(6), 1.1*GZ(6), 'GZ')
t2     = [ 'GM = ' num2str(GM) ' m' ];
text(59, GM, t2)
[ phiST,  GZ_ST ] = ginput(1);
plot([ phiST phiST ], [ 0 GZ_ST ], 'k-')
text(phiST, -0.1, '\phi_{ST}')
phiREF = 5 + 2*phiST;        % reference angle, deg
plot([ phiREF phiREF ], [ 0 max(GZ) ], 'k-')
text(phiREF, -0.1, '\phi_{REF}')
hRESm  = 0.01*phiST - 0.05;   % min required residual arm, m
resid  = GZ - (kf + kw);      % array of residual arms, m
%        find residual arm at reference angle
hRES   = spline(phi, resid, phiREF);
if hRES > hRESm
       t0 = ' greater than'
elseif hRES == hRESm
       t0 = ' equal to'
else
       t0 = ' less than'
end
fprintf(fid, '            \n')
fprintf(fid, 'The angle of static
                equilibrium is %5.1f degrees.\n',phiST);
fprintf(fid, 'The residual arm is  %5.3f m       \n', hRES);

fprintf(fid, 'at reference angle %5.1f degrees,
                          %that is\n',phiREF);
fprintf(fid, '%s the required arm %5.3f m.      \n', t0, hRESm);
hold off
fclose(fid)
```

The following example illustrates an application of the function bv1033 to a realistic ship.

Example 10.2 – An application of the wind criterion

This example is based on an undergraduate project carried out by I. Ganoni and D. Zigelman, then students at the TECHNION (Zigelman and Ganoni, 1985). The subject of the project was the reconstitution and analysis of the hydrostatic and hydrodynamic properties of a frigate similar to the Italian Navy Ship *Maestrale*. The lines and other particulars were based on the few details provided by Kehoe, Brower and Meier (1980). To distinguish our example ship from the real one, we shall call it *Maestral*; its main dimensions are: L_{pp}, 114.000 m; B, 12.900 m; D, 8.775 m. Table 10.2 contains the average of the cross-curves of stability in ten wave phases, for a volume of displacement $\nabla = 2943$ m^3.

Example 10.1 illustrates a MATLAB function that automatically checks the wind criterion of BV 1033. To run this function, the cross-curves of stability of the example ship were written to a file, maestrale.m, in the format:

```
Maestral = [
0         0
...       ...
90        5.493 ];
```

The following lines show how to prepare the input and how to invoke the function.

```
maestrale                    % load the cross-curves
cond = [ 1.03*9.81*2943 5.835 4.097 6.681 0.06 ];
sail = [ 1166.55 8.415 ];
bv1033(cond, Maestral, sail, 70)
```

Table 10.2 Frigate Maestral, average of cross-curves in ten wave phases

Heel angle (°)	w (m)	Heel angle (°)	w (m)
0	0	50	4.769
5	0.582	55	5.034
10	1.159	60	5.249
15	1.726	65	5.416
20	2.272	70	5.531
25	2.785	75	5.595
30	3.265	80	5.610
35	3.706	85	5.576
40	4.104	90	5.493
45	4.459		

The resulting diagram of stability is shown in Figure 10.3, the report, printed to file bv1033.out, appears below:

```
Stability of ship Maestral acc. to BV 1033
Displacement .................  29736.955 kN
KG ..........................      5.835 m
Metacentric height, GM .......      0.846 m
Mean draft, T ................      4.097 m
Free-surface arm ............      0.060 m
Sail area ...................   1166.550 sq m
Sail area centroid above BL ..      8.415 m
Wind pressure ...............      1.000 MPa
     Heel      Righting   Heeling
     angle       arm        arm
      deg         m          m
      0.0       0.000      0.249
      5.0       0.073      0.252
     10.0       0.146      0.251
     15.0       0.216      0.246
     20.0       0.276      0.238
     25.0       0.319      0.227
     30.0       0.348      0.214
     35.0       0.359      0.199
     40.0       0.353      0.185
     45.0       0.333      0.171
     50.0       0.299      0.158
     55.0       0.254      0.147
     60.0       0.196      0.138
     65.0       0.128      0.131
     70.0       0.048      0.126
     75.0      -0.041      0.123
     80.0      -0.136      0.122
     85.0      -0.237      0.122
     90.0      -0.342      0.122

The angle of static equilibrium is   17.0 degrees.
The residual arm is   0.168 m
at reference angle 39.1 degrees, that is
greater than the required arm 0.120 m.
```

10.5 Exercises

Exercise 10.1 – Trochoidal wave

Plot the trochoidal waves prescribed by BV 1033 for ships of 50, 100 and 200 m length. Show, on the same plots, the still-water line.

Table 10.3 *Lido 9*, cross-curves in seaway, 44.16 m³, trim −0.325 m

Heel angle (°)	Wave trough (m)	Still water (m)	Wave crest (m)
0	0.000	0.000	0.000
5	0.360	0.397	0.395
10	0.713	0.770	0.773
15	1.055	1.111	1.124
20	1.375	1.421	1.445
25	1.671	1.704	1.727
30	1.946	1.967	1.966
35	2.200	2.206	2.166
40	2.429	2.410	2.336
45	2.622	2.582	2.477
50	2.766	2.735	2.588
55	2.867	2.868	2.671
60	2.934	2.950	2.729
65	2.959	2.960	2.756
70	2.955	2.932	2.767
75	2.925	2.875	2.744
80	2.856	2.789	2.678
85	2.756	2.679	2.582
90	2.637	2.548	2.458

Exercise 10.2 – Lido 9, cross-curves in seaway
Table 10.3 contains the ℓ_k levers of the vessel *Lido 9*, for a volume of displacement equal to 44.16 m³ and the full-load trim −0.325 m. The data are calculated in wave trough, in still water, and on wave crest. According to the BV 1033 stability regulations of the German Federal Navy the wave length equals the length between perpendiculars, that is $\lambda = 15.5$ m, and the wave height is calculated from

$$H = \frac{\lambda}{10 + \lambda/20} = 1.439 \, \text{m}$$

Assuming that the height of the centre of gravity is $\overline{KG} = 2.21$ m, calculate and plot the diagrams of statical stability ($\overline{GZ}$ curves) for the three conditions: wave trough, still water, wave crest.

Using the same data as in Example 6.1 and the wind arm prescribed by the BV 1022 regulations, check the range of positive residual arms in wave trough and on wave crest. According to BV 1033, the range of positive residual arms should be at least 10°, and the maximum residual arm not less than 0.1 m.

11
Flooding and damage condition

11.1 Introduction

In the preceding chapters, we discussed the buoyancy and stability of **intact** ships. Ships, however, can suffer damages during their service. Hull damages that affect the buoyancy can be caused by collision, by grounding or by enemy action. Water can enter the damaged compartment and cause changes of draught, trim and heel. Above certain limits, such changes can lead to ship loss. We expect a ship to **survive** a reasonable amount of damage, that is an amount compatible with the size and tasks of the vessel. More specifically, we require that a ship that suffered hull damage, to an extent not larger than defined by pertinent regulations, should continue to float and be stable under moderate environmental conditions. Then, passengers and crew can be saved. Possibly the ship herself can be towed to a safe harbour.

To achieve survivability as defined above, the ship hull is subdivided into a number of watertight compartments. The lengths of the compartments should be such that after the flooding of a certain number of adjacent compartments, the waterline shall not lie above a line prescribed by relevant regulations. The same regulations specify the number of adjacent compartments that should be assumed flooded. This number depends on the size and the mission of the ship. The reason for considering adjacent compartments is simple. Collision, grounding or single enemy action usually damage adjacent compartments. Flooding of adjacent compartments also can be more dangerous than flooding of two non-adjacent compartments. Adjacent compartments situated at some distance from the midship section can cause large trim and submerge openings above the deck, leading thus to further flooding. Also, submerging part of the deck reduces the waterplane area and can cause a substantial decrease of the metacentric radius. Flooding of non-adjacent compartments, for example one in the forebody, the other in the afterbody, can produce negligible trim. Then, even with relatively large draught increases, the deck does not submerge, the waterplane area is not reduced, and the metacentric height may be sufficient. If the deck does not submerge, no openings are submerged. The need for international regulations governing the **subdivision** of the hull into watertight compartments became clear after the *Titanic* disaster, in April 1912. A meeting was convened in London leading to the adoption on 20 January 1914 of an **International Convention of the Safety of Life at Sea**. The convention is better known under its acronym, **SOLAS**. The first convention should have been applied in July 1915, but the First World War stopped the process. In 1929, a new conference was held in London. The adopted text entered into force in 1933. Technical developments made necessary a new conference; it was held in 1948. The next edition was the 1960 SOLAS Convention, organized this time by IMO (about IMO see Section 8.2). Several amendments were adopted in the following years. The 1974 SOLAS Convention was again held in London. Since then many important amendments were issued, some of them influenced by major marine disasters, such as those of the roll-on/roll-off passenger ferries *Herald of the Free Enterprise*, near Zeebrugge, in March 1987, and *Estonia*, on 28 September 1994. At the moment of this publication SOLAS 1974 together with all its amendments is the convention in force (see SOLAS 2001).

SOLAS prescriptions cover many aspects of ship safety, among them fire protection, life boats and rafts, radars, radio equipment, and emergency lighting. What interests us in this book are the prescriptions referring to subdivision and damage stability. A detailed history of SOLAS activities can be found on a website organized by *Metal Safe Sign International Ltd*, http://www.mss-int.com. A short history of damage regulations appears in Gilbert and Card (1990). A commented history of the SOLAS achievements can be read in Payne (1994). Because of the overwhelming importance of the SOLAS regulations we give here the translations of the official title in three other languages:

Fr   Convention internationale pour la sauvegarde de la vie humaine en mer

G   Internationales Übereinkommen zum Schutz des menschlichen Lebens auf See

I   Convenzione internazionale per la salvaguardia della vita umana in mare

The SOLAS regulations apply to merchant ships. Damage regulations for warships are provided in the same regulations that deal with their intact stability (see Chapters 8 and 10).

The European Commission sponsored researches on survivability in damage condition, mainly the project HARDER. The Nordic countries established a project entitled 'Safety of passenger/ro–ro vessels' (Svensen and Vassalos, 1998).

An alternative term used in damage considerations is **bilging**. Derrett and Barrass (2000) define it as follows: 'let an empty compartment be holed ... below the waterline to such an extent that the water may ... flow freely into and out of the compartment. A vessel holed in this way is said to be bilged.'

Roll-on/Roll-off ships, shortly Ro/Ro, are particularly sensitive to damage. To enable easy loading and unloading of vehicles these vessels are provided with a deck space uninterrupted by bulkheads. For the same reasons, that deck is close to the waterline. Damage can easily cause deck flooding with consequences like

$\overline{KG}$ increase, large free-surface effect and added weight. Little and Hutchinson (1995) quote, 'Over the past 14 years, 44 RO/RO vessels have capsized.' Pawlowski (1999) appreciates, 'Roll-on/roll-off (RO/RO) ships are considered by the maritime profession ... as the most unsafe ships in operation.' Statistics on loss of life due to RO/RO disasters are simply frightening. For example, Ross, Roberts and Tighe (1997) quote 193 casualties in the case on the *Herald of Free Enterprise*, 910 in the *Estonia* disaster. A few Ro/Ro's sank in one and a half minute after an accident. No wonder that many studies have been dedicated to this type of vessel. As some of them refer to constructive measures, we think that their treatment belongs to books on Ship Design, not here. We cite, however, the papers whose contents are close to the subject of this chapter.

In this chapter, we give the definitions related to flooding and explain the principles on which flooding and damage calculations are based. To illustrate these principles we apply them to box shaped vessels. We also summarize a few pertinent regulations and codes of practice. When performing calculations for real-life projects, the reader is advised to refer to the full text of the most recent edition of the regulations to be applied.

Flooding and damage stability calculations for real ship forms are rather complex and tedious. Finding the floating condition requires iterative procedures. Today, such calculations are performed on computers; therefore, we do not describe them. We also give in this chapter the translations of the most important terms introduced in it.

11.2 A few definitions

In this section, we introduce a few terms defined in the SOLAS conventions; they are also used by other regulations. The hull is subdivided into compartments by means of **watertight bulkheads**. This term is translated into three other languages as

Fr   cloisons étanches
G   Schotten
I   paratie stagne

The deck up to which these bulkheads extend is called in English **bulkhead deck**, in other languages

Fr   pont de cloisonnement
G   Schottendeck
I   ponte delle paratie

After flooding of a prescribed number of compartments the ship shall not submerge beyond a line situated at least 76 mm (3 in) below the deck at side. The said line is called in English **margin line**, in other languages

Fr	ligne de surimmersion
G	Tauchgrenze
I	linea limite

The **floodable length** at a given point of the ship length is the maximum length, with the centre at that point, that can be flooded without submerging the ship beyond the margin line. This subject is treated in more detail in Section 11.6. The term 'floodable length' is translated as

Fr	longueur envahissable
G	flutbare Länge
I	lunghezza allagabile

In Figure 11.1, we see the sketch of a ship subdivided by four bulkheads. The three waterlines WL_1, WL_2 and WL_3 are tangent to the margin line. They are examples of limit lines beyond which no further submergence of the damaged ship is admissible. If the bulkhead deck is not continuous, a continuous margin line can be assumed such as having no point at a distance less than 76 mm below the deck at side.

Let us suppose that calculating the volume of a compartment starting from its dimensions we obtain the value v. There is almost no case in which this volume can be fully flooded because almost always there are some objects in the compartment. Even in an empty tank there are usually structural members – such as frames, floors and deck beams – sounding instruments and stairs for entering the tank and inspecting it. If we deduct the volumes of such objects from the volume v, we obtain the volume of the water that can flood the compartment; let it be v_F. The ratio

$$\mu = \frac{v_F}{v} \tag{11.1}$$

is called **permeability**; it is often noted by μ. More correctly, we should talk about **volume permeability**, to distinguish it from a related notion that is the **surface permeability**. Indeed, because of the objects stored or located in a compartment, the free-surface area is smaller than that calculated from the

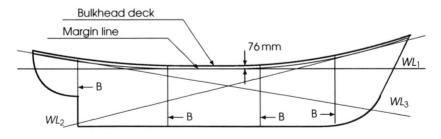

B – Watertight bulkhead

Figure 11.1 A few definitions

dimensions of the compartment. Also the moment of inertia of the free-surface area is calculated on the basis of the dimensions of the compartment. For example, if the calculations are carried out by a computer programme, they are based on an input that describes only the geometry of the tank and not its contents. The moment of inertia of the surface free to heel is smaller than the value found as above because the area considered is partially occupied by fixed objects that do not contribute to the free-surface effect. Then, it is necessary to multiply the calculated value by the surface permeability.

Typical values of volume permeability can be found in textbooks and in various regulations. Examples of the latter are given in this chapter. When the recommended values do not seem plausible, it is necessary to calculate in detail the volume of the objects found in the compartment. When there are no better data, the surface permeability can be assumed equal to the volume permeability of the same compartment.

The term 'permeability' is translated into other languages as follows

Fr	(coefficient de) perméabilité
G	Flutbarkeit
I	(coefficiente di) permeabilitá

Usually, permeabilities are given in percent, for example 85 for machinery spaces. In calculations, however, we must multiply by 0.85, and not by 85. Moreover, some computer programmes, such as ARCHIMEDES, require as input the number 0.85 and not 85. Therefore, in the following sections permeabilities are mainly given in the format 0.95, 0.85 etc., rather than as percentages.

11.3 Two methods for finding the ship condition after flooding

There are two ways of calculating the effect of flooding. One way is known as the **method of lost buoyancy**, the other as the **method of added weight**.

The method of lost buoyancy assumes that a flooded compartment does not supply buoyancy. This is what happens in reality. If we refer to Figures 2.4 and 2.5, we can imagine that if there is open communication between a compartment and the surrounding water, the water inside the compartment exercises pressures equal to and opposed to those of the external water. Then, the buoyancy force predicted by the Archimedes' principle is cancelled by the weight of the flooding water.

In the method of lost buoyancy the volume of the flooded compartment does not belong anymore to the vessel, while the weight of its structures is still part of the displacement. The 'remaining' vessel must change position until force and moment equilibria are re-established. During the process not only the displacement, but also the position of the centre of gravity remains constant. The method is also known as **method of constant displacement**. As the flooding water does not belong to the ship, it causes no free-surface effect.

In the method of added weight the water entering a damaged compartment is considered as belonging to the ship; its mass must be added to the ship displacement. Hence the term 'added weight'. Following modern practice, we actually work with masses; however, we keep the traditional name of the method, i.e. we use the word 'weight'. Another reason may be the need to avoid confusion with the term added mass mentioned in Section 6.12 and detailed in Chapter 12. The latter term does not belong to the theory of flooding and damage stability.

In the method of added weight the displacement of the flooded vessel is calculated as the sum of the intact displacement and the mass of the flooding water. The position of the centre of gravity of the damaged vessel is obtained from the sums of the moments of the intact vessel and of the flooding water. Becoming part of the vessel, the flooding water produces a free-surface effect that must be calculated and considered in all equations. For very small trim and negligible heel changes we can write

$$
\begin{aligned}
\Delta_{\mathrm{F}} &= \Delta_{\mathrm{I}} + \rho \cdot v \\
LCG_{\mathrm{F}} \cdot \Delta_{\mathrm{F}} &= LCG_{\mathrm{I}} \cdot \Delta_{\mathrm{I}} + lcg \cdot \rho \cdot v \qquad (11.2) \\
TCG_{\mathrm{F}} \cdot \Delta_{\mathrm{F}} &= tcg \cdot \rho \cdot v
\end{aligned}
$$

where the subscript F distinguishes the properties of the flooded vessel, and the subscript I those of the intact ship. By lcg we mean the longitudinal centre of gravity of the flooding water volume, v, and by tcg its transverse centre of gravity. We assume $TCG_{\mathrm{I}} = 0$. When the trim and the heel are not negligible, we must consider the vertical coordinates of the centres of gravity of the intact ship and of the flooding water volume. Example 11.1 shows how to do this for non-zero trim and zero heel.

To exemplify the above principles we follow an idea presented in *Handbuch der Werften* and later used by Watson (1998). While the latter solves algebraically the general problem, we prefer to solve it numerically and thus allow the reader to visualize the differences between methods and those between the intact and the damaged vessel. We choose the very simple example of the pontoon shown in Figure 11.2. Two transverse bulkheads subdivide the hull into three watertight compartments. In the following two sections we assume that Compartment 2 is damaged and calculate the consequences of its flooding. We choose deliberately a compartment symmetric about the midship transverse plane of symmetry of the pontoon. Thus, the flooding of Compartment 2 produces no trim. Also, the compartment extends for the full ship breadth and its flooding produces no heel. The only change of position is **parallel sinking**. Thus, the complex calculations necessary for conventional ship forms, for large trim, or for unsymmetrical flooding, do not obscure the principles and it is possible to obtain immediately a good insight of the processes involved. For the same reasons we assume that the volume and surface permeabilities are equal to 1. We leave to an exercise the

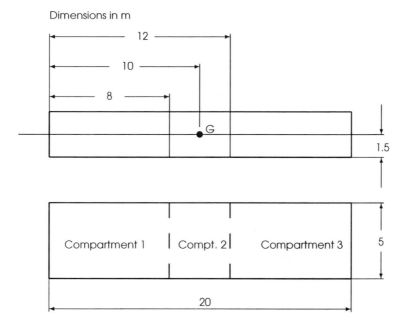

Dimensions in m

Figure 11.2 A simple pontoon – intact condition

informal proof that taking permeability into account does not change the qualitative results. Although based on different physical models, calculations by the two methods yield the same final draught, as it should be expected. Moreover, the stability properties calculated by the two methods are identical, if we compare the initial righting moments. Here, the term 'initial' has the meaning defined in Chapter 2 where we consider 'initial stability' as a property governing the behaviour of the floating body in a small heel range around the upright position. In that range the righting moment equals

$$M_{\mathrm{R}} = \Delta \overline{GM} \sin \phi$$

As we are going to see, we obtain by the two methods the same M_{R} value. In the method of lost buoyancy the displacement remains equal to that of the intact vessel. In the method of added weight the displacement increases by the mass of the flooding water. To keep the product M_{R} constant, the other factor, $\overline{GM}$, must be smaller. At a first glance it may be surprising that the two methods yield different metacentric heights. The explanation given above shows that it should be so because the considered displacements are different. What should be kept in mind, after reading the examples, is that displacement and metacentric height have different significances in the two methods. Therefore, damage stability data should include the mention of the method by which they were obtained. Computer programmes use the method of lost buoyancy.

The length of the assumed pontoon is $L = 20$ m, the beam, $B = 5$ m, and the draught in intact condition, $T_{\mathrm{I}} = 1.5$ m. Let the vertical centre of gravity be $\overline{KG_I} = 1.5\,\text{m}$. The following calculations were carried out in MATLAB, using the full precision of the software. The results are rounded off to a reasonable number of decimal digits. We first find the data of the intact pontoon. The displacement volume is

$$\nabla_I = LBT_I = 20 \times 5 \times 1.5 = 150\,\text{m}^3$$

The mass displacement equals

$$\Delta_I = \rho\nabla_I = 1.025 \times 150 = 153.75\,\text{t}$$

The moment of inertia of the waterplane area about the centreline equals

$$I_I = \frac{B^3 L}{12} = \frac{5^3 \times 20}{12} = 208.3333\,\text{m}^4$$

and the resulting metacentric radius is

$$\overline{BM}_I = \frac{I_I}{\nabla_I} = \frac{208.3333}{150} = 1.389\,\text{m}$$

For such a simple form we could have found directly the metacentric radius as

$$\overline{BM}_I = \frac{B^3 L/12}{LBT_I} = \frac{B^2}{12T_I} = \frac{5^2}{12 \times 1.5} = 1.389\,\text{m}$$

The height of the centre of buoyancy is

$$\overline{KB}_I = \frac{T_I}{2} = 0.75\,\text{m}$$

and the metacentric height is

$$\overline{GM}_I = \overline{KB}_I + \overline{BM}_I - \overline{KG}_I = 0.75 + 1.389 - 1.50 = 0.639\,\text{m}$$

For small heel angles the righting moment in intact condition is calculated as

$$M_{RI} = \Delta_I \overline{GM}_I \sin\phi = 153.75 \times 0.693 \times \sin\phi = 98.229 \sin\phi\,\text{t m}$$

11.3.1 Lost buoyancy

The translations of the term 'method of lost buoyancy' in three other languages are

Fr   La méthode des carènes perdues
G   Methode des wegfallender Verdrängung
I   Il metodo per perdita di galleggiabilità

In the method of lost buoyancy, the flooded compartment does not supply buoyancy. As shown in Figure 11.3, the buoyant hull is composed only of Compartments 1 and 3. After loosing the central compartment, the waterplane area is equal to

$$A_{\mathrm{L}} = (L - l)B = (20 - 4) \times 5 = 80\,\mathrm{m}^2$$

To compensate for the loss of buoyancy of the central compartment the draught increases to

$$T_{\mathrm{L}} = \frac{\nabla_{\mathrm{I}}}{A_{\mathrm{L}}} = \frac{150}{80} = 1.875\,\mathrm{m}$$

The height of the centre of buoyancy increases to

$$\overline{KB}_{\mathrm{L}} = \frac{T_{\mathrm{L}}}{2} = \frac{1.875}{2} = 0.938\,\mathrm{m}$$

We calculate the moment of inertia of the waterplane as

$$I_{\mathrm{L}} = \frac{B^3(L - l)}{12} = \frac{5^3(20 - 4)}{12} = 166.6667\,\mathrm{m}^4$$

and the metacentric radius as

$$\overline{BM}_{\mathrm{L}} = \frac{I_{\mathrm{L}}}{\nabla_{\mathrm{I}}} = \frac{166.6667}{150} = 1.111\,\mathrm{m}$$

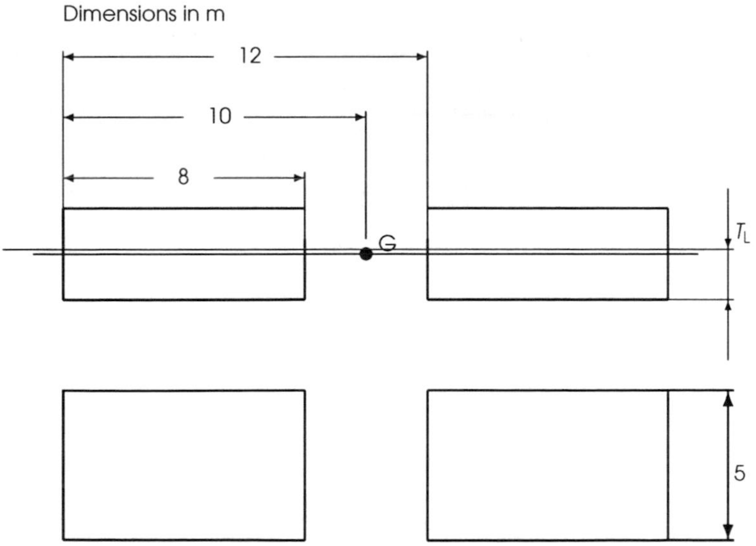

Figure 11.3 A simple pontoon – damage calculation by the method of lost buoyancy

Finally, the metacentric height is

$$\overline{GM}_{\mathrm{L}} = \overline{KB}_{\mathrm{L}} + \overline{BM}_{\mathrm{L}} - \overline{KG}_{\mathrm{I}} = 0.938 + 1.111 - 1.5 = 0.549\,\mathrm{m}$$

and the righting moment for small heel angles, in the lost-buoyancy method

$$M_{\mathrm{RL}} = \Delta_{\mathrm{I}}\overline{GM}_{\mathrm{L}} \sin\phi = 153.75 \times 0.549 \sin\phi = 84.349 \sin\phi\,\mathrm{t\,m}$$

11.3.2 Added weight

The translations of the term 'added-weight method' in three other languages are

Fr   La méthode par addition de poids
G   Methode des hinzukommenden Gewichts
I   Il metodo del peso imbarcato

For this section see Figure 11.4. Because of the added weight of the flooding water the draught of the pontoon must increase by a quantity δT. The volume of flooding water equals

$$v = lB(T_{\mathrm{I}} + \delta T) \tag{11.3}$$

The additional buoyant volume of the vessel, due to parallel sinking, is

$$\delta\nabla = LB\delta T \tag{11.4}$$

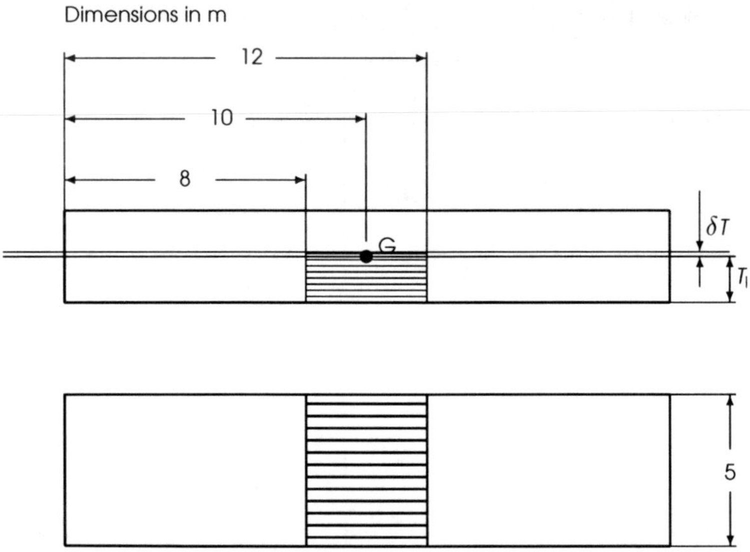

Figure 11.4 A simple pontoon – damage calculation by the method of added weight

To obtain the draught increment, δT, we equate the two volumes, that is we write $v = \delta \nabla$. Algebraic manipulation and numerical calculation yield

$$\delta T = \frac{lT_{\mathrm{I}}}{L-l} = \frac{4 \times 1.5}{20-4} = 0.375 \, \mathrm{m}$$

The draught after flooding is

$$T_{\mathrm{A}} = T_{\mathrm{I}} + \delta T = 1.500 + 0.375 = 1.875 \, \mathrm{m}$$

The volume of flooding water is calculated as

$$v = lBT_{\mathrm{A}} = 4 \times 5 \times 1.875 = 37.5 \, \mathrm{m}^3$$

and the height of its centre of gravity

$$\overline{kb} = \frac{T_{\mathrm{A}}}{2} = \frac{1.875}{2} = 0.938 \, \mathrm{m}$$

The displacement volume of the flooded pontoon is

$$\nabla_{\mathrm{A}} = LBT_{\mathrm{A}} = 20 \times 5 \times 1.875 = 187.5 \, \mathrm{m}^3$$

We consider the flooding water as an added weight; therefore, we must calculate a new centre of gravity. The calculations are shown in Table 11.1. The moment of inertia of the damage waterplane is the same as in the initial condition, that is $I_{\mathrm{A}} = 208.333 \, \mathrm{m}^4$. Then, the metacentric radius equals

$$\overline{BM}_{\mathrm{A}} = \frac{I_{\mathrm{A}}}{\nabla_{\mathrm{A}}} = \frac{208.333}{187.5} = 1.111 \, \mathrm{m}$$

In this method, the flooding water is considered as belonging to the displacement. Therefore, if there is a free surface its effect must be calculated. The moment of inertia of the free surface in the flooded compartment equals

$$i = \frac{B^3 l}{12} = \frac{5^3 \times 4}{12} = 41.667 \, \mathrm{m}^4$$

and the lever arm of the free surface effect is

$$\ell_{\mathrm{F}} = \frac{\rho i}{\rho \nabla_{\mathrm{A}}} = \frac{41.667}{187.5} = 0.222 \, \mathrm{m}$$

Table 11.1 $\overline{KG}$ by the method of added weight

	Volume	Centre of gravity	Moment
Initial	150.0	1.5	225.000
Added	37.5	0.938	35.156
Total	187.5	1.388	260.156

The height of the centre of buoyancy is yielded by

$$\overline{KB}_A = \frac{T_A}{2} = \frac{1.875}{2} = 0.938\,\mathrm{m}$$

The corresponding metacentric height is calculated as

$$
\begin{aligned}
\overline{GM_A} &= \overline{KB}_A + \overline{BM}_A - \overline{KG}_A - \ell_F \\
&= 0.938 + 1.111 - 1.388 - 0.222 \\
&= 0.439\,\mathrm{m}
\end{aligned}
$$

With the mass displacement

$$\Delta_A = \rho \nabla_A = 1.025 \times 187.5 = 192.188\,\mathrm{t}$$

we obtain the righting moment for small angles of heel, in the added-weight method

$$M_{RA} = \Delta_A \overline{GM}_A \sin\phi = 192.188 \times 0.439 \sin\phi = 84.349 \sin\phi\,\mathrm{t\,m}$$

11.3.3 The comparison

Table 11.2 summarizes the results of the preceding two sections. As expected, both the method of lost buoyancy and that of added weight yield the same draught 1.875 m, and the same initial righting moment, $84.349 \sin\phi$ t m. The displacements and the metacentric heights are different, but their products, $\Delta \overline{GM}$, are the same. As happens in most cases, the righting moment in damage condition is less than in intact condition.

Table 11.2 Flooding calculations – a comparison of methods

	Intact condition	Damaged, lost buoyancy	Damaged, by added weight
Draught, m	1.500	1.875	1.875
∇, m^3	150.000	150.000	187.500
Δ, t	153.750	153.750	192.188
$\overline{KB}$, m	0.750	0.938	0.938
$\overline{BM}$, m	1.389	1.111	1.111
$\overline{KG}$, m	1.500	1.500	1.388
$\overline{GM}$, m	0.639	0.549	0.439
$\Delta\overline{GM}$, tm	98.229	84.349	84.349

11.4 Details of the flooding process

The free surface in a compartment open to the sea behaves differently than that in an intact tank. In Figure 11.5(a), $W_I L_I$ is the waterline in upright position and $W_\phi L_\phi$, the waterline in a heeled position. We assume that the water level in the side tank is the same as the external water level. In the heeled position the water surface in the tank changes to FS, a line parallel to $W_\phi L_\phi$. The volume of water in the tank remains constant. In Figure 11.5(b) the side tank is damaged and in open communication with the sea. If the waterline in the heeled position is $W_\phi L_\phi$, this is also the water level in the damaged tank. The water volume is no longer constant, but varies with the heel angle. For the case shown in the figure, the volume increases by the slice comprised between the lines $W_\phi L_\phi$ and FS. This change of volume must be taken into account in the added-weight method. Figure 11.5(b) shows a case of **unsymmetrical flooding**. This kind of flooding can easily submerge the deck. The consequences may be a drastic reduction of stability and the submergence of openings such as vents. Therefore, care must be exercised when placing longitudinal bulkheads. Sometimes, to compensate unsymmetrical flooding it is necessary to open a connection between the damaged tank and a tank situated symmetrically on the other side of the ship. This action is called **cross-flooding**. The UK-Navy document SSP 24 warns against the potential danger presented by longitudinal bulkheads.

Cross-flooding takes some time and can cause a slow change of the ship position. Söding (2002) lists other slow-flooding processes such as occurring 'through open or non-watertight doors, hatches with non-watertight or partly open hatch covers, through pipes, ventilation ducts...'. In his paper, Söding

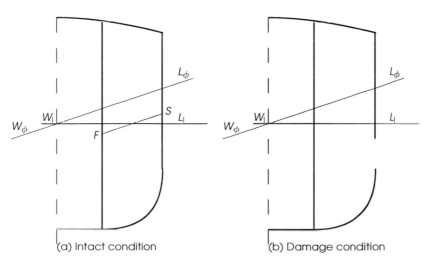

(a) Intact condition   (b) Damage condition

Figure 11.5 Free surface in intact and in damaged tank

describes the mathematics of such water flows. Air can be trapped above the flooding-water surface. If the top envelope of the compartment is airtight flooding is stopped. If not, it is only slowed down.

Between the position of intact condition and the final damage position (provided that an equilibrium position can be found) the vessel can pass through intermediate positions more dangerous than the final one. It is necessary to check if such positions exist and if the ship can survive them.

11.5 Damage stability regulations

11.5.1 SOLAS

Regulation 5 of the convention specifies how to calculate the permeabilities to be considered. Thus, the permeability, in percentage, throughout the machinery space shall be

$$85 + 10 \left(\frac{a - c}{v} \right)$$

where a is the volume of passenger spaces situated under the margin line, within the limits of the machinery space, c is the volume of between-deck spaces, in the same zone, appropriated to cargo, coal, or store, and v, the whole volume of the machinery space below the margin line.

The percent permeability of spaces forward or abaft of the machinery spaces should be found from

$$63 + 35 \frac{a}{v}$$

where a is the volume of passenger spaces under the margin line, in the respective zone, and v, the whole volume, under the margin line, in the same zone.

The **maximum permissible length** of a compartment having its centre at a given point of the ship length is obtained from the floodable length by multiplying the latter by an appropriate number called **factor of subdivision**. For example, a factor of subdivision equal to 1 means that the margin line should not submerge if one compartment is submerged, while a factor of subdivision equal to 0.5 means that the margin line should not submerge when two compartments are flooded.

Regulation 6 of the convention shows how to calculate the factor of subdivision as a function of the ship length and the nature of the ship service. First, SOLAS defines a factor, A, applicable to ships primarily engaged in cargo transportation

$$A = \frac{58.2}{L - 60} + 0.18$$

For $L = 131$, $A = 1$. Another factor, B, is applicable for ships primarily engaged in passenger transportation

$$B = \frac{30.3}{L - 42} + 0.18$$

For $L = 79$, $B = 1$.

A **criterion of service numeral**, C_s, is calculated as function of the ship length, L, the volume of machinery and bunker spaces, M, the volume of passenger spaces below the margin line, P, the number of passengers for which the ship is certified, N, and the whole volume of the ship below the margin line, V. There are two formulas for calculating C_s; their choice depends upon the product $P_1 = KN$, where $K = 0.056L$. If P_1 is greater than P,

$$C_s = 72 \frac{M + 2P_1}{V + P_1 - P}$$

otherwise

$$C_s = 72 \frac{M + 2P}{V}$$

For ships of length 131 m and above, having a criterion numeral $C_s \leq 23$, the subdivision abaft the forepeak is governed by the factor A. If $C_s \geq 123$ the subdivision is governed by the factor B. For $23 < C_s < 123$, the subdivision factor should be interpolated as

$$F = A - \frac{(A - B)(C_s - 23)}{100}$$

If $79 \leq L < 131$, a number S should be calculated from

$$S = \frac{3.754 - 25L}{13}$$

If $C_s = S$, $F = 1$. If $C_s \geq 123$, the subdivision is governed by the factor B. If C_s lies between S and 123, the subdivision factor is interpolated as

$$F = 1 - \frac{(1 - B)(C_s - S)}{123 - S}$$

If $79 \leq L < 131$ and $C_s < S$, or if $L < 79$, $F - 1$.

Regulation 7 of the convention contains special requirements for the subdivision of passenger ships. Regulation 8 specifies the criteria of stability in the final condition after damage. The heeling arm to be considered as the one that results from the largest of the following moments:

- crowding of all passengers on one side;
- launching of all fully loaded, davit-operated survival craft on one side;
- due to wind pressure.

We call **residual righting lever arm** the difference

$$\overline{GZ} - \text{heeling arm}$$

The range of positive residual arm shall be not less than $15°$. The area under the righting-arm curve should be at least $0.015\,\text{m rad}$, between the angle of static equilibrium and the smallest of the following:

- angle of progressive flooding;
- $22°$ if one compartment is flooded, $27°$ if two or more adjacent compartments are flooded.

The moment due to the crowding of passengers shall be calculated assuming 4 persons per m^2 and a mass of $75\,\text{kg}$ for each passenger. The moment due to the launching of survival craft shall be calculated assuming all lifeboats and rescue boats fitted on the side that heeled down, while the davits are swung out and fully loaded. The wind heeling moment shall be calculated assuming a pressure of $120\,\text{N m}^{-2}$.

11.5.2 Probabilistic regulations

Wendel (1960a) introduces the notion of **probability of survival after damage**. A year later, a summary in French appears in Anonymous (1961). This paper mentions a translation into French of Wendel's original paper (in Bulletin Technique du Bureau Veritas, February 1961) and calls the method 'une nouvelle voie', that is 'a new way'. Much has been written since then on the probabilistic approach; we mention here only a few publications, such as Rao (1968), Wendel (1970), Abicht and Bakenhus (1970), Abicht, Kastner and Wendel (1977), Wendel (1977). Over the years Wendel used new and better statistics to improve the functions of probability density and probability introduced by him. The general idea is to consider the probability of occurrence of a damage of length y and transverse extent t, with the centre at a position x on the ship length. Statistics of marine accidents should allow the formulation of a function of probability density, $f(x, y, t)$. The probability itself is obtained by triple integration of the density function. The IMO regulation A265 introduces probabilistic regulations for passenger ships, and SOLAS 1974, Part B1, defines probabilistic rules for cargo ships. Concisely, Regulation 25 of the SOLAS convention defines a **degree of subdivison**

$$R = (0.002 + 0.0009L^3)^{1/3}$$

where L is measured in metres. An **attained subdivision index** shall be calculated as

$$A = \Sigma p_i s_i$$

where p_i represents the probability that the ith compartment or group of compartments may be flooded, and s_i is the probability of survival after flooding the ith compartment or group of compartments. The attained subdivision index, A, should not be less than the required subdivision index, R.

Early details of the standard for subdivision and damage stability of dry cargo ships are given by Gilbert and Card (1990). A critical discussion of the IMO 1992 probabilistic damage criteria for dry cargo ships appears in Sonnenschein and Yang (1993). The probabilistic SOLAS regulations are discussed in some detail by Watson (1998) who also exemplifies them numerically. Ravn *et al.* (2002) exemplify the application of the rules to Ro–Ro vessels.

Serious criticism of the SOLAS probabilistic approach to damage can be found in Björkman (1995). Quoting from the title of the paper, 'apparent anomalies in SOLAS and MARPOL requirements'. Watson (1998) writes, 'There would seem to be two main objections to the probabilistic rules. The first of these is the extremely large amount of calculations required, which although acceptable in the computer age, is scarcely to be welcomed. The other objection is the lack of guidance that it gives to a designer, who may be even driven to continuing use of the deterministic method in initial design, changing to the probabilistic later – and hoping this does not entail major changes!'

The 'CORDIS RTD PROJECTS' database of the European Communities, 2000, defines as follows the objective of project HARDER:

> 'The process of harmonisation of damage stability regulations according to the probabilistic approach is undergoing scrutiny . . . before being proposed for adoption by IMO . . . However, ongoing investigations started revealing serious lack of robustness and consistency and more importantly a worrying lack of rationale in the choice of parameters that are likely to affect the evolution of the overall design and safety of ships.
>
> A recent application of existing tools by a committee of the relevant IMO working group . . . revealed that, before confidence in the whole process is irreversibly affected, concerted effort at European level must address the thorough validation of calculations, the proper choice of parameters and the definition of levels of acceptance . . . '

A report on the progress of the project HARDER is contained in the IMO document SLF 45/3/3 of 19 April 2002. The report covers 'Investigations and proposed formulations for the factor "S": the probability of survival after flooding'. The approach adopted in the project HARDER is explained by Rusås (2002). As the probabilistic regulations are bound to change, we do not detail them in this book.

11.5.3 The US Navy

The regulations of the US Navy are contained in a document known as DDS-079-1. Part of the regulations are classified, part of those that are not classified can be found in Nickum (1988) or Watson (1998). For a ship shorter than 30.5 m (100 ft) the flooding of any compartment should not submerge her beyond the margin line. Ships longer than 30.5 m and shorter than 91.5 m (300 ft) should meet the same submergence criterion with two flooded compartments. Ships longer than 91.5 m should meet the submergence criterion with a damage extent of $0.15L$ or 21 m, whichever is greater.

When checking stability under wind, the righting arm, $\overline{GZ}$, should be reduced by $0.05 \cos \phi$ to account for unknown unsymmetrical flooding or transverse shift of loose material. As for intact condition (see Figure 8.4), the standard identifies two areas between the righting-arm and the wind-arm curves. The area A_1 is situated between the angle of static equilibrium and the angle of downflooding or 45°, whichever is smaller. The area A_2 is situated to the left, from the angle of static equilibrium to an angle of roll. The wind velocity and the angle of roll should be taken from DDS-079-1. As in the intact condition, the standard requires that $A_1/A_2 \geq 1.4$.

The US Navy uses the concept of **V lines** to define a zone in which the bulkheads must be completely watertight. We refer to Figure 11.6. Part (a) of the figure shows a longitudinal ship section near a bulkhead. Let us assume that after checking all required combinations of flooded compartments, the highest

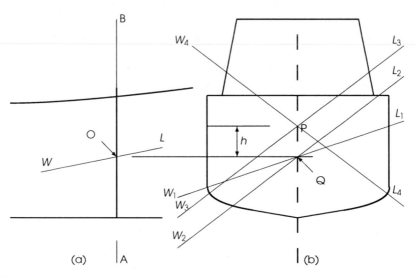

Figure 11.6 V lines

waterline on the considered bulkhead is WL; it intersects the bulkhead at O. In part (b) of the figure, we show the transverse section AB that contains the bulkhead. The intersection of WL with the bulkhead passes though the point Q. The standard assumes that unsymmetrical flooding can heel the vessel by $15°$. The waterline corresponding to this angle is W_1L_1. Rolling and transient motions can increase the heel angle by a value that depends on the ship size and should be taken from the standard. We obtain thus the waterline W_2L_2. Finally, to take into account the relative motion in waves (that is the difference between ship motion and wave-surface motion) we draw another waterline translated up by $h = 1.22$ m (4 ft); this is waterline W_3L_3. Obviously, unsymmetrical flooding followed by rolling can occur to the other side too so that we must consider the waterline W_4L_4 symmetrical of W_3L_3 about the centreline. The waterlines W_3L_3 and W_4L_4 intersect at the point P. We identify a V-shaped limit line, W_4PL_3, hence the term 'V lines'. The region below the V lines must be kept watertight; severe restrictions refer to it and they must be read in detail.

11.5.4 The UK Navy

The standard of damage stability of the UK Navy is defined in the same documents NES 109 and and SSP 24 that contain the prescriptions for intact stability (see Section 8.4). We briefly discuss here only the rules referring to vessels with a military role. The degree of damage to be assumed depends on the ship size, as follows:

Waterline length	Damage extent
$L_{WL} < 30$ m	any single compartment
$30 \leq L_{WL} \leq 92$	any two adjacent main compartments, that is compartments of minimum 6-m length
> 92 m	damage anywhere extending 15% of L_{WL} or 21 m, whichever is greater.

The permeabilities to be used are

Watertight, void compartment and tanks	0.97
Workshops, offices, operational and accommodation spaces	0.95
Vehicle decks	0.90
Machinery compartments	0.85
Store rooms, cargo holds	0.60

The wind speeds to be considered depend on the ship displacement, Δ, measured in tonnes, that is metric tons of weight.

Displacement Δ, tonnes	Nominal wind speed, knots
$\Delta \leq 1000$	$V = 20 + 0.005\Delta$
$1000 < \Delta \leq 5000$	$V = 5.06 \ln \Delta - 10$
$5000 < \Delta$	$V = 22.5 + 0.15\sqrt{\Delta}$

The following criteria of stability should be met (see also Figure 8.4):

1. Angle of list or loll not larger than $30°$;
2. Righting arm $\overline{GZ}$ at first static angle not larger than 0.6 maximum righting arm;
3. Area A_1 greater than A_{min} as given by

    $\Delta \leq 5000\,\text{t}$ $A_{min} = 2.74 \times 10^{-2} - 1.97 \times 10^{-6}\Delta\,\text{m rad}$

    $5000 < \Delta < 50000\,\text{t}$ $A_{min} = 0.164\Delta^{-0.265}$

    $\Delta > 50000\,\text{t}$ consult Sea Technology Group
4. $A_1 > A_2$;
5. Trim does not lead to downflooding;
6. $\overline{GM}_L > 0$

Like the US Navy, the UK Navy uses the concept of V lines to define a zone in which the bulkheads must be completely watertight; some values, however, may be more severe. We refer again to Figure 11.6. Part (a) of the figure shows a longitudinal ship section near a bulkhead. Let us assume that after checking all required combinations of flooded compartments, the highest waterline on the considered bulkhead is WL; it intersects the bulkhead at O. In part (b) of the figure, we show the transverse section AB that contains the bulkhead. The intersection of WL with the bulkhead passes though the point Q. The standard assumes that unsymmetrical flooding can heel the vessel by $20°$. The waterline corresponding to this angle is W_1L_1. Rolling and transient motions can increase the heel angle by $15°$, leading to the waterline W_2L_2. Finally, to take into account the relative motion in waves (that is the difference between ship motion and wave-surface motion) we draw another waterline translated up by $h = 1.5\,\text{m}$; this is waterline W_3L_3. Obviously, unsymmetrical flooding followed by rolling can occur to the other side too so that we must consider the waterline W_4L_4. The waterlines W_3L_3 and W_4L_4 intersect at the point P. Thus, we identify a V-shaped limit line, W_4PL_3, hence the term 'V lines'. The region below the V lines must be kept watertight; severe restrictions refer to it and they must be read in detail.

11.5.5 The German Navy

The BV 1003 regulations are rather laconic about flooding and damage stability. The main requirement refers to the extent of damage. For ships under 30 m length, only one compartment should be assumed flooded. For larger ships a damage length equal to

$0.18L_{WL} + 3.6\,\text{m},$

but not exceeding 18 m, should be considered. Compartments shorter than 1.8 m should not be taken into account as such, but should be attached to the adjacent compartments. The leak may occur at any place along the ship, and all compartment combinations that can be flooded in the prescribed leak length should be considered. The damage may extend transversely till a longitudinal bulkhead, and vertically from keel up to the bulkhead deck.

Damage stability is considered sufficient if

- the deck-at-side line does not submerge;
- without beam wind, and if symmetrically flooded, the ship floats in upright condition;
- in intermediate positions the list does not exceed 25° and the residual arm is larger than 0.05 m;
- under a wind pressure of 0.3 kN m^{-2} openings of intact compartments do not submerge, the list does not exceed 25° and the residual lever arm is larger than 0.05 m.

If not all criteria can be met, the regulations allow for decisions based on a probabilistic factor of safety.

11.5.6 A code for large commercial sailing or motor vessels

The code published by the UK Maritime and Coastguard Agency specifies that the free flooding of any one compartment should not submerge the vessel beyond the margin line. The damage should be assumed anywhere, but not at the place of a bulkhead. A damage of the latter kind would flood two adjacent compartments, a hypothesis not to be considered for vessels under 85 m. Vessels of 85 m and above should be checked for the flooding of two compartments.

In the damaged condition the angle of equilibrium should not exceed 7° and the range of positive righting arms should not be less than 15° up to the flooding angle. In addition, the maximum righting arm should not be less than 0.1 m and the area under the righting-arm curve not less than 0.015 m rad. The permeabilities to be used in calculations are

stores	0.60
stores, but not a substantial amount of them	0.95
accommodation	0.95
machinery	0.85
liquids	0.95 or 0, whichever leads to worse predictions

The expression 'not a substantial amount of them' is not detailed.

11.5.7 A code for small workboats and pilot boats

The code published by the UK Maritime and Coastguard Agency contains damage provisions for vessels up to 15 m in length and over, certified to carry 15 or more persons and to operate in an area up to 150 miles from a safe haven. The regulations are the same as those described for sailing vessels in Subsection 11.5.6, except that there is no mention of the two-compartment standard for lengths of 85 m and over.

11.5.8 EC regulations for internal-water vessels

The following prescriptions are taken from a proposal to modify the directive 82/714 CEE, of 4 October 1982, issued by the European Parliament. The intact-stability provisions of the same document are summarized in Chapter 8.

A collision bulkhead should be fitted at a distance of minimum $0.04L_{WL}$ from the forward perpendicular, but not less than 4 m and no more than $0.04L_{WL}+2$ m. Compartments abaft of the collision bulkhead are considered watertight only if their length is at least $0.10L_{WL}$, but not less than 4 m. Special instructions are given if longitudinal watertight bulkheads are present.

The minimum permeability values to be considered are:

passenger and crew spaces	0.95
machinery spaces, including boilers	0.85
spaces for cargo, luggage, or provisions	0.75
double bottoms, fuel tanks	either 0.95 or 0

Following the flooding of any compartment the margin line should not submerge. The righting moment in damage condition, M_R, should be calculated for the downflooding angle or for the angle at which the bulkhead deck submerges, whichever is the smallest. For all flooding stages, it is required that

$$M_R > 0.2M_P = 0.2 \times 1.5bP$$

where M_P is the moment due to passenger crowding on one side, b is the maximum available deck breadth at 0.5 m above the deck, and P is the total mass of the persons aboard. The regulations assume 3.75 persons per m^2, and a mass of 75 kg per person. The document explains in detail how to calculate the available deck area, that is the deck area that can be occupied by crowding persons.

11.5.9 Swiss regulations for internal-water vessels

The following prescriptions are extracted from a decree of the Swiss Federal Council (Schweizerische Bundesrat) of 9 March 2001, that modifies a Federal Law of 8 November 1978. This is the same document that is quoted in Chapter 8 for its intact-stability prescriptions.

A ship should be provided with at least one collision bulkhead and two bulkheads that limit the machinery space. If the machinery space is placed aft, the second machinery bulkhead can be omitted. The distance between the collision bulkhead and the intersection of the stem (bow) with the load waterline should lie between $L_{WL}/12$ and $L_{WL}/8$. If this distance is shorter, it is necessary to prove by calculations that the fully loaded ship continues to float when the two foremost compartments are flooded. In no intermediary position should the deck-at-side line submerge. This proof is not necessary if the ship has on both sides watertight compartments extending longitudinally $L_{WL}/8$ from the intersection of the stem with the load waterline, and transversely at least $B/5$.

11.6 The curve of floodable lengths

Today computer programmes receive as input the descriptions of the hull surface and of the internal subdivision. In the simplest form, the input can consist of off-sets, bulkhead positions and compartment permeabilities. Then, it is possible to check in a few seconds what happens when certain compartment combinations are flooded. If the results do not meet the criteria relevant to the project, we can change the positions of bulkheads and run flooding and damage-stability calculations for the newly defined subdivision. Before the advent of digital computers the above procedure took a lot of time; therefore, it could not be repeated many times. Just to give an idea, manual flooding calculations for one compartment combination could take something like three hours. Usually, the calculations were not purely manual because most Naval Architects used slide rules, adding machines and planimetres. Still it was not possible to speed up the work. To improve efficiency, Naval Architects devised ingenious, very elegant methods; one of them produces the **curve of floodable lengths**. To explain it we refer to Figure 11.7. In the lower part of the figure, we show a ship outline with four transverse bulkheads; above it we show a curve of floodable lengths and how to use it.

Let us consider a point situated a distance x from the aftermost point of the ship. Let us assume that we calculated the maximum length of the compartment having its centre at x and that will not submerge the margin line, and that length is L_F. In other words, if we consider a compartment that extends from $x - L_F/2$ to $x + L_F/2$, this is the longest compartment with centre at x that when flooded will not submerge the ship beyond the margin line.

Now, we plot a point with the given x-coordinate and the y-coordinate equal to L_F measured at half the scale used for x values. For example, if the ship outline is drawn at the scale 1:100, we plot y values at the scale 1:200. There were Naval Architects who used the same scale for both coordinates; however, the reader will discover that there is an advantage in the procedure preferred by us. Plotting in this way all (x, L_F) pairs, we obtain the curve marked 1; this is the curve of floodable lengths.

Now, let us check if the middle compartment meets the submergence-to-the-margin-line requirement. Counting from aft forward, we talk about the compartment limited by the second and the third bulkhead. Let us assume that this is

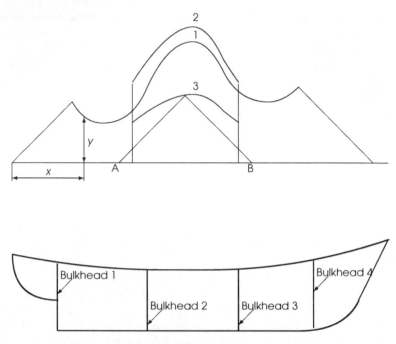

Figure 11.7 The curve of floodable lengths

a machinery compartment with permeability $\mu = 0.85$. Therefore, within the limits of this compartment we can increase the floodable lengths by dividing them by 0.85. The resulting curve is marked 2. Let us further assume that we are dealing with a ship subject to a 'two-compartment' standard (factor of subdivision $F = 0.5$). Then, we divide by 2 the ordinates of the curve 2, obtaining the curve marked 3. This is the curve of permissible lengths. On the curve 3, we find the point corresponding to the centre of the machinery compartment and draw from it two lines at $45°$ with the horizontal. The two lines intercept the base line at A and B. Both A and B are outside the bulkheads that limit the machinery compartment. We conclude that the length of this compartment meets the submergence criterion. Indeed, as the y-coordinate of the curve of floodable lengths is equal to half the length L_F, we obtain on the horizontal axis a length $\overline{AB} = L_\mathrm{F}/(\mu F)$, that is the permissible length. It is larger than the length of the compartment. To draw the lines at $45°$ we can use commercially available set squares (triangles). If we plot both x and y values at the same scale, we must draw check lines at an angle equal to $\arctan 2$; there are no set squares for this angle.

In Figure 11.7, we can identify the properties common to all curves of floodable lengths and give more insight into the flooding process.

1. At the extremities, the curve turns into straight-line segments inclined 45° with respect to the horizontal. Let us choose any point of the curve in that region. Drawing from it lines at 45°, that is descending along the first or the last curve segment, we reach the extremities of the ship. These are indeed the limits of the floodable compartments at the ship extremities because there is no vessel beyond them.
2. The straight lines at the ship extremities rise up to local maxima. Then the curve descends until it reaches local minima. Usually the ship breadth decreases toward the ship extremities and frequently the keel line turns up. Thus, compartment volumes per unit length decrease toward the extremities. Therefore, floodable lengths in that region can be larger and this causes the local maxima.
3. As we go towards the midship the compartment volumes per unit length increase, while still being remote from the midship. Flooding of such compartments can submerge the margin line by trimming the vessel. Therefore, they must be kept short and this explains the local minima.
4. The curve has an absolute maximum close to the midship. Flooding in that region does not cause appreciable trim; therefore, floodable lengths can be larger.

The term 'curve of floodable lengths' is translated as

Fr   Courbe des longueurs envahissable
G   Kurve der flutbaren Längen
I   curva delle lunghezze allagabili

A very elegant method for calculating points on the curve of floodable lengths was devised by Shirokauer in 1928. A detailed description of the method can be found in Nickum (1988), Section 4. A more concise description is given by Schneekluth (1988), Section 7.2. The procedure begins by drawing a set of waterlines tangent to the margin line. For each of these lines the Naval Architect calculates the volume and the centre of the volume of flooding water that would submerge the vessel to that waterline. The calculations are based on Equations such as (11.2). The boundaries of the compartment are found by trial-and-error using the curve of sectional areas corresponding to the given waterlines.

11.7 Summary

Ships can be damaged by collision, grounding, or enemy action. A vessel can survive damage of some extent if the hull is subdivided into watertight compartments by means of watertight bulkheads. The subdivision should be designed to make sure that after the flooding of a given number of compartments the ship can float and be stable under moderate environmental conditions. The subdivision of merchant ships should meet criteria defined by the international Convention on the Safety of Life at Sea, shortly SOLAS. The first SOLAS conference was convened in 1914, following the *Titanic* disaster. It was followed by the 1929, 1948, 1960 and 1974 conventions. The latter conference, completed with many important amendments, is in force at the time of this writing. Warships are subject to damage regulations defined by the respective navies.

The SOLAS convention defines as bulkhead deck, the deck reached by the watertight bulkheads. The margin line is a line passing at least 76 mm (3 in) below the side of the bulkhead deck. If the bulkhead deck is not continuous, the margin line should be defined as a continuous line that is everywhere at least 76 mm below the bulkhead deck. The term floodable length refers to a function of the position along the ship length. For a given position, say P, the floodable length is the maximum length of a compartment with the centre at P and whose flooding will not submerge the vessel beyond the margin line.

Let v be the volume of a compartment calculated from its geometrical dimensions. Almost always there are some objects in the compartment: therefore, the net volume that can be flooded, v_F, is less than v. We call the ratio $\mu = v_F/v$ volume permeability. The same objects that reduce the volume that can be flooded, reduce also the free surface area that contributes to the free-surface effect. We define a surface permeability as the ratio of the net free surface to the total free surface calculated from the geometric dimensions of the compartment. The moment of inertia of the free-surface calculated from the geometry of the compartment should be multiplied by the surface permeability.

There are two methods of calculating the properties of a flooded vessel: the method of lost buoyancy and the method of added weight. In the method of lost buoyancy we assume that a damaged compartment does not provide buoyancy. The displacement of the vessel and the centre of gravity do not change. The ship must change position until the undamaged compartments provide the buoyancy force and moments that balance the weight of the vessel. As the flooding water does not belong to the vessel, but to the surrounding environment, it does not cause a free-surface effect. This method corresponds to what happens in reality; it is the method used by computer programmes. In the method of added weight we consider the flooding water as a weight added to the displacement. The displacement and the centre of gravity change until the equilibrium of forces and moments is established and the level of flooding water is equal to that of the surrounding water. As the flooding water is now part of the vessel, it causes a free-surface effect. The two methods yield the same final equilibrium position and the same righting moment, $\Delta \overline{GM} \sin \phi$, in damage condition. As the displacements are different, the metacentric heights, $\overline{GM}$ also are different so as to yield the same product $\Delta \overline{GM}$.

SOLAS and other codes of practice also prescribe damage-stability criteria. For example, some criteria specify minimum value and range of positive residual arms and of areas under the righting-arm curve. Flooding and damage stability can be studied on ship models, in test basins, or by computer simulation. A paper dealing with the former approach is that of Ross, Roberts and Tighe (1997); it refers to ro–ro ferries. A few papers dealing with the latter approach are quoted in Chapter 13.

11.8 Examples

Example 11.1 – Analysis of the flooding calculations of a simple barge
This example is taken from Schatz (1983). We consider the box-shaped barge
shown in Figure 11.8, assuming as initial data $\nabla_I = 1824\,\mathrm{m}^3$, $\overline{KG} = 3.0\,\mathrm{m}$, and
$LCG = 0\,\mathrm{m}$. These values were fed as input to the programme ARCHIMEDES,
together with the information that Compartments 2.1 and 2.2 are flooded. The
permeabilities of the two compartments are 1. Using various run options of the
programme, we calculate the properties of the intact hull, of the flooded hull,
and of the flooded volume. The results are shown in Table 11.3.
   The programme ARCHIMEDES uses two systems of coordinates. A system
xyz is attached to the ship. The ship offsets, the limits of compartments, the
displacement and the centre of gravity are input in this system. The programme is
invoked specifying the numbers of the flooded compartments. The calculations
are run in the lost-buoyancy method and the results are given in a system of
coordinates, $\xi\eta\zeta$, fixed in space. In this example, only the trim changed. A
sketch of the coordinate systems involved is shown in Figure 11.9. The data of
the damaged hull and of the flooded compartments, columns 3 and 4 in Table 11.3
are given in the $\xi\zeta$ system. To get more insight into the process let us check if
the results fulfill the equations of equilibrium (11.2). To do this we must use data
expressed in the same system of coordinates. For example, we transform the
coordinates of the centre of gravity using an equation deduced from Figure 11.9:

$$\xi_G = x_G \cos\psi + z_G \sin\psi = LCG \cos\psi + \overline{KG} \sin\psi \qquad (11.5)$$

First, we calculate

$$\psi = \arctan \frac{\mathrm{trim}}{L_{\mathrm{pp}}} = \arctan \frac{-1.092}{76} = 0.823°$$

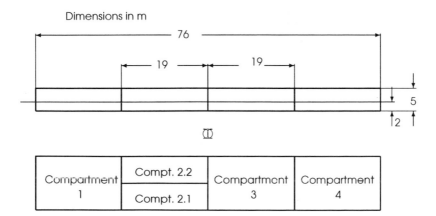

Dimensions in m

Figure 11.8 A simple barge – damage calculation

Table 11.3 Simple barge – Compartments 2.1 and 2.2 flooded

	Intact condition	Damaged, hull	Flooded compartment
Draught, m	1.999	2.711	2.711
∇, m^3	1824.000	2472.682	649.294
Δ, t	1869.600	2534.500	665.5628
$\overline{KG}$, m	3.000		1.285
LCG, m, from midship	0.000		−9.671
LCB, m, from midship	0.000	2.670	
Trim, m	0.000	−1.092	−1.092
$\overline{KB}$, m	0.750	1.337	0.915
$\overline{BM}$, m	1.389	4.427	1.139
$\overline{GM}$, m	4.001		0.454
FS moment of inertia			2736.276

The moment of the intact-displacement volume about the midship section, in the trimmed position, is

$$\nabla_{\mathrm{I}}(LCG \cos \psi + \overline{KG} \sin \psi) = 1824(0 \times \cos(-0.823°) + s \sin(-0.823°)$$
$$= -78.616 \ \mathrm{m}^4$$

The moment of the flooded compartment equals

$$v \cdot lcg = 649.294 \times (-9.671) = -6279.322 \ \mathrm{m}^4$$

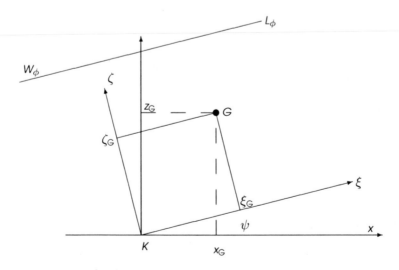

Figure 11.9 A simple barge – coordinate systems used in calculations

The moment of the flooded barge resulting directly from hydrostatic calculations is

$$\nabla_F \cdot LCF_F = 2472.682 \times (-2.570) = -6354.793 \text{ m}^4$$

The deviation between the two moments is less than 0.05%; the equilibrium of moments is fulfilled. As to the equilibrium of forces, we can easily see that $1824 + 649.294$ is practically equal to 2472.682.

The programme ARCHIMEDES, like other computer programmes, carries out calculations by the lost-buoyancy method. Then the final displacement volume remains equal to the intact volume, 1824 m^3, while the calculated metacentric height, $\overline{GM}$, is 2.858 m. The righting moment for small heel angles, in the lost-buoyancy method, is

$$M_{RL} = 1.025 \times 1824 \times 2.858 \sin \phi = 5342.3 \sin \phi \, \text{t m}$$

As an exercise let us compare this moment with that predicted by the added-weight method. Hydrostatic calculations for the damaged barge yield $\overline{KM} = 5.764$ m. Capacity calculations for the compartments 2.1 and 2.2 give a total volume of flooding water equal to 649.294 m^3, with a height of the centre of gravity at 1.286 m. In Table 11.4, we calculate the damage displacement and the coordinates of its centre of gravity in the added-weight method. Capacity calculations for the flooded compartments yield a total moment of inertia of the free surfaces, $i = 2736.276 \text{ m}^3$. The corresponding lever arm of the free surface is

$$l_F = \frac{i}{\nabla_A} = \frac{2736.276}{2473.294} = 1.106 \, \text{m}$$

The resulting metacentric height is

$$\overline{GM}_A = \overline{KM} - \overline{KG}_A - l_F = 5.764 - 2.550 - 1.106 = 2.107 \, \text{m}$$

and the righting moment for small heel angles, in the added-weight method

$$\begin{aligned} M_{RA} = \rho \nabla_A \overline{GM}_A \sin \phi &= 1.025 \times 2473.294 \times 2.107 \sin \phi \\ &= 5341.7 \sin \phi \, \text{t m} \end{aligned}$$

Due to errors of numerical calculations the values of M_{RL} and M_{RA} differ by 0.03%; in fact they are equal, as expected.

Table 11.4 Simple barge – added-weight calculations

	Volume (m^3)	kg (m)	Moment (m^4)	lcg (m)	Moment (m^4)
Intact hull	1824.000	3.000	5472.000	0.000	0.000
Flooding water	649.294	1.286	834.992	−9.671	−6279.322
Flooded hull	2473.294	2.550	6306.992	−2.539	−6279.322

Table 11.5 Flooding calculations – a comparison of methods considering permeability

	Intact condition	Damaged, lost buoyancy	Damaged, by added weight
Draught, m	1.500	1.829	1.829
∇, m^3	150.000	150.000	182.927
Δ, t	153.750	153.750	187.500
$\overline{KB}$, m	0.750	0.915	0.915
$\overline{BM}$, m	1.389	1.139	1.139
$\overline{KG}$, m	1.500	1.500	1.395
$\overline{GM}$, m	0.639	0.554	0.454
$\Delta\overline{GM}$, tm	98.229	85.104	85.104

11.9 Exercise

Exercise 11.1 – Comparison of methods while considering permeability
In Subsections 11.3.1 and 11.3.2, we compared the lost-buoyancy method to the added-weight-method, but, to simplify things, we did not consider permeabilities. This exercise is meant to show the reader that even if we consider permeabilities, the two methods yield the same draught and the same righting moment in damage condition. The reader is invited to redo the calculations in the mentioned sections, but under the assumption that the volume and surface permeabilities of the flooded compartment equal 0.9.

A hint for using the method of lost buoyancy is that the waterplane area, LB, is reduced by the floodable area of Compartment 2, μBl. The hint for the method of added weight is that the volume of flooding water equals $\mu l B T_{\mathrm{A}}$, where T_{A} is the draught in damage condition. The results should be those shown in Table 11.5.

12
Linear ship response in waves

12.1 Introduction

The title of the book is 'Ship hydrostatics and stability'. This chapter describes processes that are not hydrostatic, but can affect stability. We elaborate here on some reservations expressed in Section 6.12 and sketch the way towards more realistic models. First, we need a wave theory that can be used in the description of real seas. Therefore, we introduce the theory of **linear waves**. Next, we show how real seas can be described as a **superposition** of regular waves. This leads to the introduction of **sea spectra**. A floating body moves in six degrees of freedom. The oscillating body generates waves that absorb part of its energy. The integration of pressures over the hull surface yields the forces and moments acting on the body. We return here, without detailing, to the notions of added mass and **damping coefficients** introduced in Section 6.12. A full treatment would go far beyond the scope of the book; therefore, we limit ourselves to mentioning a few important results.

The problems of mooring and anchoring deserve special treatment and their importance has grown with the development of offshore structures. We cannot discuss here the behaviour of **compliant floating structures,** that is moored floating structures, but give an example of how the mooring can change the natural frequencies of a floating body. We mention in this chapter a few methods of reducing ship motions, mainly the roll. This allows us to show that under very particular conditions, free water surfaces can help, a result that seems surprising in the light of the theory developed in Chapter 6.

The models introduced in this chapter are too complex to yield explicit mathematical expressions that can be directly applied in engineering practice. It is only possible to implement the models in computer programmes that yield numerical results. The input to such programmes is a statistical description of the sea considered as a **random process**. Correspondingly, the output, that is the ship **response,** is also a random process.

This chapter assumes the knowledge of more mathematics than the rest of the book. Mathematical developments are concise, leaving to the interested reader the task of completing them or to refer to specialized books. The reader who cannot follow the mathematical treatment can find in the summary a non-mathematical description of the main subjects.

12.2 Linear wave theory

In Subsection 10.2.3, we introduced the theory of trochoidal waves. Trochoidal waves approximate well the shape of swells and are prescribed by certain codes of practice for stability and bending-moment calculations. Another wave theory is preferred for the description of real seas and for the calculation of ship motions; it is the theory of linear waves. The basic assumptions are

1. the sea water is **incompressible**;
2. there is no viscosity, i.e. the sea water is **inviscid**;
3. there is no surface tension;
4. no fluid particle turns around itself, i.e. the motion is irrotational;
5. the wave amplitude is much smaller than the wave length.

The first assumption, that of incompressibility, is certainly valid at the small depth and the wave velocities experienced by surface vessels. This is a substantial difference from phenomena experienced in aerodynamics. Excepting roll damping, the second assumption, the lack of viscous phenomena, leads to results confirmed by experience. For roll, certain corrections are necessary; often they are done by empiric means. Surface tension plays a role only for very small waves, such as the ripple that can be seen on the surface of a swell. We shall see how the fourth hypothesis, that is irrotational flow, makes possible the development of an elegant **potential theory** that greatly simplifies the analysis. The fifth hypothesis, low-amplitude waves, is not very realistic; surprisingly, it leads to realistic results.

We consider two-dimensional waves, that is waves with parallel crests of infinite length, such as shown in Figure 12.1. The crests are parallel to the y direction and we are only interested in what happens in the x and z directions. Let u be the horizontal and w the vertical velocity of a water particle. We note by ρ the water density. The theory of fluid dynamics shows that the rate of change of the mass of a unit volume of water is

$$\frac{\partial(\rho u)}{\partial x} + \frac{\partial(\rho w)}{\partial z}$$

The density of an incompressible fluid, ρ, is constant. Then, the condition that the mass of unit volume of water does not change is expressed as

$$\frac{\partial u}{\partial x} + \frac{\partial w}{\partial z} = 0 \tag{12.1}$$

Equation (12.1) is known as the **equation of continuity**; it states that the **divergence** of the vector with components u, w is zero. The assumption of irrotational

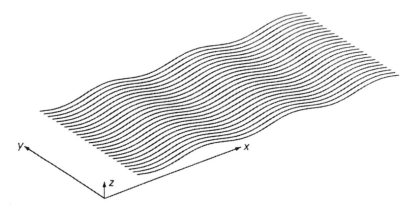

Figure 12.1 Two-dimensional waves: swell

motion is expressed by the condition that the **curl** of the vector with components u, w is zero. In two dimensions this is

$$\frac{\partial w}{\partial x} - \frac{\partial u}{\partial z} = 0 \qquad (12.2)$$

We define a **velocity potential**, Φ, such that

$$u = \frac{\partial \Phi}{\partial x}, \qquad w = \frac{\partial \Phi}{\partial z} \qquad (12.3)$$

These expressions verify, indeed, Eq. (12.2). Substituting Eq. (12.3) into Eq. (12.1) yields the **Laplace equation**

$$\frac{\partial^2 \Phi}{\partial x^2} + \frac{\partial^2 \Phi}{\partial z^2} = 0 \qquad (12.4)$$

This equation must be solved together with a set of **boundary conditions**. Let $\zeta(x, z, t)$ be the **elevation** of the free surface and z the vertical coordinate measured from the mean water surface upwards. In simple terms, ζ represents the wave profile. The **kinematic condition**

$$\frac{\partial \zeta}{\partial t} = \frac{\partial \Phi}{\partial z}, \qquad \text{at} \quad z = 0 \qquad (12.5)$$

states that the vertical velocity of the wave surface equals the vertical velocity of a water particle at the mean water level. This is an approximation acceptable for small wave amplitudes.

The **dynamic free-surface condition** states that the water pressure on the wave surface is equal to the atmospheric pressure

$$\frac{\partial \Phi}{\partial t} + g\zeta + \frac{1}{2}\left[\left(\frac{\partial \Phi}{\partial x}\right)^2 + \left(\frac{\partial \Phi}{\partial z}\right)^2\right] = 0 \quad \text{on} \quad z = \zeta(x, y, z) \qquad (12.6)$$

Assuming small wave amplitudes we can neglect the squares of particle velocities and thus we remain with the condition

$$g\zeta + \frac{\partial \Phi}{\partial t} = 0, \quad \text{at} \quad z = 0 \tag{12.7}$$

From Eqs. (12.5) and (12.7) we obtain the linearized free-surface condition

$$\frac{\partial^2 \Phi}{\partial t^2} + g\frac{\partial \Phi}{\partial z} = 0 \tag{12.8}$$

Additional boundary conditions must be written for the sea bottom, for walls that limit the water domain, and for the surfaces of bodies floating in that domain. As the water does not pass through such boundaries, the velocity components normal to such boundaries should be zero.

Let the wave length be λ, and the **wave number** $k = 2\pi/\lambda$. The vertical coordinate of a water particle is $z = 0$ at the mean sea level, and $z = -d$, at the depth d. We give the results of the theory for infinite-depth water as these are the most interesting for sea-going ships. We leave to an exercise the proof that these results fulfill the Laplace equation and the boundary conditions. The solution that interests us is the potential

$$\Phi = \frac{g\zeta_0}{\omega}e^{kz}\cos(\omega t - kx) \tag{12.9}$$

The equation of the sea surface is

$$\zeta = \zeta_0 \sin(\omega t - kx) \tag{12.10}$$

The following relationship exists between the wave length, λ, and the **wave period**, T,

$$\lambda = \frac{g}{2\pi}T^2 \tag{12.11}$$

Figure 12.2 shows the propagation of the wave described by Eq. (12.10). The wave period is $T = 6.5$ s, and the wave length given by Eq. (12.11) is $\lambda = 65.965$ m. The wave height, $H = 2\zeta_0$, equals $\lambda/20$, a ratio often used in Naval Architecture.

The speed of propagation of the wave shape is called **celerity**, a term that comes from the Latin 'celeritas', speed. From Eq. (12.11) we find the celerity

$$c = \frac{\lambda}{T} = \sqrt{\frac{g\lambda}{2\pi}} \tag{12.12}$$

We immediately see that long waves propagate faster than short waves. Therefore, we say that water waves are **dispersive**. Acoustic waves, for example, are not dispersive.

The components of the water-particle velocity are

$$u = \omega\zeta_0 e^{kz}\sin(\omega t - kx) \tag{12.13}$$
$$w = \omega\zeta_0 e^{kz}\cos(\omega t - kx) \tag{12.14}$$

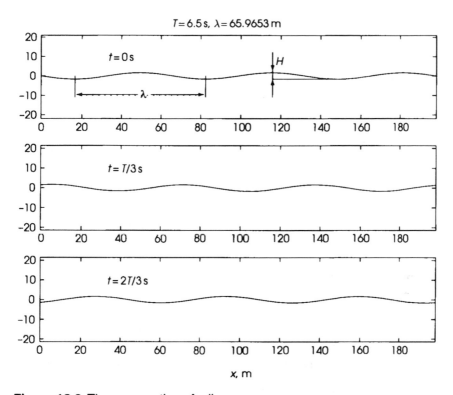

$T = 6.5\,\mathrm{s},\ \lambda = 65.9653\,\mathrm{m}$

$x,\ \mathrm{m}$

Figure 12.2 The propagation of a linear wave

We invite the reader to use the latter equations and prove that in infinite-depth water, the particles move on circular orbits whose radii decrease with depth. At a depth equal to about one-half wave length the orbital motion becomes negligible.

Figure 12.3 shows the orbit of a water particle at the surface of the wave represented in Figure 12.2. The orbital velocities, u and v, are shown at two time instants, i.e. $t = 1$ s and $t = 4$ s.

12.3 Modelling real seas

We can register the elevation of the sea at a given point and obtain a function of time $\zeta = f(t)$. Alternatively, we can consider the sea surface at a given time instant, t_0, and a given coordinate y_0. Then, we can register the elevations along the x-axis and obtain a function $\zeta = g(x)$. Both representations have an **irregular aspect** in the sense that there is no pattern that repeats itself. The linear wave theory allows us to represent the sea surface as the superpositon of a large number of sine waves, that is

$$\zeta = \sum_{i=1}^{N} A_i \sin(\omega_i t - k_i x + \epsilon_i) \tag{12.15}$$

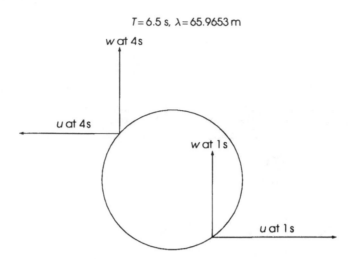

Figure 12.3 Orbital velocities at the sea surface

where A_i is the wave amplitude, ω_i the angular frequency, k_i the wave number, and ϵ_i the phase of the ith wave. We assume that the numbers ϵ_i are random and uniformly distributed between 0 and 2π. To explain how the superposition of sine waves can produce an irregular sea we refer to Figure 12.4. The lower

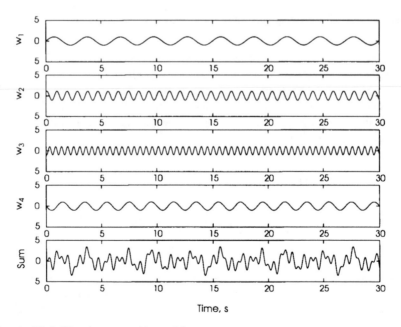

Figure 12.4 The superposition of four waves

curve represents the sum of the four sine waves plotted above it. A periodical pattern can still be detected; however, as the number of components increases, any periodicity disappears and there is no pattern that repeats itself.

As the wave phases, ϵ_i, are random, the sea surface is a random process. Let us consider a segment of a wave record, such as in Figure 12.5. We distinguish two types of trough-to-crest heights. When measuring the height H_1, the trough and the crest lie on two sides of the mean sea level, while H_2 is measured between two points on the same side of the mean sea level. Experience shows that heights of the first type, H_1, follow approximately the **Raleigh distribution**

$$f(H) = \frac{H}{4m_0} e^{-\frac{H^2}{8m_0}} \tag{12.16}$$

The **mean height** is

$$H_\mathrm{m} = \int_0^\infty H f(H) \mathrm{d}H = \sqrt{2\pi m_0} \tag{12.17}$$

An important characteristic is the **significant wave height** defined as the **mean of the highest third of the wave heights**

$$H_{1/3} = \int_{H_0}^\infty H f(H) \mathrm{d}h \tag{12.18}$$

where H_0 is defined by

$$\int_{H_0}^\infty f(H) \mathrm{d}h = \frac{1}{3} \tag{12.19}$$

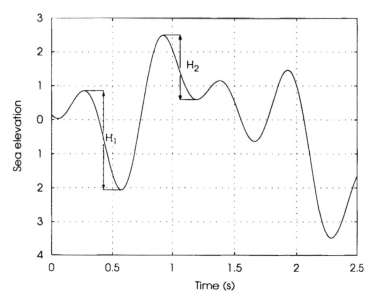

Figure 12.5 For the definition of the significant wave height

The significant height allows the calculation of other characteristics, for example the sea spectrum. A natural question arises: given the significant height, $H_{1/3}$, what is the maximum wave height, H_{max}, that can be expected? It appears that the larger the number of waves considered, the higher the maximum wave height that can be expected. Using data in Bonnefille (1992) we find that $H_{max}/H_{1/3}$ varies from 1.2 for a sample of ten waves, to 1.92 for 1000 waves.

Let us return now to Eq. (12.15). It can be shown that the total energy of N wave components, per unit sea area, equals

$$E_0 = \frac{1}{2}\sum_{i=1}^{N} A_i^2 \qquad (12.20)$$

To define the **wave spectrum**, $S(\omega)$, we consider a band extending from ω_j to $\omega_j + \Delta\omega$ and write

$$S(\omega_j)\Delta\omega = \frac{1}{2}A_j^2 \qquad (12.21)$$

where A_j is the amplitude of the wave component in the frequency band considered by us. For example, in Figure 12.6 we consider the band of breadth $\Delta\omega$ centred around 0.8 rad s^{-1}. In this case $A_j^2/2 = 0.08$ m^{2}s. The area of this band, like the whole area under the spectrum curve, is measured in m^2.

The wave spectrum describes the distribution of wave energy versus wave angular frequency. At the end of Section 12.4 we shall find an important use of this concept. Wave spectra can be obtained from measurements. A number of

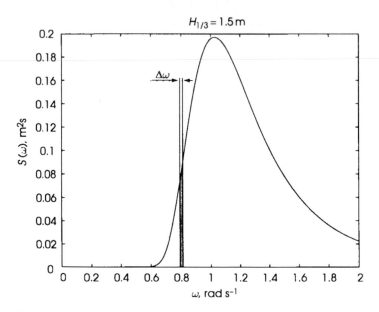

Figure 12.6 A Pierson–Moskovitz spectrum

formulae have been proposed for calculating **standard spectra** on the basis of a few given or measured sea characteristics. We shall give only one example, the **Pierson–Moskovitz spectrum** as described by Fossen (1994)

$$S = A\omega^{-5}\,e^{-B\omega^{-4}}\,\mathrm{m^2 s} \tag{12.22}$$

where

$$
\begin{aligned}
A &= 8.1 \times 10^{-3} g^2 \,\mathrm{m^2\,s^{-4}} \\
B &= 0.0323 \left(\frac{g}{H_{1/3}}\right)\mathrm{s^{-4}}
\end{aligned}
$$

This spectrum corresponds to fully developed seas recorded in the North Atlantic; an example is shown in Figure 12.6. The theory of linear waves exposed in this section is a first-order approximation in which the wave shape moves, but there is no mass transport. This approximation is sufficient for moving ships as their speed is usually larger than the 'drift' caused by waves. For stationary structures it may be necessary to consider higher order approximations that predict a drift.

12.4 Wave induced forces and motions

Like any other free body, a ship moves in six degrees of motion; we describe them with the aid of Figure 12.7. The six motions of a ship have traditional names that were adopted in the previous century also for planes and cars. We follow the notation of Faltinsen (1993). Three motions are linear; they are described below.

1. **Surge**, along the x-axis; we note it by η_1.
2. **Sway**, in the direction of the y-axis; we use the notation η_2.
3. **Heave**, along the z-axis; we note it by η_2.

The other three degrees of freedom define angular motions, as detailed below.

1. **Roll**, around the x-axis; we note it by η_3.
2. **Pitch**, around the y-axis; we use the notation η_5.
3. **Yaw**, around the z-axis; it is noted by η_6.

The motion of any point on a floating body is the resultant of all six motions

$$s = \eta_1 \mathbf{i} + \eta_2 \mathbf{j} + \eta_3 \mathbf{k} + \omega \times \mathbf{r} \tag{12.23}$$

where $\mathbf{i}$ is the unit vector on the x-axis, $\mathbf{j}$, the unit vector on the y-axis, $\mathbf{k}$, the unit vector on the z-axis, and $\times$ denotes the vector product. The rotation vector is

$$\omega = \eta_4 \mathbf{i} + \eta_5 \mathbf{j} + \eta_6 \mathbf{k}$$

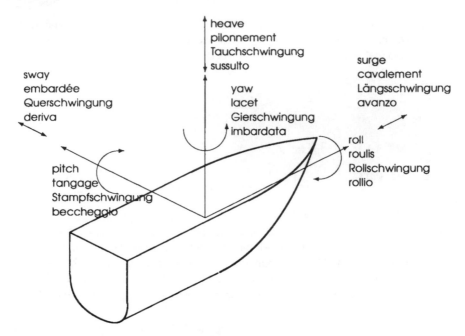

Figure 12.7 Ship motions – definitions

and the position vector of a point with coordinates x, y, z is

$$\mathbf{r} = x\mathbf{i} + y\mathbf{j} + z\mathbf{k} \qquad (12.24)$$

For example, the vertical motion of the point with coordinates x, y, z is the resultant of the heave, roll and pitch motions

$$(\eta_3 + y\eta_4 - x\eta_5)\mathbf{k}$$

For particular purposes we can write an equation of motion in one degree of freedom, without considering the influence of the motions in the other degrees of freedom. We say that such equations describe **uncoupled** motions. Thus, in Section 6.7 we developed a non-linear equation of roll, the non-linear term being $\rho\Delta\overline{GZ}$. In Section 6.8 we linearized the equation for small roll angles. We neglected the damping term that for roll is non-linear. An example of an uncoupled roll equation with linear damping and a forcing term due to a trochoidal wave is given by Schneekluth (1988)

$$\frac{\mathrm{d}^2\eta_4}{\mathrm{d}t^2} + 2n\frac{\mathrm{d}\eta_4}{\mathrm{d}t} + \omega_{n4}^2\eta_4 = \omega_{n4}^2\frac{2\pi\zeta_0}{\lambda}\sin(\omega_\mathrm{W}t) \qquad (12.25)$$

where n is a linear damping coefficient, ω_{n4} is the ship natural angular frequency in roll, ζ_0 is the wave amplitude, and ω_W, the wave angular frequency.

Equations for uncoupled pitch motion can be developed in the same way as those of roll, substituting $\overline{GM_\mathrm{L}}$ for $\overline{GM}$. For example, Schneekluth (1988) gives the following equation for undamped pitch

$$\frac{d^2\eta_5}{dt^2} + \frac{g\overline{GM_L}}{i_{55}^2}\eta_5 - \frac{g\overline{GM_L}}{i_{55}^2}\gamma \sin\frac{2\pi t}{T_E} = 0 \tag{12.26}$$

where i_{55} is the radius of inertia of the ship mass about the Oy axis, γ is the maximum pitch amplitude, and T_E is the period of encounter. Obviously

$$\omega_{n5} = \frac{\sqrt{g\overline{GM_L}}}{y_{55}} \tag{12.27}$$

is the ship natural, angular frequency in pitch, and

$$\omega_E = \frac{2\pi}{T_E}$$

is the angular frequency of encounter.
We use Figure 12.8 to develop an equation of the uncoupled heave motion

$$(m + A_{33})\ddot{\eta}_3 + b\dot{\eta}_3 + \rho g A_W \eta_3 = \rho g A_W \zeta_0 \cos \omega_E t \tag{12.28}$$

Above, we assumed that the wave length is large compared to the dimensions of the waterplane. In Figure 12.8(b) we see a mass-dashpot-spring analogy of the heaving body. This analogy holds only for the form of the governing equations. In Figure 12.8(b) the damping coefficient, b, is a constant. In Figure 12.8(a) the added mass in heave, A_{33}, and the damping coefficient, b, are functions of the

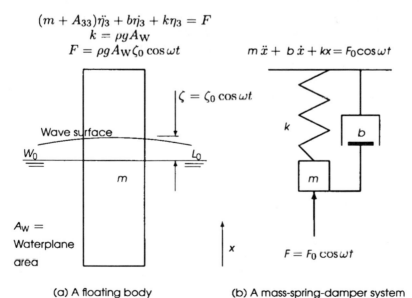

(a) A floating body   (b) A mass-spring-damper system

Figure 12.8 A heaving, floating body as a second-order dynamic system

frequency of oscillation. After the extinction of transients, that is in steady state, the frequency of oscillation is equal to the exciting frequency, that is the wave frequency, ω_W, for a body that does not move, and the frequency of encounter, ω_E, for a moving, floating body.

Let us return to the terms that are proportional to motion in the equations of roll, pitch and heave

$$g\Delta\overline{GM}\eta_4, \qquad g\Delta\overline{GM_L}\eta_5, \qquad \rho g A_W \eta_3$$

These terms represent two hydrostatic moments and one hydrostatic force that oppose the motion and tend to return the floating body to its initial position. The collective name for those moments and force is **restoring forces**. Only the roll, pitch and heave motion are opposed by hydrostatic restoring forces. There are no hydrostatic restoring forces that oppose surge, sway or yaw.

The equations of uncoupled motions are simplified models that allow us to reach a few important conclusions. In reality, certain couplings exist between the various motions. Thus, we already know that during roll the centre of buoyancy moves along the ship causing pitch. As pointed out by Schneekluth (1988), the combination of roll and pitch motions causes an oscillation of the roll axis and induces yaw. Also, the combination of roll and pitch induces heave. Moreover, one motion can influence the added masses and the damping coefficients of other motions. The most complete model of **coupled motions** is

$$(\mathbf{M} + \mathbf{A})\ddot{\eta} + \mathbf{B}\dot{\eta} + \mathbf{C}\eta = \mathrm{Re}(\mathbf{F}e^{-i\omega_E t}) \qquad (12.29)$$

Above, $\mathbf{M}$ is a 6-by-6 matrix whose elements are the ship mass and its moments of inertia about the three axes of coordinates, and $\mathbf{A}$ is a 6-by-6 matrix of added masses (general term including added masses and added moments of inertia). The vectors of motions, speeds and accelerations are

$$\eta = \begin{bmatrix} \eta_1 \\ \eta_2 \\ \eta_3 \\ \eta_4 \\ \eta_5 \\ \eta_6 \end{bmatrix}, \qquad \dot{\eta} = \begin{bmatrix} \dot{\eta}_1 \\ \dot{\eta}_2 \\ \dot{\eta}_3 \\ \dot{\eta}_4 \\ \dot{\eta}_5 \\ \dot{\eta}_6 \end{bmatrix}, \qquad \ddot{\eta} = \begin{bmatrix} \ddot{\eta}_1 \\ \ddot{\eta}_2 \\ \ddot{\eta}_3 \\ \ddot{\eta}_4 \\ \ddot{\eta}_5 \\ \ddot{\eta}_6 \end{bmatrix}$$

The expression $\mathrm{Re}(\mathbf{F}e^{-i\omega_E t})$ means the real part of the vector of sinusoidal exciting forces and moments.

For a ship displaying port-to-starboard symmetry a part of the elements of the matrix $\mathbf{M}$ are zero, and another part are symmetric. The system of six ordinary differential equations can be simplified in many practical situations. Thus, for a floating structure presenting symmetry about the xOz plane, and with the centre of gravity in the position $(0, 0, z_G)$, Faltinsen (1993) shows that the matrix of inertias becomes

$$\mathbf{M} = \begin{bmatrix} M & 0 & 0 & 0 & Mz_{\mathrm{G}} & 0 \\ 0 & M & 0 & -Mz_{\mathrm{G}} & 0 & 0 \\ 0 & 0 & M & 0 & 0 & 0 \\ 0 & -Mz_{\mathrm{G}} & 0 & I_4 & 0 & -I_{46} \\ Mz_{\mathrm{G}} & 0 & 0 & 0 & I_5 & 0 \\ 0 & 0 & 0 & -I_{46} & 0 & I_6 \end{bmatrix}$$

where M is the mass of the floating body, I_4, the moment of inertia about the x-axis, I_{46}, the product of inertia about the x- and z-axis, and I_6, the moment of inertia about the z-axis. Certain symmetries also can appear in the matrices of added masses, $\mathbf{A}$, and damping coefficients, $\mathbf{B}$. Remember, added masses and damping coefficients are functions of the frequency of oscillation. For a structure symmetric about the xOz plane the motions of surge, heave and pitch (vertical-plane motions) can be uncoupled from those of sway, roll and yaw.

The equations shown above are linear. Then, if for a wave amplitude equal to 1 the resulting motion amplitude is η_{a}, for a wave amplitude equal to A the motion amplitude will be $A\eta_{\mathrm{a}}$. Further, the principle of superposition applies to motions as it applies to waves. The response to the sum of several waves is the sum of the responses to the individual waves. Then, if we characterize the exciting waves by their spectrum, we can characterize the resulting motion by a motion spectrum.

In Subsection 6.9.5 we introduced the concept of transfer function for a simple case of roll motion. The transfer function obtained from a differential equation such as those shown in this chapter is a function of frequency. Let the transfer function of the ith motion be $Y_i(\omega)$. The spectrum of the respective motion, $S_{\eta_i}(\omega)$, is related to the wave spectrum, $S_w(\omega)$, by the relationship

$$S_{\eta_i}(\omega) = [Y_i(\omega)Y_i(-\omega)]\,S_w(\omega) \tag{12.30}$$

The expression between square brackets is called **response amplitude operator**, shortly **RAO**. The response amplitude operators of the various motions can be obtained from the coupled equations of all motions. All motions occur at the frequency of the exciting force, but have different phases.

12.5 A note on natural periods

If a linear mass-dashpot-spring system, such as that shown in Figure 12.8(b), is excited by a force whose period is close to that of the system, the response amplitude can be very large; we talk about **resonance**. Theoretically, at zero damping the response is unbounded. In practice any physical system is damped to a certain extent and this limits the response to bounded values. Large-amplitude oscillations reduce the performance of the crew and the equipment and, therefore, they should be avoided.

A very efficient means of avoiding resonance is to ensure that the natural period of the floating body is remote from that of the waves prevailing in the region of operation. In general, it is not possible to change the natural periods of ships because their designs must meet other important requirements. It is possible to change the natural periods of moored platforms, such as those used in offshore technology. To show an example let us refer to Figure 12.8(a). The natural period of the undamped and uncoupled heave motion of the shown body is

$$T_{n3} = 2\pi \sqrt{\frac{M + A_{33}}{\rho g A_{\mathrm{W}}}}$$

Let us assume that the floating body is moored as shown in Figure 12.9. The mooring cable is tensioned; it pulls the floating body down increasing its draught beyond the value corresponding to its mass, M. Thus, if we note by V the submerged volume, and by T_{c} the tension in the cable, we can write

$$\rho g V = g M + T_{\mathrm{c}}$$

If the floating body is an offshore platform, we call it **tension leg platform**, shortly **TLP**. When the floating body oscillates vertically, the hydrostatic force that opposes the heave motion is that predicted in Figure 12.8. An additional force develops in the cable; its value, according to the theory of elasticity, is

$$\frac{AE}{\ell} \eta_3 \tag{12.31}$$

where A is the sectional area of the cable, E, the Young modulus of the material of the cable, and ℓ, the cable length. This second force is usually much larger than

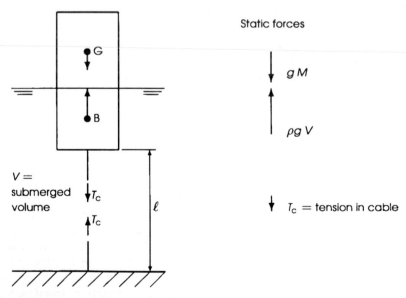

Figure 12.9 A tension-leg floating body

the hydrostatic force. Then, a good approximation of the uncoupled, undamped natural period of heave is

$$T_{n3} = 2\pi\sqrt{\frac{M + A_{33}}{AE/\ell}} \qquad (12.32)$$

and it can differ much from that of the unmoored platform. Lateral mooring lines act like non-linear springs and can change the periods of other motions.

Natural periods can change temporarily when a ship enters confined waters. The added masses are influenced by close vertical walls and by a close bottom. Schneekluth (1988) cites the case of a barge with a B/T ratio equal to 2. When performing the roll test in a depth equal to $1.25T$, the added mass in roll was found to be 2.7 times larger than in deep water. The measured roll period appeared larger than in deep water, leaving the impression that the stability was worse than in reality. Schneekluth appreciates that the added mass in roll, A_{33}, is approximately 15% of the ship mass, M, and that bilge keels increase the added mass by approximately 6%.

12.6 Roll stabilizers

There are many systems of reducing roll amplitude; their aim is to produce forces whose moment can be added to the righting moment. The simplest and cheapest system is represented by the **bilge keels**; they are steel profiles assembled on part of the ship length, close to the bilge. Bilge keels act in two ways. First, a hydrodynamic resistance force develops on them; it is opposed to the roll motion. Second, bilge keels cause vortexes that increase the viscous damping of the roll motion. As shown in the previous chapters, some codes of stability acknowledge the contribution of bilge keels and provide for corresponding corrections of some requirements. Bilge keels are **passive devices**.

Roll fins are wing-shaped bodies that extend transversely; usually they can be rotated by a control system that receives as input the roll angle, velocity and acceleration. The forward ship velocity causes hydrodynamic forces on the wings, forces that oppose the roll motion. No helpful forces are produced at low ship speeds. Rudders can be used as active anti-roll devices. Their action is coupled with other motions and influences manoeuvering.

We do not expand on the devices mentioned above, but prefer to concentrate on another possibility because its relation to stability is evident and because it contradicts to some extent the theory that any liquid free surface endangers stability. We mean **anti-roll tanks**. To explain their action we use a simple mechanical analogy. We consider a classical oscillating system composed of a mass, a spring and a dashpot. If a smaller mass is attached to the main mass by a spring, and if the second mass and spring are properly dimensioned, their vibration damps the oscillations of the main mass. This is the principle of the **Frahm vibration absorber**. In a similar mode, if two tanks, one on starboard, the other on the port side, are connected by a pipe, and water flows between them in a certain phase to the roll motion, this cross-flow opposes the roll motion. The main mass-spring-damper system above is the analogue of the ship and the small mass-spring system is the analogue of the anti-roll tanks.

We consider in Figure 12.10(a) a system composed of the mass m_1, the linear spring k_1, and the viscous damper (dashpot) c, and an auxiliary system composed of the mass m_2 and the linear spring k_2. A sinusoidal force, $F_0 \sin \omega t$, acts on the main mass, m_1. The position of the mass m_1 is measured by the variable x_1, that of the mass m_2 by the variable x_2. If properly 'tuned', the auxiliary system (k_2, m_2), 'absorbs' the forced vibrations of the main system. To show this we first write the equations that govern the behaviour of the composed system. The first Eq. (12.33) describes the forces that act on the mass m_1, and the second equation refers to the forces acting on the mass m_2,

$$m_1 \frac{d^2 x_1}{dt^2} + c \frac{dx_1}{dt} + k_1 x_1 + k_2(x_1 - x_2) = F_0 \sin \omega t$$

$$m_2 \frac{dx_2}{dt} + k_2(x_2 - x_1) = 0 \qquad (12.33)$$

We assume that the initial conditions are all zero, that is $x_1 = 0$, $dx_1/dt = 0$, $x_2 = 0$, $dx_2/dt = 0$. Taking Laplace transforms and noting with s the Laplace-transform variable, with $X_1(s)$ the Laplace transform of $x_1(t)$, and with $X_2(s)$ that of x_2, we obtain

$$[m_1 s^2 + cs + k_1 + k_2]X_1(s) - k_2 X_2(s) = \frac{F_0 \omega}{s^2 + \omega^2}$$

$$k_2 X_1(s) - (m_2 s^2 + k_2)X_2(s) = 0 \qquad (12.34)$$

Eliminating $X_2(s)$ from Eq. (12.34) we arrive at

$$X_1(s) = \frac{F_0 \omega}{s^2 + \omega^2} \frac{m_2 s^2 + k_2}{(m_2 s + k_2)(m_1 s^2 + cs + k_1 + k_2) - k_2^2} \qquad (12.35)$$

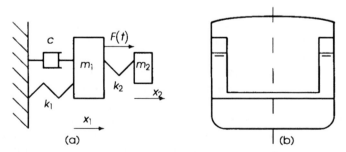

Figure 12.10 (a) A Frahm vibration absorber, (b) Flume tanks

Let us choose $k_2/m_2 = \omega^2$, i.e. we tune the auxiliary system to the exciting frequency ω. Then, the Laplace transform of the amplitude of oscillation of the main mass, m_1, becomes

$$X_1(s) = \frac{F_0 m_2 \omega}{(m_1 s^2 + k_2)(m_1 s^2 + cs + k_1 + k_2) - k_2^2} \qquad (12.36)$$

Churchill (1958) shows that the roots of the denominator (poles) have negative real parts so that the oscillation $x_1(t)$ is damped. A simulation of a system with a Frahm vibration absorber is shown in Example 12.1.

In Figure 12.10, we sketch a section through a ship equipped with **flume tanks**. A transverse pipe connects the two tanks. The flow of water between the two sides can be controlled by throttling the pipe or by acting on the outflow of air above the free surfaces. The water in the flume tanks causes a free-surface effect. Therefore, a tradeoff is necessary between the benefits of roll stabilizing and the disadvantage of reducing the effective metacentric height.

A friend of this author, Shimon Lipiner, described years ago an experiment carried out at the University of Glasgow. Tests on the model of a Ro/Ro ship were meant to show how disastrous can be the effect of water on the uninterrupted car deck. For the particular parameters involved in that experiment, the observed effect was a reduction instead of an increase of the roll amplitude. The water on deck acted then as a Frahm stabilizer. Figure 12.11 reproduced from McGeorge (2002) by courtesy of Butterworth-Heinemann describes the action of a passive tank stabilizer.

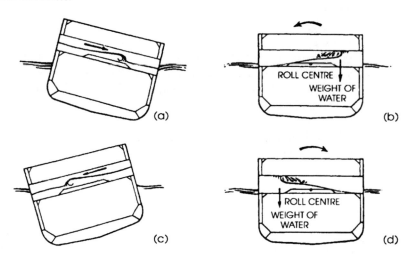

Figure 12.11 Brown-NPL pasive tank stabiliser: (a) Stern view of ship with passive tank rolled to starboard. The water is moving in the direction shown, (b) Ship rolling to port. The water in the tank on the starboard side provides a moment opposing the roll velocity, (c) Ship at the end of its roll to port. The water is providing no moment to the ship, (d) Ship rolling to starboard. The water in the tank on the port side provides a moment opposing the roll velocity

12.7 Summary

To calculate the motion of a floating body in real waves we need an adequate description of a real sea. Therefore, we consider the real sea as the result of the superposition of a large number of linear waves. The theory of linear waves is based on the following assumptions:

1. the sea water is incompressible;
2. the sea water is inviscid (no viscous effects);
3. surface tension plays no role;
4. no water particle turns around itself (irrotational motion);
5. the wave amplitude is small compared to the wave length.

The above assumptions allow the development of an elegant theory in which the velocities of water particles can be derived from a velocity potential. The record of sea elevations in a fixed point is a function of time in which we cannot find any pattern that repeats itself. We can, however, characterize the sea by statistical quantities. One important example is the significant wave height defined as the mean of the highest third of trough-to-crest heights. The heights are measured between trough and crests situated on different sides of the sea level.

Another statistical characteristic of the sea is the wave spectrum, actually the distribution of wave energy as function of the wave frequency. Sea spectra can be measured or can be calculated on the basis of sea characteristics, such as the significant wave height. Formulae for standard spectra have been proposed for various ocean or sea regions.

Floating bodies move in six degrees of freedom. Three motions are linear: surge along the x-axis, sway along the y-axis, and heave along the z-axis, where the axes of coordinates are those defined in Chapter 1. The other three motions are angular: roll around the x-axis, pitch around the y-axis, and yaw around the z-axis.

We can write a differential equation for one particular motion without considering the influence of other motions. We say then that the motion is uncoupled. In reality certain couplings exist between motions. For example, we know from Chapter 2 that roll induces pitch. Moreover, one motion can influence the added masses and damping coefficients of other motions. The most general representation of motions in six degrees of freedom is by a system of six ordinary differential equations. The port-to-starboard symmetry of many floating structures simplifies the matrices of inertia, added masses and damping coefficients and allow the decoupling of equations. Then, for example, we can write a system of three equations for the vertical-plane motions, heave, surge and pitch, and another system for sway, roll and yaw.

Moorings can change the natural frequencies of motions. An example is that of tension-leg platforms. As the name says, the mooring 'tendons' are tensioned so that they pull down the platform and increase its draught beyond that corresponding to the platform mass. An elastic force develops in the tensioned tendons; it opposes heave and is much larger than the hydrostatic force developed by the added submerged volume in heave. The natural period in heave is changed so that it is remote from that of the waves prevailing in the region of operation. Natural periods of ships can change in confined waters because of the proximity of vertical walls and bottom. This effect must be avoided when performing roll tests.

The roll amplitude can be reduced by passive devices, such as bilge keels, or by active devices, such as roll fins. A frequently used roll stabilizer employs two tanks (flume tanks) connected by a transversal pipe. When properly tuned, the cross-flow between the two tanks opposes the roll motion. This is a case in which a free surface helps. However, a tradeoff must be done between the good effect on roll and the reduction of effective metacentric height due to the free-surface effect of the water in the flume tanks.

12.8 Examples

Example 12.1 – Simulating a Frahm vibration absorber
Let us simulate the behaviour of a system provided with a vibration absorber, such as described in Section 12.6. Dividing both sides of the first Eq. (12.33) by m_1 and both sides of the second equation by m_2, we obtain

$$\ddot{x}_1 + \frac{c}{m_1}\dot{x}_1 + \frac{k_1}{m_1}(x_1 - x_2) = \frac{F_0}{m_1}\sin\omega t$$
$$\ddot{x}_2 + \frac{k_2}{m_2}(x_2 - x_1) = 0$$

$$(12.37)$$

We note by $\omega_0^2 = k_1/m_1$ the square of the natural angular frequency of the undamped main system. According to the theory developed in Section 12.6 we set $k_2/m_2 = \omega^2$, that is the square of the exciting frequency. We transform the factor k_2/m_1 as follows

$$\frac{k_2}{m_1} = \frac{k_2}{m_2}\frac{m_2}{m_1} = \omega^2\frac{m_2}{m_1}$$

With the above notations we rewrite Eq. (12.37) as

$$\ddot{x}_1 + \frac{c}{m_1}\dot{x}_1 + \omega_0^2 x_1 + \omega^2\frac{m_2}{m_1}(x_1 - x_2) = \frac{F_0}{m_1}\sin\omega t$$
$$\ddot{x}_2 + \omega^2(x_2 - x_1) = 0$$

$$(12.38)$$

For numerical integration we must convert the above system of two second-order differential equations into a system of four first-order differential equations. To do so we define the four variables

$$y_1 = \dot{x}_1 \qquad \text{the speed of mass } m_1$$
$$y_2 = x_1 \qquad \text{the motion of mass } m_1$$
$$y_3 = \dot{x}_2 \qquad \text{the speed of mass } m_2$$
$$y_4 = x_2 \qquad \text{the motion of mass } m_2$$

Using these notations the system of first-order differential equations becomes

$$
\dot{y}_1 = -\frac{c}{m_1}y_1 - \omega_0^2 y_2 - \omega^2 \frac{m_2}{m_1}(y_2 - y_4) - \frac{F_0}{m_1}\sin\omega t
$$

$$
\dot{y}_2 = y_1
$$

$$
\dot{y}_3 = -\omega_0^2(y_4 - y_2)
$$

$$
\dot{y}_4 = y_3
$$

(12.39)

As shown, for example, in Biran and Breiner (2002), Chapter 14, we write the model as the following function Frahm

```
%FRAHM Model of a Frahm vibration absorber.

    function  yd = Frahm(t, y, rm)

% Input arguments:  t  time,  y  variable,  rm m2-to-m1 ratio

% meaning of derivatives
% yd(1)        speed of main mass m1
% yd(2)        displacement of main mass m1
% yd(3)        displacement of absorbing mass m2
% yd(4)        displacement of absorbing mass m2

w0   = 2*pi/14.43;        % natural frequency of main system
w    = 2*pi/7;            % wave frequency, rad/s
c_m  = 0.1;               % damping coefficient, c-to-m1 ratio
F_m  = 1;                 % exciting amplitude, F-to-m1 ratio

yd = zeros(size(y));      % allocate space for y

% derivatives
yd(1) = -c_m*y(1)- w0^2*y(2)  - w^2*rm*(y(2) - y(4)) - F_m*sin(w*t);
yd(2) = y(1);
yd(3) = -w^2*(y(4) - y(2));
yd(4) = y(3); }
```

The ratio m_2/m_1 appears as an input argument, rm. Thus, it is possible to play with the rm value and visualize its influence. To call the function Frahm we write a script file, call_Frahm; its beginning may be

```
%CALL_FRAHM Calls ODE23 with Frahm derivatives.
%          Integrates the model of the Frahm damper.
t0 = 0.0; % initial time, s
tf = 100; % final integration time
```

```
y0 = [ 0; 0; 0; 0 ] % initial conditions
% call integration function for system
% without absorber
[ t, y ] = ode23(@Frahm, [ t0, tf ], y0, [], 0);
subplot(3, 1, 1), plot(t, y(:, 2))
    axis([ 0 100 -5 5 ])
    Ht = text(80, 3.5, 'r_m = 0');
    set(Ht, 'FontSize', 12)
    Ht = title('Displacement of main mass');
    set(Ht, 'FontSize', 14)
% call integration function with mass ratio 1/10
...
```

The results of the simulation are shown in Figure 12.12. The larger the rm ratio, the more effective the absorber is. On a ship, however, large flume tanks mean a serious reduction of the effective metacentric height and of the cargo. Hence the need for a tradeoff between advantages and disadvantages.

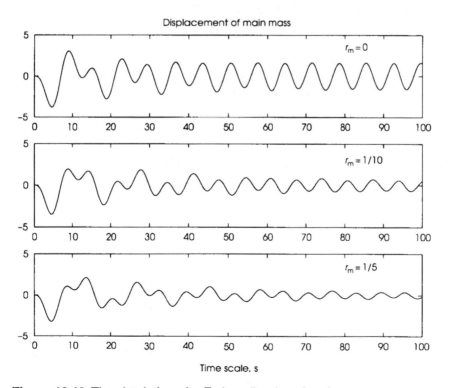

Figure 12.12 The simulation of a Frahm vibration absorber

12.9 Exercises

Exercise 12.1 – Potential wave theory
Prove that Eqs. (12.13) and (12.14) fulfill Eq. (12.4).

Exercise 12.2 – Vertical motion
Draw a sketch to prove that the vertical motion of a ship point with coordinates x, y, z is, indeed, as shown on page 278. In other words, show that the vector of the vertical motion is the resultant of three vectors produced by heave, roll and pitch.

Exercise 12.3 – A Frahm vibration absorber
Referring to Example 12.1, change the value of c_m in function Frahm and study the influence of the damping value.

Exercise 12.4 – A Frahm vibration absorber
Referring to Example 12.1 modify the file call_Frahm so as to plot also the motion of the absorbing mass m_2.

12.10 Appendix – The relationship between curl and rotation

In Figure 12.13, we consider an infinitesimal square whose sides are dx and dz. The horizontal speed of the lower left corner is u, and the vertical speed w. Then, the horizontal velocity of the upper left corner is

$$u + \frac{\partial u}{\partial z} dz$$

and the vertical velocity of the lower right corner is

$$w + \frac{\partial w}{\partial x} dx$$

The difference of velocities between the lower left and the lower right corner of the square causes a counter-clockwise rotation around the y-axis with the angular speed

$$\frac{\partial w}{\partial x}$$

The difference of horizontal speeds between the lower left and the upper left corners causes a clockwise rotation with the angular speed around the y-axis

$$\frac{\partial u}{\partial z}$$

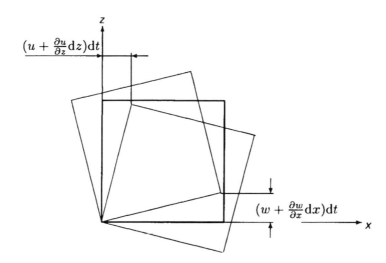

Figure 12.13 The relationship between curl and rotational motion

The resulting mean angular speed is

$$\frac{1}{2}\left(\frac{\partial w}{\partial x} - \frac{\partial u}{\partial z}\right)$$

In three-dimensional space the curl of the vector of velocities $[u, \ v, \ w]$ is calculated from the determinant

$$\mathrm{curl}([u, v, w]) = \begin{bmatrix} \mathbf{i} & \mathbf{j} & \mathbf{k} \\ \frac{\partial}{\partial x} & \frac{\partial}{\partial y} & \frac{\partial}{\partial z} \\ u & v & w \end{bmatrix} \tag{12.40}$$

where **i, j, k** are the unit vectors in the x, y, and z directions, respectively. One can see immediately that Eq. (12.2) says that there is no rotation around the y-axis.

The terms corresponding to 'curl' in continental Europe are different, for example

Fr   roteur
G   Rotor
I   rotore

13
Computer methods

13.1 Introduction

The large amount of multiplications, summations and integrations required in hydrostatic calculations made necessary a systematic approach and the use of mechanical computing devices. Amsler invented in 1856 the **planimeter**, a mechanical instrument that yields the area enclosed by a given curve. The planimeter is an **analogue computer**. Other examples of mechanical, analogue computers once widely used in Naval Architecture are the **integraph** and the **integrator**. The integraph draws the integral curve, $\int_{x_0}^{x} f(\xi)d\xi$, of a given curve, $y = f(x)$ (see Section 3.4). The integrator yields the area, the first and second moments of the area bounded by a closed curve. When digital computers appeared, they gradually replaced the mechanical instruments. To our knowledge, the first publication of a digital computer programme for Naval Architecture is due to Kantorowitz (1958). More programmes for hydrostatic calculations appeared in the following years. Today, digital computers are used extensively in modern Naval Architecture and computer programmes are commercially available. With the arrival of computer graphics, Naval Architects understood that they can apply the new techniques to solve the problems of hull definition. Today, some of the most sophisticated software packages are used for this purpose. Moreover, once the hull surface is defined, the programmes use this definition to perform hydrostatic and other calculations.

In this chapter we discuss concisely a few ways of using computers for the treatment of the subjects described in the book. A detailed treatment would require a dedicated book (for Naval Architectural graphics see Nowacki, Bloor and Oleksiewicz, 1995). Besides this, computer software changes so rapidly that it would be necessary to update the book at short intervals.

One of the first subjects treated in the book is the definition of the hull surface. It is natural to begin this chapter by showing how computers are used for this definition. To do so we first introduce a few elementary concepts of computer graphics, and afterwards we give a few simple examples of application to hull-surface definition.

The next subjects discussed in the book are hydrostatic and weight calculations. Correspondingly, we give in this book a few examples of computer implementations of these matters.

Towards the end of the book we describe simple models of dynamic ship behaviour. We end this chapter by explaining what simulation is and give a simple example that uses SIMULINK, a powerful toolbox that extends the capabilities of MATLAB.

13.2 Geometric introduction

13.2.1 Parametric curves

The ellipse shown in Figure 13.1 can be described by the Equation

$$\frac{x^2}{a^2} + \frac{y^2}{b^2} = 1 \tag{13.1}$$

where $2a$ is called **major axis** and $2b$ **minor axis**. In the particular case shown in Figure 13.1, $a = 3$ and $b = 2$. We cannot use this **implicit equation** to draw the curve by means of a computer. We can, however, derive the **explicit equation**

$$y = \pm b\sqrt{1 - (x/a)^2} \tag{13.2}$$

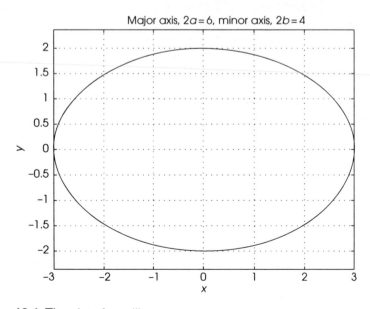

Figure 13.1 The plot of an ellipse

Now, we can draw the ellipse in MATLAB using the following commands

```
a  = 3; b = 2; x = -3: 0.01: 3;
y1 = b*(1 - (x/a).^2).^(1/2); y2 = -y1;
plot(x, y1, 'k-', x, y2, 'k-'), axis equal
```

There is another way of plotting the ellipse, namely by using a **parametric equation** of the curve. An easy-to-understand example is

$$
\begin{aligned}
x &= a \cos t \\
y &= b \sin t
\end{aligned}
\tag{13.3}
$$

where t is a **parameter** running from 0 to 2π. We invite the reader to show that Eq. (13.1) can be obtained from Eq. (13.3). The MATLAB commands that implement Eq. (13.3) are

```
a = 3; b = 2; t = 0: pi/60: 2*pi;
x = a*cos(t);
y = b*sin(t);
plot(x, y, 'k-'), axis equal
```

The parameter t identifies any point on the curve and defines the orientation of the curve – that is, the sense in which the parameter t increases. It is usual to normalize it to lie in the interval $[0, 1]$. For example, we can rewrite Eq. (13.3) as

$$
\begin{aligned}
x &= a \cos 2\pi t \\
y &= b \sin 2\pi t
\end{aligned}
\tag{13.4}
$$

where $0 \le t \le 1$.

The concepts described in this section can be easily extended to three-dimensional curves. Thus, the equations

$$
x = r \cos 2\pi t, \qquad y = r \sin 2\pi t, \qquad z = pt, \quad t = [0, 1]
$$

describe a helix with radius r and pitch p.

13.2.2 Curvature

An important characteristic of a curve is its curvature. We refer to Figure 13.2 for a formal definition. Let us consider the curve passing through the points A, B and C. The angle between the tangents at the points A and B is α, and the length of the arc AB is s. Then

$$
k = \frac{d\alpha}{ds}
\tag{13.5}
$$

is the curvature at the point A. In words, the curvature is the rate of change of the curve slope.

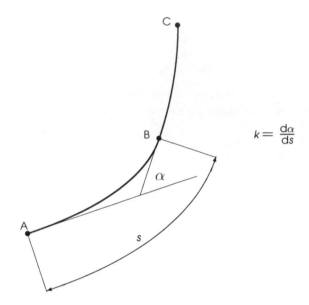

Figure 13.2 The definition of curvature

For a curve defined in the explicit form $y = f(x)$ the curvature is given by

$$k = \frac{\frac{d^2y}{dx^2}}{\left[1 + \left(\frac{dy}{dx}\right)^2\right]^{3/2}} \qquad (13.6)$$

We see that the curvature is directly proportional to the second derivative of y with respect to x. The curvature of a circle with radius r is constant along the whole curve and equal to $1/r$. For other curves the curvature may vary along the curve. The **radius of curvature** is the inverse of curvature, that is $1/k$. A most important example is the metacentric radius, $\overline{BM}$, defined in Subsection 2.8.2; it is the radius of curvature of the curve of centres of buoyancy.

The curvature has a strong influence on the shape of the curve. Fairing the lines of a ship means in a large measure taking care of curvatures. For a three-dimensional curve we have to define a second quantity, **torsion**, which is a measure of how much it bends outside of a plane. More details can be found in books on differential geometry.

13.2.3 Splines

In Naval Architecture, the term **spline** designs a wood, metal or plastic strip used to draw the curved lines of the ship. According to the *Webster's Ninth New Collegiate Dictionary*, the origin of the word is unknown and it first appeared in

1756. It can be shown that, when forced to pass through a set of given points, a spline bends so that its shape can be described by a cubic polynomial. According to Schumaker (1981) Schoenberg adopted in 1946 the term **spline functions** to describe a class of functions that approximate the behaviour of 'physical splines'.

Spline functions use polynomials to describe curves. It is easy to calculate, differentiate or integrate polynomials. On the other hand, it may be difficult to fit a single polynomial to a large number of points. A set of n points defines a polynomial of degree $n - 1$. When $n = 3$, the fitted curve is a parabola that connects the three points without oscillating. For $n = 4$, the curve may show a point of inflection and as n increases the curve may oscillate wildly between the given points. Runge (German, 1856–1927) described the phenomenon of **polynomial inflexibility**; an example in MATLAB is shown in Biran and Breiner (2002: 428–9). The general idea of the spline functions is to solve the problem by subdividing the given set of points into several subsets, to fit a polynomial to each subset, and to ensure certain continuity conditions at the junction of two polynomials. For example, let us suppose that we have to fit a spline over the interval $[x_a, \ x_b]$, and we subdivide it into two at x_i, where, by definition, $x_a < x_i < x_b$. Let $y_1(x)$ be the polynomial fitted over the interval $[x_a, \ x_i]$ and $y_2(x)$ the polynomial fitted over the interval $[x_i, \ x_b]$. Obviously, we impose the condition

$$y_1(x_i) = y_2(x_i)$$

For slope continuity, we also require that

$$\left[\frac{dy_1}{dx}\right]_{x=x_i} = \left[\frac{dy_2}{dx}\right]_{x=x_i}$$

A nicer curve is obtained when the curvature too is continuous, that is

$$\left[\frac{d^2y_1}{dx^2}\right]_{x=x_i} = \left[\frac{d^2y_2}{dx^2}\right]_{x=x_i}$$

Additional conditions can be imposed on the slopes of the curve at the beginning and the end of the interval $[x_a, \ x_b]$. The set of conditions makes possible the writing of a system of linear equations that yields all the coefficients of the two polynomials. The extension to more subintervals is straightforward.

Let us consider in Figure 13.3 a set of points arranged along a ship station. If the curve passes through all given points, as in Figure 13.4, we say that the curve is an **interpolating spline**. Figure 13.4 was drawn with the MATLAB spline function. In ship design, we may be less interested in passing the curve through all the given points, than in obtaining a fair curve. The fitted curve is then an **approximating spline**. An example obtained with the MATLAB polyfit and polyval functions is shown in Figure 13.5. In this case the curve is a single cubic polynomial fitted over seven points so that the sum of the squares of

Figure 13.3 Points along a ship station

deviations is minimal, that is a least-squares fit. The two solutions described in this paragraph do not allow the user to intervene in the fit; other solutions enable this and they are introduced in the following sections.

13.2.4 Bézier curves

Working at Citröen, Paul de Faget de Casteljau (French, born 1930, see Bieri and Prautzsch, 1999 and De Casteljau, 1999) developed a kind of curves that were further developed at Renault by Pierre Bézier (French, 1910–99). These curves, called now **Bézier curves**, are defined by a set of **control points**, B_0, B_1, ..., B_n, so that the coordinates of any point, $P(t)$, on the curve, are weighted averages of the coordinates of the control points. On the other hand, the coordinates are functions of a parameter $t = [0, 1]$. The curve begins at $t = 0$ and ends at $t = 1$.

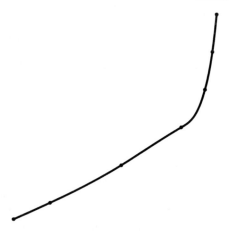

Figure 13.4 An interpolating spline

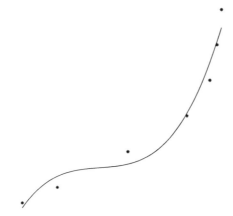

Figure 13.5 An approximating spline

The simplest Bézier curve is a straight line that connects the two points

$$\mathbf{B_0} = \begin{bmatrix} x_0 \\ y_0 \end{bmatrix}, \qquad \mathbf{B_1} = \begin{bmatrix} x_1 \\ y_1 \end{bmatrix} \qquad\qquad (13.7)$$

The coordinates of a point on the segment $\overline{B_0 B_1}$ are given as functions of the parameter t

$$\mathbf{P(t)} = \begin{bmatrix} x \\ y \end{bmatrix} = (1-t)\mathbf{B_0} + t\mathbf{B_1}, \quad t = [0,\ 1] \qquad (13.8)$$

The above equation is in fact a formula for linear interpolation. A second-degree curve is defined by three points, $\mathbf{B_0}$, $\mathbf{B_1}$, $\mathbf{B_2}$, and its equation is

$$\mathbf{P}(t) = (1-t)^2\mathbf{B_0} + 2(1-t)t\mathbf{B_1} + t^2\mathbf{B_2} \qquad (13.9)$$

It can be shown that Eq. (13.9) describes a parabola.

A cubic Bézier curve is defined by four control points, $\mathbf{B_0}, \ldots, \mathbf{B_3}$, and its equation is

$$\mathbf{P}(t) = (1-t)^3\mathbf{B_0} + 3(1-t)^2t\mathbf{B_1} + 3(1-t)t^2\mathbf{B_2} + t^3\mathbf{B_3} \qquad (13.10)$$

An example is shown in Figure 13.6. We concentrate on cubic polynomials for the simple reason that cubics are the lowest-degree curves that display inflection points. Thus, cubic curves can reproduce the change of curvature sign present in some ship lines. Increasing the degree of polynomials above 3 can cause fluctuations (see above 'polynomial inflexibility') and make computation more complex. In Example 13.1, we give the listing of a MATLAB function that plots a cubic Bézier curve.

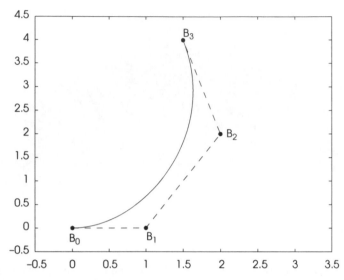

Figure 13.6 A cubic Bézier spline

The following properties of Bézier curves are given here without proof.

Property 1   The curve passes through the first and the last control point only. In Figure 13.6, the curve passes, indeed, through the points $\mathbf{B_0}$ and $\mathbf{B_3}$ only.

Property 2   The curve is tangent to the first and last segment of the control polygon. In Figure 13.6, the curve is tangent to the segments $\overline{\mathbf{B_0B_1}}$ and $\overline{\mathbf{B_2B_3}}$.

Property 3   The sum of the coefficients that multiply the coordinates of the control points equals 1. In spline theory, the functions that produce these coefficients are called **blending** or **basis functions**.

Property 4   Moving one control point influences the shape of the whole curve. Thus, in Figure 13.7, the point $\mathbf{B_3}$ was moved horizontally until it lies on the line $\overline{\mathbf{B_1B_2}}$. We see that the curve eventually becomes a straight line. As the point $\mathbf{B_3}$ is moved further to the right, a point of inflexion appears as in Figure 13.8.

The property of the tangents at the ends of a Bézier curve allows us to join two Bézier curves so that the continuity of the first derivative is achieved. For example, in Figure 13.9, two Bézier curves are joined at point $\mathbf{B_3}$, while the point $\mathbf{B_4}$ lies on the straight line defined by the points $\mathbf{B_2}$ and $\mathbf{B_3}$.

The general form of a Bézier curve of degree n is

$$\mathbf{P}(t) = \sum_{i=0}^{n} \mathbf{B_i} J_{n,i}(t), \quad 0 \le t \le 1 \tag{13.11}$$

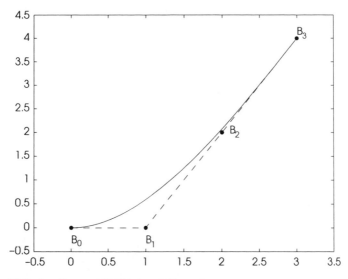

Figure 13.7 Another cubic Bézier spline

where the blending function is

$$J_{n,i}(t) = \frac{n!}{i!(n-i)!} t^i (1-t)^{n-i} \qquad (13.12)$$

and $0^0 = 1$, $0! = 1$. The blending functions of Bézier curves are also known as
Bernstein polynomials.

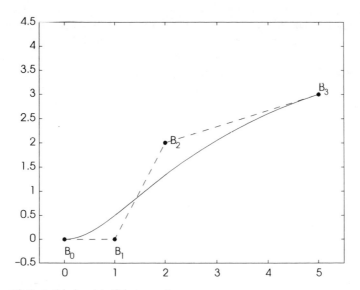

Figure 13.8 A third cubic Bézier spline

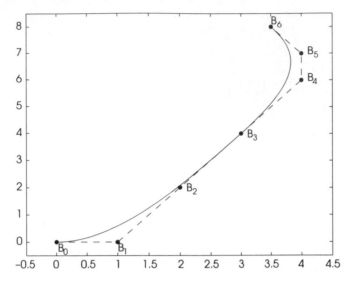

Figure 13.9 Combining two cubic Bézier splines

More degrees of freedom can be obtained by using **rational Bézier curves** defined by

$$\mathbf{P}(t) = \frac{\sum_{i=0}^{n} \mathbf{B_i} W_i J_{n,i}(t)}{\sum_{i=0}^{n} W_i J_{n,i}(t)} \qquad (13.13)$$

The numbers W_i are called **weights**. We assume that all the weights are positive so that all denominators are positive. The numerator is a vector, while the denominator is a scalar. When all $W_i = 1$, the rational curves become the non-rational Bézier curves described in this section. Rational Bézier curves can describe accurately conic sections. As these sections, that is the circle, the ellipse, the parabola and the hyperbola are second-degree curves, three control points are necessary. The kind of curve depends on the chosen weights. An application of rational Bézier curves to hull-surface design is given by Kouh and Chen (1992).

Examples of earlier uses of cubic or rational cubic splines to ship design can be found in Kouh (1987), Ganos (1988) and Söding (1990). Jorde (1997) poses a 'reverse' problem, how to define the ship lines to achieve given sectional area curves and coefficients of form.

13.2.5 B-splines

It is easy to calculate points along Bézier curves. On the other hand, moving a control point produces a **global** change of the curve. Another class of more sophisticated curves, the **B-spline curves**, do not have this disadvantage. Moving a point on the latter curves causes only a **local** change, that is a change that affects only the curve segment that neighbours the moved point. We give below the **recursive** definition of a B-spline. Given $n + 1$ control points, $B_1, \ldots, B_{n+1}$, the position vector is

$$\mathbf{P}(t) = \sum_{i=1}^{n+1} \mathbf{B_i} N_{i,k}(t), \quad t_{\min} \leq t \leq t_{\max}, \quad 2 \leq k \leq n + 1 \qquad (13.14)$$

Here k is the **order** of the B-spline, and $k - 1$, the degree of the polynomials in t. The basis functions are

$$N_{i,t}(t) = \begin{cases} 1 & \text{if } t_i \leq t \leq t_{i+1} \\ 0 & \text{otherwise} \end{cases} \qquad (13.15)$$

and

$$N_{i,k}(t) = \frac{(t - t_i)N_{i,k-1}(t)}{t_{i+k-1} - t_i} + \frac{(t_{i+k-t} - t)N_{i+1,k-1}(y)}{t_{i+k} - t_{i+1}} \qquad (13.16)$$

The set of t_i values is called **knot vector**. If the knot values are not equally spaced the B-spline is called **non-uniform**, otherwise it is called **uniform**. The sum of the basis functions is

$$\sum_{i=1}^{n+1} N_{i,k}(t) = 1 \qquad (13.17)$$

for all t.

The calculation of points along B-spline curves requires rather complex algorithms that are beyond the scope of this chapter.

The **NURBS**, or **non-uniform rational B-splines** are an extension of the B-splines; their definition is

$$\mathbf{P}(t) = \frac{\sum_{i=0}^{n+1} \mathbf{B_i} W_i M_{i,k}(t)}{\sum_{i=1}^{n=1} W_i N_{i,k}(t)}$$

As in Eq. (13.13), W_i are the weights. The basis functions, $N_{i,k}$, are defined by Eqs. (13.15) and (13.16). A book on splines that includes historical and biographical notes is that of Rogers (2001).

13.2.6 Parametric surfaces

Surfaces can be defined by implicit equations such as

$$f(x, y, z) = 0$$

This form is not suitable for computer plots; a helpful form is an explicit equation like

$$z = f(x, y)$$

However, as for curves, the preferred form in computer graphics is a parametric representation of the form

$$x = x(u, w), \qquad y = y(u, w), \qquad z = z(u, w)$$

Two parameters are sufficient, indeed, to define any point on a given surface. As an example let us consider the upper half of an ellipsoid whose parametric equations are

$$
\begin{aligned}
x &= a \cos \pi \frac{u}{2} \cos 2\pi w \\[2mm]
y &= b \cos \pi \frac{u}{2} \sin 2\pi w \\[2mm]
z &= c \sin \pi \frac{u}{2}, \quad u = [0,\ 1], \quad w = [0,\ 1]
\end{aligned}
\tag{13.18}
$$

When $a = b = c$ the ellipsoid becomes a sphere with centre in the origin of coordinates and radius 1. Then $\pi u/2$ is the analogue of what is called in geography **latitude**, and πw is the analogue of **longitude**.

Figure 13.10 shows a **wireframe view** of a surface obtained with Eq. (13.18). The curve that bounds the surface at its bottom corresponds to $u = 0$. A net composed of two **isoparametric** curve families is shown. The constant-u curves are marked $u = 0,\ 0.1,\ \ldots,\ 1$. The curve corresponding to $u = 1$ condenses to a single point, the Northern Pole in the case of a sphere. For the sake of visibility only part of the constant-w curves are marked: $w = 0.5,\ \ldots,\ 0.9$. As $\cos 0 = \cos 2\pi$, and $\sin 0 = \sin 2\pi$, the curves $w = 0$ and $w = 1$ coincide.

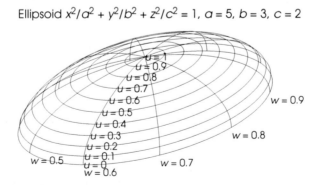

Ellipsoid $x^2/a^2 + y^2/b^2 + z^2/c^2 = 1,\ a = 5,\ b = 3,\ c = 2$

Figure 13.10 The u and w nets on a parametric ellipsoidal surface

Figure 13.10 shows that a surface can be described by a net of isoparametric curves. One procedure for generating a surface can begin by defining a family of plane curves, for example ship stations, with the help of Bézier curves, non-rational or rational B-splines, or NURBS, with the parameter u. Taking then the points $u = 0$ on all curves, we can fit them a spline of the same kind as that used for the first curves. Proceeding in the same manner for the points $u = 0.1, \ldots, u = 1$, we obtain a net of curves. Plane curves can be properly described by breaking them into spline segments and imposing continuity conditions at the junction points. Similarly, surfaces can be broken into **patches** with continuity conditions at their borders. The expressions that define the patches can be direct extensions of plane curves equations such as those described in the preceding sections. For example, a **tensor product Bézier patch** is defined by

$$\mathbf{P}(\mathbf{u}, \mathbf{w}) = \sum_{i=0}^{m} \sum_{j=0}^{n} \mathbf{B}_{ij} J_{i,m}(u) J_{j,n}(w), \quad u = [0, \ 1], \quad w = [0, \ 1]$$

where the control points, $\mathbf{B}_{ij}$ define a control polyhedron, and $J_{i,m}(u)$ and $J_{j,n}(w)$ are the basis functions we met in the section on Bézier curves. There are more possibilities and they are described in detail in the literature on geometric modelling.

13.2.7 Ruled surfaces

A particular case is that in which corresponding points on two space curves are joined by straight-line segments. For example, in Figure 13.11 we consider three of the constant-w curves shown in Figure 13.10. Then, we draw a straight line from a $u = i$ point on the curve $w = 0.6$ to the $u = i$ point on the curve $w = 0.7$, for $i = 0, \ 0.1, \ \ldots, \ 1$. The surface patch bounded by the $w = 0.6$ and the $w = 0.7$ curves is a **ruled surface**. A second ruled-surface patch is shown between the curves $w = 0.7$ and $w = 0.8$. Ruled surfaces are characterized by the fact that it is possible to lay on them straight-line segments.

13.2.8 Surface curvatures

In Figure 13.12, let $\mathbf{N}$ be the normal vector to the surface at the point P, and $\mathbf{V}$, one of the tangent vectors of the surface at the same point P. The two vectors, $\mathbf{N}$ and $\mathbf{V}$, define a plane, π_1, normal to the surface. The intersection of the plane π_1 with the given surface is a planar curve, say C. The curvature of C at the point P is the **normal curvature of C at the point P in the direction of** V. We note it by k_n. A theorem due to Euler states that there is a direction, defined by the tangent vector $\mathbf{V}_{\min}$, for which the normal curvature, $k_{\min}$, is minimal, and

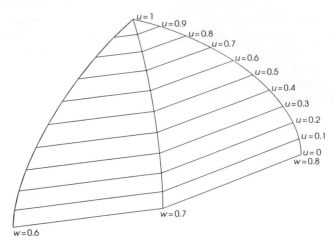

Figure 13.11 Two ruled surfaces

another direction, defined by the tangent vector $\mathbf{V}_{max}$, for which the normal curvature, k_{max}, is maximal. Moreover, the directions $\mathbf{V}_{min}$ and $\mathbf{V}_{max}$ are perpendicular. The curvatures k_{min} and k_{max} are called **principal curvatures**. For example, in Figure 13.12 the planes π_1 and π_2 are perpendicular one to the other and their intersections with the ellipsoidal surface yields curves that have the principal curvatures at the point from which starts the normal vector $\mathbf{N}$. The two curves are shown in Figure 13.13.

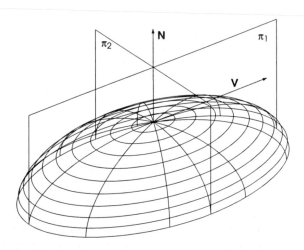

Figure 13.12 Normal curvatures

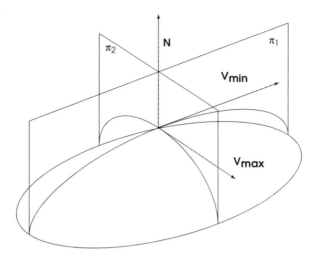

Figure 13.13 Principal curvatures

The product of the principal curvatures is known as **Gaussian curvature**:

$$K = k_{min} \cdot k_{max} \tag{13.19}$$

and the mean of the principal curvatures is known as **mean curvature**:

$$H = \frac{k_{min} + k_{max}}{2} \tag{13.20}$$

In Naval Architecture, curvatures are used for checking the fairness of surfaces.

A surface with zero Gaussian curvature is developable. By this term we understand a surface that can be unrolled on a plane surface without stretching. In practical terms, if a patch of the hull surface is developable, that patch can be manufactured by rolling a plate without stretching it. Thus, a developable surface is produced by a simpler and cheaper process than a non-developable surface that requires pressing or forging. A necessary condition for a surface to be developable is for it to be a ruled surface. Cylindrical surfaces are developable and so are cone surfaces. The sphere is not developable and this causes problems in mapping the earth surface. Readers interested in a rigorous theory of surface curvatures can refer to Davies and Samuels (1996) and Marsh (1999). The literature on splines and surface modelling is very rich. To the books already cited we would like to add Rogers and Adams (1990), Piegl (1991), Hoschek and Lasser (1993), Farin (1999), Mortenson (1997) and Piegl and Tiller (1997).

13.3 Hull modelling

13.3.1 Mathematical ship lines

De Heere and Bakker (1970) cite Chapman (Fredrik Henrik af Chapman, Swedish Vice-Admiral and Naval Architect, 1721–1808) as having described ship lines as early as 1760 by parabolae of the form

$$y = 1 - x^n$$

and sections by

$$y = 1 - z^n$$

In 1915, David Watson Taylor (American Rear Admiral, 1864–1940) published a work in which he used 5-th degree polynomials to describe ship forms. Names of later pioneers are Weinblum, Benson and Kerwin. More details on the history of mathematical ship lines can be found in De Heere and Bakker (1970), Saunders (1972, Chapter 49) and Nowacki *et al.* (1995). Kuo (1971) describes the state of the art at the beginning of the 70s. Present-day Naval Architectural computer programmes use mainly B-splines and NURBS.

13.3.2 Fairing

In Subsection 1.4.3, we defined the problem of fairing. A major object of the developers of mathematical ship lines was to obtain fair curves. Digital computers enabled a practical approach. Some early methods are briefly described in Kuo (1971), Section 9.3. A programme used for many years by the Danish Ship Research Institute is due to Kantorowitz (1967a,b). Calkins *et al.* (1989) use one of the first techniques proposed for fairing, namely differences. Their idea is to plot the 1st and the 2nd differences of offsets. In addition, their software allows for the rotation of views and thus greatly facilitates the detection of unfair segments.

As mentioned in Subsections 13.2.2 and 13.2.8, plots of the curvature of ship lines can help fairing. Surface-modelling programmes like MultiSurf and SurfaceWorks (see next section) allow to do this in an interactive way. More about curvature and fairing can be read in Wagner, Luo and Stelson (1995), Tuohy, Latorre and Munchmeyer (1996), Pigounakis, Sapidis and Kaklis (1996) and Farouki (1998). Rabien (1996) gives some features of the Euklid fairing programme.

13.3.3 Modelling with MultiSurf and SurfaceWorks

In this section, we are going to describe a few steps of the hull-modelling process performed with the help of MultiSurf and SurfaceWorks, two products of Aero-Hydro. We like these surface modellers for their excellent visual interface, the possibilities of defining and capturing many relationships between the various elements of a design, and the wide range of useful point, curve and surface types. A recent possibility is that of connecting SurfaceWorks to SolidWorks.

The programmes described in this section are based on a concept developed by John Letcher; he called it **relational geometry** (see Letcher, Shook and Shepherd, 1995 and Mortenson, 1997, Chapter 12). The idea is to establish a hierarchy of dependencies between the elements that are successively created when defining a surface or a hull surface composed of several surfaces. To model a surface one has to define a set of control, or **supporting** curves. To define a supporting curve, the user has to enter a number of supporting points; they are the control points of the various kinds of curves. Points can be entered giving their absolute coordinates, or the coordinate-differences from given, **absolute** points. Moreover, it is possible to define points constrained to stay on given curves or surfaces. When the position of a supporting point or curve is changed, any **dependent** points, curves or surfaces are automatically updated. Relational geometry considerably simplifies the problems of intersections between surfaces and the modification of lines.

Both MultiSurf and SurfaceWorks use a system of coordinates with the origin in the forward perpendicular, the x-axis positive towards aft, the y-axis positive towards starboard, and the z-axis positive upwards. When opening a new model file, a dialogue box allows the user to define an axis or plane of symmetry, and the units. For a ship the plane of symmetry is $y = 0$.

We begin by 'creating' a set of points that define a desired curve, for example a station. Thus, in MultiSurf, a first point, $p01$, is created with the help of the dialogue box shown in Figure 13.14. The last line is highlighted; it contains

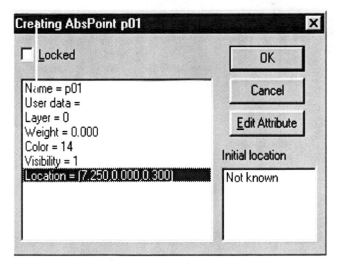

Figure 13.14 MultiSurf, the dialogue box for defining an absolute three-dimensional point

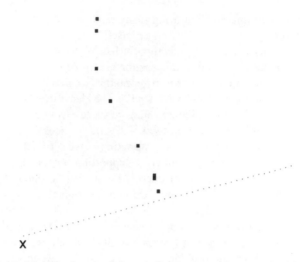

Figure 13.15 MultiSurf, points that define a control curve, in this case a transverse section

the coordinates of the point, $x = 17.250$, $y = 0.000$, $z = 3.000$. There is a quick way of defining a set of points, such as shown in Figure 13.15. In this example all the points are situated along a station; they have in common the value $x = 17.250$ m.

To 'create' the curve defined by the points in Figure 13.15 the user has to select the points and specify the curve kind. A Bcurve (this is the MultiSurf terminology for B-splines) uses the **support** points as a control polygon (see Subsection 13.2.4), while a Ccurve (MultiSurf terminology for cubic splines) passes through all support points. Figure 13.16 shows the Bcurve defined by

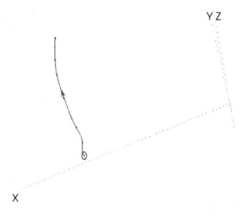

Figure 13.16 MultiSurf, a curve that defines a transverse section

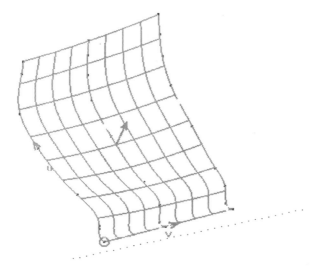

Figure 13.17 MultiSurf, a surface defined by control curves such as those in Figure 13.16

the points in Figure 13.15. The display also shows the point in which the curve parameter has the value 0, and the positive direction of this parameter.

Several curves, such as the one shown in Figure 13.16, can be used as support of a surface. To 'create' a surface the user selects a set of curves and then, through pull-down menus, the user choses the surface kind. An example of surface is shown in Figure 13.17. Any point on this surface is defined by the two parameters u and v. The display shows the origin of the parameters, the direction in which the parameter values increase, and a normal vector.

To exemplify a few additional features, we use this time screens of the SurfaceWorks package. In Figure 13.18 we see a set of four points along a station. The window in the lower, left corner of Figure 13.18 contains a list of these points. Figure 13.19 shows the B-spline that uses the points in Figure 13.18 as control points. At full scale it is possible to see that the curve passes only through the first and the last point, but very close to the others. The display shows again the origin and the positive sense of the curve parameter.

Figure 13.19 is an axonometric view of the curve. Figure 13.20 is an orthographic view normal to the x-axis. In Figure 13.21, we see the same station and below it a plot of its curvature. In this case we have a simple third-degree B-spline; the plot of its curvature is smooth. In other cases the curve we are interested in can be a **polyline** composed of several curves. Then, the curvature plot can help in fairing the composed curve. Usually, it is not possible to define a single surface that fits the whole hull of a ship. Then, it is necessary to define several surfaces that can be joined together along common edges. A surface is defined by a set of supporting curves, for example, the bow profile, some transverse curves, etc.

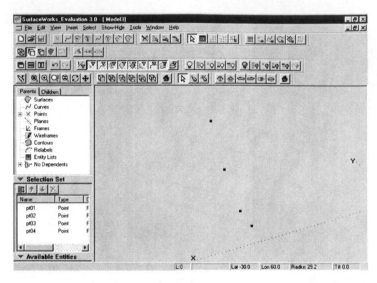

Figure 13.18 SurfaceWorks, points that define a control (supporting) curve for a surface

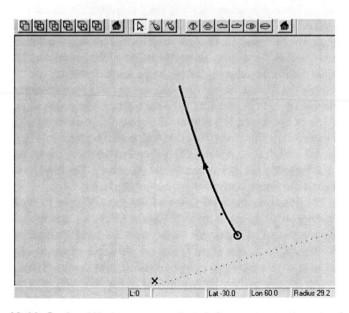

Figure 13.19 SurfaceWorks, a curve that defines a transverse section

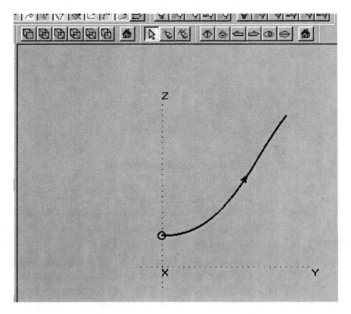

Figure 13.20 The same curve projected in the direction of the *x*-axis

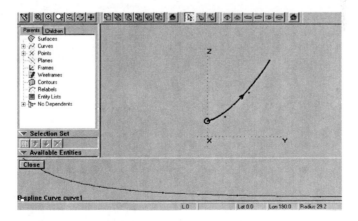

Figure 13.21 The curvature of the same curve

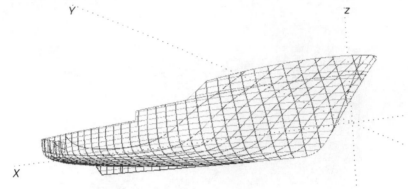

Figure 13.22 The wireframe view of a powerboat

Figure 13.22 shows a wireframe view of a powerboat. The hull surface is composed of the following surfaces: bow round, bulwark, bulwark round, hull, keel forward, keel aft, and transom.

The software enables the user to view the hull from any angle, for example as in Figure 13.23. Other views can be used to check the appearance and the fairing of the hull. The **rendered** view may be very helpful; we do not show an example because it is not interesting in black and white.

Three plots of surface curvature are possible: normal, mean or Gaussian. We have chosen the plot of normal curvature shown in Figure 13.24. The

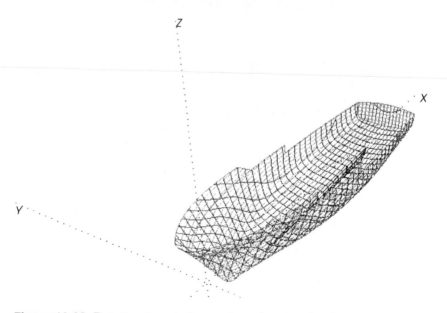

Figure 13.23 Rotating the wireframe view of a powerboat

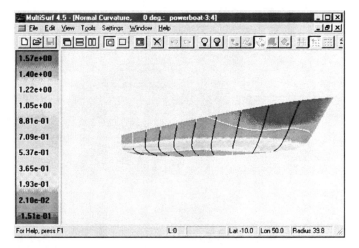

Figure 13.24 A plot of normal curvatures

view - offsets option displays the transverse stations to be used for hydro-static calculations. The display for our powerboat is shown in Figure 13.25. A drawing of the ship lines produced by the programme is shown in Figure 13.26. A dialogue box enables the user to choose the display. For example, Figure 13.26 is produced with the option Body plan top right, same scale.

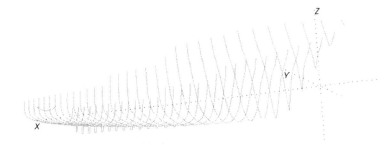

Figure 13.25 A plot of the offsets of a powerboat

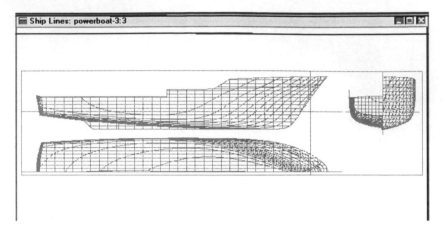

Figure 13.26 The lines of a powerboat

13.4 Calculations without and with the computer

Before the era of computers, the Naval Architect prepared a documentation that was later used for calculating the data of possible loading cases. The documentation included:

- hydrostatic curves;
- cross-curves of stability;
- capacity tables that contained the filled volumes and centres of gravity of holds and tanks, and the moments of inertia of the free surfaces of tanks.

For a given load case, the Naval Architect, or the ship Master, performed the weight calculations that yielded the displacement and the coordinates of the centre of gravity. The data for holds and tanks were based on the tables of capacity. The next step was to find the draught, the trim and the height $\overline{KM}$ by interpolating over the hydrostatic curves. Finally, the curve of static stability was calculated and drawn after interpolating over the cross-curves of stability. It is in this way that **stability booklets** were prepared; they contained the calculations and the curves of stability for several pre-planned loadings. The same method was employed by the ship Master for checking if it is possible to transport some unusual cargo.

The above procedure is still followed in many cases, with the difference that the basic documentation is calculated and plotted with the help of digital computers, and the weight and $\overline{GZ}$ calculations are carried out with the aid of hand calculators, possibly with the help of an electronic spreadsheet. However, since the introduction of personal computers and the development of Naval Architectural software for such computers, it is possible to proceed in a more efficient way. Thus, it is sufficient to store in the computer a description of the hull and of its subdivision into holds and tanks. The model can be completed with a description of the sail area necessary for calculating the wind arm. Then, the user can define a loading case by entering for each hold or tank a measure of its filling, for example the filling height, and the specific gravity of the cargo. The computer programme calculates almost instantly the parameters of the floating conditions and the characteristics of stability, and it does so without rough approximations and interpolations. For example, in a manual, straightforward trim calculation one has to use the moment to change trim, MCT, read from the hydrostatic curves. Hydrostatic curves are usually calculated for the ship on even keel; therefore, using the MCT value read in them means to assume that this value remains constant within the trim range. Computer calculations, on the other hand, do not need this assumption. The floating condition is found by successive iterations that stop when the conditions of equilibrium are met with a given tolerance.

The ship data stored in the computer constitute a **ship model**; it can be organized as a **data base**. In this sense, Biran and Kantorowitz (1986) and Biran, Kantorowitz and Yanai (1987) describe the use of relational data bases. Johnson, Glinos, Anderson *et al.* (1990), Carnduf and Gray (1992) and Reich (1994) discuss more types of data bases. Many modern ships are provided with board computers that contain the data of the ship and a dedicated computer programme. Moreover, the computer can be connected to sensors that supply on line the tank and hold filling heights.

13.4.1 Hydrostatic calculations

Some hydrostatic calculations are straightforward in the sense that we can perform them in a single iteration. For example, if we want to calculate hydrostatic curves we must perform integrations for a draught T_0, then for a draught T_1, and so on. Chapter 4 shows how to carry out such calculations. Other calculations can be carried out only by iterations. For example, let us assume that we want to calculate the righting arm of a given ship, for a given displacement volume, ∇_0, and the heel angle ϕ_i. We do not know the draught, T_0, corresponding to the given parameters. We must start with an initial guess, T_{init}, draw the waterline, $W_0 L_0$, corresponding to this draught and the heel angle ϕ_i, and calculate the actual displacement volume. If the guess T_{init} was not based on previous calculations, almost certainly we shall find a displacement volume $\nabla_1 \neq \nabla_0$. If the deviation is larger than an acceptable value, ϵ, we must try another waterplane, $W_1 L_1$, parallel to the initial guess waterline, $W_0 L_0$. This time we proceed in a more 'educated' manner. Readers familiar with the Newton–Raphson procedure may readily understand why we use the derivative of the displacement volume with respect to the draught, that is the waterplane area, A_W. We calculate a draught correction

$$\delta T = \frac{\nabla_0 - \nabla_1}{A_W}$$

and we start again with a corrected draught

$$T_1 = T_{\text{init}} + \delta T$$

We continue so until the **stopping condition**

$$|\nabla_0 - \nabla_N| \le \epsilon$$

is met.

A much more difficult, but frequent problem is that of finding the floating condition of a ship for a given loading. The input is composed of the displacement volume and the coordinates of the centre of gravity. The output is the triple of parameters that define the floating condition, that is the draught, the heel and the trim. To solve the problem we can think of a Newton-like procedure in three variables. Such a procedure implies the calculation of a Jacobian whose elements are nine partial derivatives. Not less difficult is the problem of finding the floating condition of a damaged ship, provided the ship can still float. The Naval Architect has to find the draught, trim and heel for which the conditions described in Section 11.3 are met. In physical terms, the Naval Architect must find the ship position in which the water level in the flooded compartments is the same as that of the surrounding water and the centres of buoyancy and gravity lie on a common vertical. Some details of the above problems can be found in Söding (1978). The calculations of hydrostatic data from surface patches is discussed by Rabien (1985).

Many ingenious methods for solving the above problems have been devised; by elegant procedures they ensured satisfactory precision in reasonable calculating times. The methods based on mechanical computers are particularly interesting. Details can be found in older books. For example, an original publication of a method for calculating lever arms at large heel angles is due to Leparmentier (1899). Other methods for calculating cross-curves of stability are described by Rondeleux (1911), Dankwardt (1957), Attwood and Pengelly (1960), Krappinger (1960), Semyonov-Tyan-Shansky (no year given), De Heere and Bakker (1970), Hervieu (1985), Rawson and Tupper (1996). Methods of flooding calculations are explained, for example, in Semyonov-Tyan-Shansky (no year given) and De Heere and Bakker (1970).

As mentioned, the first publication about a computer programme for Naval Architectural calculations is that of Kantorowitz (1958); it contains also an analysis of calculation errors. The first computer programmes worked in the **batch mode**; an input had to be submitted to the computer, the computer produced an output. For many years the input was contained in a set of punched cards, later it could be written on a file. An example of such a programme is ARCHIMEDES written at the University of Hannover (see Poulsen, 1980). The input consists of several sequences of numbers. One sequence defines the calculations to be performed, a second sequence describes the hull surface, a third sequence defines the subdivision into compartments and tanks, a fourth the longitudinal distribution of masses, a fifth defines run parameters such as the draught, trim, the wave characteristics, and the identifiers of the compartments to be considered flooded.

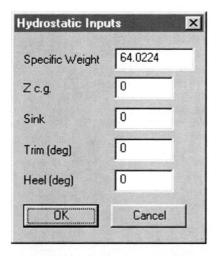

Figure 13.27 The MultiSurf dialogue box for entering the input for hydrostatic calculations

The programme ARCHIMEDES could be run for hydrostatic calculations, capacity calculations (compartment and tank volumes, centres of gravity, and free surfaces), cross-curves of stability, damage stability, and longitudinal bending. Many examples in this book were obtained with the ARCHIMEDES programme. A newer version of the software, ARCHIMEDES II, is described by Söding nd Tonguc (1989).

Recent programmes have a graphic interface that enables the user to build and change interactively the ship model, to define run parameters and run calculations. The output consists of tables and graphs.

Hydrostatic calculations can be performed in MultiSurf or SurfaceWorks after obtaining the offsets (see Figure 13.25). Figure 13.27 shows the dialogue box in which the user has to input the height of the centre of gravity, under $Z.c.g$, the draught, under $Sink$, and the trim and the heel. A rich output is produced; Figure 13.28 shows only a fragment. A disadvantage of this implementation is that each draught-trim-$\overline{KG}$ combination requires a separate run. Aerohydro supplies another programme, Hydro, that enables a more convenient operation and yields also graphs. So do several packages marketed by other companies.

13.5 Simulations

The term **simulation** is frequently used in modern technical literature. The word derives from the Latin 'simulare', which means to imitate, pretend, counterfeit. In our context, by simulation we understand computer runs that yield an

```
34 stations,   6036 points
     Inputs
Sink              4.00        Spec. Wt.         64.02
Trim, deg.        0.00        Z c.g.            -3.00
Heel, deg.        0.00
     Dimensions
W.L. Length      18.50        W.L. Beam          5.11
W.L. Fwd. X      -1.80        Draft              6.00
W.L. Aft  X      16.70
     Displacement
Volume          681.8         Ctr.Buoy. X       14.68
Displ't.      43653.1         Ctr.Buoy. Y       -0.00
LCB (% w.l.)     89.1         Ctr.Buoy. Z       -3.14
     Waterplane
W.P. Area       12.52         Ctr.Flotn. X       0.33
LCF (% w.l.)    11.5
     Wetted Surface
Wetd.Area       613.29        Ctr. W.S. X       15.30
Ctr. W.S. Z      -3.07
     Lateral Plane
L.P. Area       132.81        Ctr. L.P. X       13.23
Ctr. L.P. Z      -3.43
     Initial Stability
Trans. GM        3.89         Trans.RMPD      2963.2
```

Figure 13.28 A fragment of the output of hydrostatic calculations carried out in MultiSurf

approximation of the behaviour of a real-life system we are interested in. The steps involved in this activity are described below:

1. The building of a physical model that describes the most important features of the real-life system.
2. The translation of the physical model into a mathematical model. Many mathematical models are composed of ordinary differential equations that describe the evolution of physical quantities as functions of time.
3. The translation of the mathematical model into a computer programme.
4. The running of the computer programme and the output of results.

For several good reasons the physical model cannot describe all features of the real-life system. First, we may not be aware of some details of the phenomenon under study. Next, to use manageable mathematics we must accept simplifying assumptions. Last but not least, we must keep the computation time within reasonabe limits and to achieve this we may be forced to accept more simplifying assumptions.

It follows that computer simulations do not exactly reproduce the behaviour of real-life systems; they only 'simulate' part of that behaviour. Better results can be certainly obtained by experiments, especially at full scale. It is easy to imagine that full-scale experiments on ships may be very expensive so that they cannot be carried out frequently. Dangerous experiments that can lead to ship loss may not be possible at all. Such tests can be performed only on reduced-scale models. Still, basin tests too are expensive and their extent is usually limited by the available budget. Simulations may replace dangerous experiments, basin tests can be completed by simulations. Then, part of the possible cases can be simulated, part tested on basin models. The basin tests can be used to correct or validate the computer model.

It is possible to measure the motions of a ship model in a test basin equipped with a wave maker. Then, the motions are recorded as functions of time. It is also possible to simulate ship motions as functions of time, that is to simulate in the **time domain**. However, such measurements or simulations in the time domain have limitations. As explained in Chapter 12, the sea surface is a random process; therefore, ship motions are also random processes. To simulate a given spectrum in the basin or in a computer programme, it is necessary to draw a number of random phases. The resulting motions do not describe all possible situations, but are only an example of such possibilities. We say that we obtain a **realization** of the random process. Moreover, for practical reasons, the duration of a basin test is limited. Then, the time span may not be sufficient for the worst event to happen. Although we may afford simulation times longer than basin tests, they still may be insufficient for obtaining the worst events.

More results can be obtained by calculating motions as functions of frequency, that is calculating in the **frequency domain**. Programmes that perform such calculaitons are available both through universities and on the market. The software calculates the added masses and damping coefficients, for a series of frequencies, by using potential theory and certain simplifying assumptions. Next, the software calculates the response amplitude operators, RAOs, of various motions or **events**. For a wave frequency component, and given ship heading and speed, the programme calculates the frequency of encounter and transforms the spectra from functions of wave frequency to functions of the frequency of encounter. Response spectra are obtained as products of the spectra of encounter and RAOs. Statistics can be extracted from the spectra, for instance **root mean square**, shortly **RMS** values of the motions.

Taking into consideration the motion of the sea surface, the heave and the pitch, the programme yields the motion of a deck point relative to the sea surface and calculates the probability of having waves on deck. Other events whose probability can be calculated are slamming and propeller racing, while the motions, velocities and accelerations of given ship points are obtained as combinations of motions in the various degrees of freedom. An example of ship motions simulated in the time domain can be found in Elsimillawy and Miller (1986). Examples of studies of capsizing in the time domain are in Gawthrop, Kountzeris and Roberts (1988) and Kat and Paulling (1989). An example of simulation in frequency domain is given by Kim, Chou and Tien (1980).

13.5.1 A simple example of roll simulation

Subsection 9.3.2 shows how to implement in MATLAB a Mathieu equation and simulate the roll motion produced by parametric excitation. More complicated models can be simulated in a similar manner by writing the governing equations as systems of first-order differential equations and calling an integration routine. The more complex the system becomes, the more difficult it is to proceed in this way. The programmer must write more lines and arrange them in the order in which information must be passed from one programme line to another. Software packages have been written to make simulation easier. The common feature of the various packages is that the programmer does not have to care about the order in which information must be passed. Also, routines and functions frequently used in simulations are available in libraries from which the user can readily call them. The programmer has only to describe the various relationships, the software will detail the equations and arrange them in the required order. In this section we give one very simple example of the capabilities of modern simulation software. As we give in the book examples in MATLAB, it is natural to use here the related simulation package, SIMULINK. Let us consider the following roll equation

$$\Delta i^2 \ddot{\phi} + g\Delta \overline{GZ} = M_\mathrm{H} \tag{13.21}$$

where Δ is the displacement mass, i, the mass radius of inertia, $\overline{GZ}$, the righting arm, and M_H, a heeling moment. We rewrite Eq. (13.21) as

$$\ddot{\phi} + \frac{g}{i^2}\overline{GZ} = \frac{M_\mathrm{H}}{\Delta i^2} \tag{13.22}$$

In this example we neglect added mass and damping, but use a non-linear function for $\overline{GZ}$ and can accept a variety of heeling moments. To represent this equation in SIMULINK we draw the **block diagram** shown in Figure 13.29 by putting in blocks taken from the libraries of the software and connecting them by lines that define the relationships between blocks. At the beginning we put two blocks representing heeling moments, M_H. For the wind moment we use a **step function**. Initially the moment is zero, at a given moment it jumps to a prescribed value that remains constant in continuation. For the wave moment we use a sine function, but it is not difficult to input a sum of sines.

The next block to the right is a **switch**; it is used to select one of the heeling moments, M_H. The block called `Heeling arm` performs the division of the heeling moment by the displacement value supplied by the block called `displacement`. Follows a summation point. At this point the value $g\overline{GZ}$ is subtracted from the heeling arm. The output of the summation block is

$$\frac{M_\mathrm{H}}{\Delta} - g\overline{GZ}$$

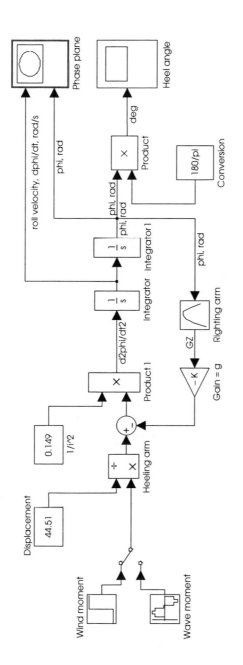

roll

Figure 13.29 Simulating roll in SIMULINK

Continuing to the right, we find a block that multiplies by $1/i^2$ the output of the summation block; the result is

$$\left(\frac{M_H}{\Delta} - g\overline{GZ}\right)\frac{1}{i^2}$$

We immediately see from Eq. (13.22) that the output of the block called `Product1` is the roll acceleration, $\ddot{\phi}$. This acceleration is the input to an integrator. The symbol

$$\frac{1}{s}$$

that marks the integrator block reminds the integration of Laplace transforms. The output of the integrator is the roll velocity, $\dot{\phi}$, in radians per second. The roll velocity is supplied as input to two output blocks. One block, above at right, is an oscilloscope, shortly scope, marked `Phase plane`. The other block, an integrator marked `Integrator 1`, outputs the roll angle, ϕ.

Following a path to the left, the roll angle becomes the input of a block called `Righting arm`. This block contains $\overline{GZ}$ values as functions of ϕ. In a **gain** block the $\overline{GZ}$ value is multiplied by the acceleration of gravity, g, and at the summation point, the product is subtracted from the heeling arm. Following rightward paths, the roll angle is supplied directly to the scope `Phase plane`, while converted to degrees is input to the scope `Heel angle`. The scope `phase plane` displays the roll velocity versus the roll angle. The scope `angle` displays the roll angle versus time.

13.6 Summary

Ship projects require the drawing of lines that cannot be described by simple mathematical expressions, and also extensive calculations, mainly iterated integrations. Interesting attempts have been made to use mathematical ship lines, but until the second half of the last century the procedures for drawing and fairing ship lines remained manual. As to calculations, many elegant methods were devised, not a few of them based on mechanical, analogue computers, such as planimeters, integrators and integraphs. As in other engineering fields, in the domain of Naval Architecture the advent of digital computers greatly improved the techniques and made possible important advances. Naval Architects were among the first engineers to use massive computer programmes.

The development of computer graphics has made possible the use of computers in the design of hull surfaces. In computer graphics, curves are defined parametrically

$$x = x(t), \qquad y = y(t), \qquad z = z(t)$$

where the parameter, t, is frequently normalized so as to vary from 0 to 1.

The central idea in computer graphics is to define curves by piecewise polynomials. In simple words, the interval over which the whole curve should be defined is subdivided into subintervals, a polynomial is fitted over each subinterval and conditions of continuity are ensured at the junction of any two intervals. The conditions of continuity include the equality of coordinates at the junction point and the equality of the first, possibly also the second derivative at that point. The latter conditions mean continuity of tangent and curvature.

The simplest examples of curves used in computer graphics are the Bézier curves. The coordinates of a point on a Bézier curve are weighted means of the coordinates of n control points that form a control polygon. The degree of the polynomial representing the Bézier curve is $n - 1$. An extension of the Bézier curves are the rational Bézier curves; they can describe more curve kinds than the non-rational Bézier curves.

Moving a control point of a Bézier curve produces a general change of the whole curve. B-splines avoid this disadvantage by using a more complicated scheme in which the polynomials change between control points. Moving a control point of a B-spline produces only a local change of the curve. A powerful extension of the B-splines are the non-uniform rational B-splines, shortly NURBS. Computer programmes for ship graphics use mainly B-splines and NURBS.

Naval Architectural calculations involve many integrations. The calculations for hydrostatic curves can be performed straightforward. Other calculations can be carried out only by iterations, e.g. for finding the cross-curves of stability or the floating condition of a ship for a given loading, possibly also a given damage. Systematic and elegant methods were devised for performing the calculations with acceptable precision, in a reasonable time. Many methods used mechanical, analogue computers. When digital computers became available it was possible to write computer programmes that performed the calculations in a faster and more versatile way. The first programmes worked in the batch mode. The input was first introduced on punched cards, later on files. The programme was run and the output printed on paper. Present-day programmes are interactive and graphic user interfaces facilitate the input and yield a better and pleasant output. The interface enables the user to build and change interactively the ship model. This model includes the definitions of the hull surface, of the subdivision into compartments, holds and tanks, the materials in holds and tanks, and the sail area required for the calculation of wind arms.

Another use of computer programmes is in the simulation of the behaviour of ships and other floating structures in waves or after damage. Thus, it is possible to study situations that would be too dangerous to experiment them on real ships. Simulations can be carried out in the time domain or in the frequency domain. In the latter approach, one input is a sea spectrum, the output consists of spectra of motions and probability of events such as deck wetness, slamming or propeller racing. Simulations are used also for studying the stability of ships in the presence of parametric excitation. When the model used in simulation consists of ordinary differential equations the work can be greatly facilitated by using special simulation software. Then, the user employs a graphical interface to build the model with blocks dragged from libraries. The software produces the governing equations and arranges them in the order required for a correct information flow.

13.7 Examples

Example 13.1 – Cubic Bézier curve

```
%BEZIER Produces the position vector of a cubic
%Bezier spline

    function P = Bezier(B0, B1, B2, B3)

% Input arguments are the four control points
% B0, B1, B2, B3 whose coordinates are given
% in the format [ x; y ]. Output is the
% position vector P with coordinates given in
% the same format.

% calculate array of coefficients, in fact
% Bernstein polynomials
t  = [ 0: 0.02: 1 ]';                    % parameter
C0 = (1 - t).^3;
C1 = 3*t.*(1 - t).^2;
C2 = 3*t.^2.*(1 - t);
C3 = t.^3;
C = [ C0 C1 C2 C3 ];
% form control polygon and separate coordinates
B  = [ B0 B1 B2 B3 ];
xB = B(1, :); yB = B(2, :)
% calculate points of position vector
xP = C*xB'; yP = C*yB';
P = [ xP'; yP' ]
```

13.8 Exercises

Exercise 13.1 – Parametric ellipse
Write the MATLAB commands that plot an ellipse by means of Eq. (13.4).

Exercise 13.2 – Bézier curves
Show that the sum of the coefficients in Eq. (13.9) equals 1 for all t values.

Bibliography

Abicht, W. (1971). Die Sicherheit der Schiffe im nachlaufenden unregelmäßigen Seegang. *Schiff und Hafen*, **23**, No. 12, 988–90.

Abicht, W. (1973). Unterteilung und Lecksicherheit. *Hansa*, **110**, No. 15/16, 1380–2.

Abicht, W. and Bakenhus, J. (1970). Berechnung der Wahrscheinlichkeit des Überstehens von Verletzungen quer- und längsunterteilter Schiffe. In *Handbuch der Werften*, Vol. X, pp. 38–49.

Abicht, W., Kastner, S. and Wendel, K. (1976). *Stability of ships, safety from capsizing, and remarks on subdivision and freeboard*, Rept. 19. Hannover: Technical University Hannover.

Abicht, W., Kastner, S. and Wendel, K. (1977). Stability of ships, safety from capsizing, and remarks on subdivision and freeboard. In *Proceeding of the Second West European Conference on Marine Technology*, 23–27 May, Paper No. 9, pp. 95–118.

Anonymous (1872). The stability of the 'Captain', 'Monarch', and some other iron-clades. *Naval Science* (E.J. Reed, ed.), **1**, 26 and following.

Anonymous (1961). Sécurité et compartimentage. *Bulletin du Bureau Veritas*, **45**, No. 5, May, 91.

Appel, P. (1921). *Traité de Mécanique Rationelle*, 3d edition, Vol. 3. Paris: Gauthier-Villars et Cie.

Arndt, B. (1960A). Ermittlung von Mindestwerten für die Stabilität. *Schiffstechnik*, **7**, No. 35, 35–46.

Arndt, B. (1960B). Die stabilitätserprobung des Segelschulschiffs "Gorch Fock". *Schiffstechnik*, **7**, No. 39, 177–90.

Arndt, B. (1962). Einige Berechnungen der Seegangsstabilität. *Schiffstechnik*, **9**, No. 48, 157–60.

Arndt, B. (1964). Systematische Berechnungen der Seegangsstabilität für ein Frachtschiff mit einer Völligkeit von 0,63. *Hansa*, No. 24, 2479–91.

Arndt, B. (1965). Ausarbeitung einer Stabilitätsvorschrift für die Bundesmarine. *Jahrbuch der Schiffbautechnischen Gesellschaft*, **59**, 594–608. English translation 'Elaboration of a stability regulation for the German Federal Navy'. BSRA translation no. 5052.

Arndt, B. (1968). Schüttgut und Kentersicherheit. *Hansa*, **105**, Nov., 2013–26.

Arndt, B. and Roden, S. (1958). Stabilität bei vor- und achterlichem Seegang. *Schiffstechnik*, **29**, No. 5, 192–9.

Arndt, B., Brandl, H. and Vogt, K. (1982). 20 years of experience – Stability regulations of the West-German Navy. *Second International Conference on Stability of Ships and Ocean Vehicles*, Tokyo, 111–21.

Arndt, B., Kastner, S. and Roden, S. (1960). Die Stabilitätserprobung des Segelschulschiffs "Gorch Fock". *Schiffstechnik*, **7**, No. 39, 177–90.

Arscott, F.M. (1964). *Periodic Differential Equations – An Introduction to Mathieu, Lamé and Allied Functions*. Oxford: Pergamon Press.

ASTM (2001). *Guide F1321-92 Standard Guide for Conducting a Stability Test (Lightweight Survey and Inclining Experiment) to Determine the light Ship Displacement and Centers of Gravity of a Vessel*. http://www..astm.org/DATABASE.CART/PAGES/F1321.htm.

Attwood, E.L. and Pengelly, H.S. (1960). *Theoretical Naval Architecture*, new edition expanded by Sims, A.J. London: Longmans.

Bieri, H.P. and Prautzsch, H. (1999). Preface. *Computer Aided Geometric Design*, **16**, 579–81.

Biran, A. and Breiner, M. (2002). *MATLAB 6 for Engineers*. Harlow, England: Prentice Hall. Previous edition translated into German as: *MATLAB 5 für Ingenieure – Systematische und praktische Einführung* (1999). Bonn: Addison-Wesley. Greek translation 1999, Ekdoseis Tziola.

Biran, A. and Kantorowitz, E. (1986). Ship design system integrated around a relational data base. In *CADMO 86* (G.A. Keramidas, and T.K.S. Murthy, eds), pp. 85–94, Berlin: Springer-Verlag.

Biran, A., Kantorowitz, E. and Yanai, J. (1987). A Relational Data Base for Naval Architecture. In *International Symposium on Advanced Research for Ships and Shipping in the Nineties, CETENA's 25th Anniversary*, S. Margherita Ligure, Italy, 1–3 October.

Birbănescu-Biran, A. (1979). A system's theory approach in Naval Architecture. *International Shipbuilding Progress*, **26**, No. 295, March, 55–60.

Birbănescu-Biran, A. (translated and extended) (1982). *User's Guide for the Program System "Arhimedes 76" for hydrostatic calculations*. Technion. See also original, Poulsen (1980).

Birbănescu-Biran, A. (1985). *User's guide for the program STABIL for intact stability of naval vessels*, release 2. Haifa: Technion.

Birbănescu-Biran, A. (1988). Classification systems for ship items: a formal approach and its application. *Marine Technology*, **15**, No. 1, 67–73.

Björkman, A. (1995). On probabilistic damage stability. *The Naval Architect*, Oct., E484–5.

Bonnefille, R. (1992). *Cours d'Hydraulique Maritime*, 3rd edition. Paris: Masson.

Borisenko, A.I. and Tarapov, I.E. (1979). *Vector and Tensor Analysis with Applications*, translated and edited by Silverman, R.A. New York: Dover Publications.

Bouteloup, J. (1979). *Vagues, Marées, Courants Marins*. Paris: Presses Universitaires de France.

Bovet, D.M., Johnson, R.E. and Jones, E.L. (1974). Recent Coast Guard research into vessel stability. *Marine Technology*, **11**, No. 4, 329–39.

Brandl, H. (1981). Seegangsstabilität nach der Hebelarmbilanz auf Schiffen der Bundeswehr–Marine. *Hansa*, **118**, No. 20, 1497–503.

Burcher, R.K. (1979). The influence of hull shape on transverse stability. In *RINA Spring Meetings*, paper No. 9.

Calkins, D.E., Theodoracatos, V.E., Aguilar, G.D. and Bryant, D.M. (1989). Small craft hull form surface definition in a high-level computer graphics design environment. *SNAME Transactions*, **97**, 85–113.

Cardo, A., Ceschia, M., Francescutto, A. and Nabergoj, R. (1978). Stabilità della nave e movimento di rollio: caso di momento sbandante non variabile, *Tecnica Italiana*, No. 1, 1–9.

Carnduff, T.W. and Gray, W.A. (1992). Object oriented computing techniques in ship design. In *Computer Applications in the Automation of Shipyard Operation and Ship Design IV*, pp. 301–14.

Cartmell, M. (1990). *Introduction to Linear, Parametric and Nonlinear Vibrations*. London: Chapman and Hall.

Cesari, L. (1971). *Asymptotic Behaviour and Stability Problems in Ordinary Differential Equations*. Berlin: Springer-Verlag.

Chantrel, J.M. (1984). *Instabilités Paramétriques dans le Mouvement des Corps Flottants – Application au cas des bouées de chargement*, Thèse presentée à "Ecole Nationale Supérieure de Mécanique pour l'obtention du diplôme de Docteur-Ingénieur". Paris: Éditions Technip.

Churchill, R.V. (1958). *Operational Mathematics*, 2nd edition. New York: McGraw-Hill Book Company.

Cleary, C., Daidola, J.C. and Reyling, C.J. (1966). Sailing ship intact stability criteria. *Marine Technology*, **33**, No. 3, 218–32.

Comstock, J.P. (ed.) (1967). *Principles of Naval Architecture*. N.Y.: SNAME.

Costaguta, U.F. (1981). *Fondamenti di Idronautica*. Milano: Ulrico Hoepli.

Cunningham, W.J. (1958). *Introduction to Nonlinear Analysis*. New York: McGraw-Hill Book Company.

Dahle, A. and Kjærland, O. (1980). The capsizing of M/S Helland-Hansen. The investigation and recommendations for preventing similar accidents. *Norwegian Maritime Research*, **8**, No. 3, 2–13.

Dahle, A. and Kjærland, O. (1980). The capsizing of M/S Helland-Hansen. The investigation and recommendations for preventing similar accidents. *The Naval Architect*, No. 3, March, 51–70.

Dahle, A.E. and Myrhaug, D. (1996). Capsize risk of fishing vessels. *Schiffstechnik/Ship Technology Research*, **43**, 164–71.

Dankwardt, E. (ed.) (1957) Schiffstheorie. In *Schiffbautechnisches Handbuch*, Vol. I, Part 1 (W. Henschke, ed.). VEB Verlag Technik Berlin.

Davies, A. and Samuels, P. (1996). *An introduction to Computational Geometry for Curves and Surfaces*. Oxford: Clarendon Press.

Deakin, B. (1991). The development of stability standards for U.K. sailing vessels. *The Naval Architect*, Jan., 1–19.

De Casteljau, P. de F. (1999). De Casteljau's autobiaiography: My time at Citroën. *Computer Aided Geometric Design*, **16**, 583–6.

De Heere, R.F.S. and Bakker, A.R. (1970). *Buoyancy and Stability of Ships*. London: George G. Harrap & Co.

Den Hartog, J.P. (1956). *Mechanical Vibrations*, 4th edition. New York: McGraw-Hill Book Company.

Derrett, D.R., revised by Barrass, C.B. (2000). *Ship Stability for Masters and Mates*, 5th edition. Oxford: Butterworth-Heinemann.

Devauchelle, P. (1986). *Dynamique du Navire*. Paris: Masson.

DIN 81209-1 (1999). *Geometrie und Stabilität von Schiffen – Formelzeichen, Benennungen, Definitionen. Teil 1: Allgemeines, Überwasser-Einrumpfschiffe* (Geometry and stability of ships – Symbols for formulae, nomenclature, deffinitions. Part 1. General, surface monohull ships).

Dorf, R.C. and Bishop, R.H. (2001). *Modern Control Systems*, 9th edition. Upper Saddle River, N.J.: Prentice-Hall.

Douglas, J.F., Gasiorek, J.M. and Swaffield, J.A. (1979). *Fluid Mechanics*. London: Pitman Publishing Limited.

Doyère, Ch. (1927). *Théorie du Navire*. Baillière.

Dunwoody, A.B. (1989). Roll of a ship in astern seas – Metacentric height spectra. *Jr. of Ship Research*, **33**, No. 3, Sept., 221–8.

Elsimillawy, N. and Miller, N.S. (1986). Time simulation of ship motions: a guide to the factors degrading dynamic stability. *SNAME Transactions*, **94**, 215–40.

Faltinsen, O.M. (1993). *Sea Loads on Ships and Offshore Structures*. Cambridge: Cambridge University Press.

Farin, G. (1999). *NURBS from Projective Geometry to Practical Use*. Natick, Ma: A.K. Peters.

Farouki, R.T. (1998). On integrating lines of curvature. *Computer Aided Geometric Design*, **15**, 187–92.

Fog, N.G. (1984). Creative definition of ship hulls using a B-spline surface. *Computer-Aided Design*, **16**, No. 4, July, 225–9.

Fossen, T. (1994). *Guidance and Control of Ocean Vehicles*. Chichester: John Wiley and Sons.

Furttenbach, J. (1968). *Architectura Navalis*, 2nd edition. Hamburg: Germanischer Lloyd.

Ganos, G.G. (1988). Methodical series of traditional Greek fishing boats. Dissertation for the degree of Doctor of Engineering, Loukakis, Th. supervisor. National Technical University of Athens, June.

Gawthrop, P.J., Kountzeris, A. and Roberts, J.B. (1988). Parametric excitation of nonlinear ship roll motion from forced roll data. *Journal of Ship Research*, **32**, No. 2, 101–11.

Gilbert, R.R. and Card, J.C. (1990). The new international standard for subdivision and damage stability of dry cargo ships. *Marine Technology*, **27**, No. 2, 117–27.

Gray, A. (1993). *Modern Differential Geometry of Curves and Surfaces*. Boca Raton, Florida: CRC Press.

Grimshaw, R. (1990). *Nonlinear Ordinary Differential Equations*. Oxford: Blackwell Scientific Publications.

Grochowalski, S. (1989). Investigation into the physics of ship capsizing by combined captive and free-running model tests. *SNAME Transactions*, **97**, 169–212.

Hansen, E.O. (1985). An analytical treatment of the accuracy of the results of the inclining experiment. *Naval Engineers Journal*, **97**, No. 4, May, 97–115; discussion in No. 5, July, 82–4.

Harpen, N.T. van (1971). Eisen te stellen aan de stabiliteit en het reserve-drijfvermogen van bovenwaterscheppen der Koninklijke Marine en het Loodswezen. *S. en W.*, **38**, No. 4, 1972.

Helas, G. (1982). Stabilität von Schiffen in nachlaufenden Seegang. *Seewirtschaft*, **14**, No. 9, 440–1.

Hendrickson, R. (1997). *Encyclopedia of Word and Phrase Origins*, expanded edition. New York: Facts on File.

Henschke, W. (ed.) (1957). *Schiffbautechnisches Handbuch*, Vol. 1, 2nd edition. Berlin: VEB Verlag Technik.

Hervieu, R. (1985). *Statique du Navire*. Paris: Masson.

Hoschek, J. and Lasser, D. (1993). *Fundamentals of Computer Aided Geometric Design*, translated by Schumaker, L.L. Wellesley Ma: A.K. Peters.

Hsu, C.S. (1977). On nonlinear parametric excitation problems. In *Advances in Applied Mechanics* (Chia-Shun Yih, ed.), **17**, 245–301.

Hua, J. (1996). A theoretical study of the capsize of the ferry "Herald of Free Enterprise". *International Shipbuilding Progress.* **43**, No. 435, 209–35.

Ilie, D. (1974). *Teoria Generală a Plutitorilor.* Bucharest: Editura Academiei Republicii Socialiste România.

IMO (1995). *Code on Intact Stability for All Types of Ships Covered by IMO Instruments – Resolution A749(18).* London: International Maritime Organization.

INSEAN (1962). *Carene di Pescherecci,* Quaderno n. 1. Roma: INSEAN (Vasca Navale).

INSEAN (1963). *Carene di Petroliere,* Quaderno n. 2. Roma: INSEAN (Vasca Navale).

ISO 7460 (1983). International standard: *Shipbuilding – Shiplines – Identification of Geometric Data.*

ISO 7462 (1985). International standard: *Shipbuilding – Principal dimensions – Terminology and Definitions for Computer Applications,* 5th edition, English and French.

ISO 7463 (1990). International standard: *Shipbuilding and Marine Structures – Symbols for Computer Applications.*

Jakić, K. (1980). A new theory of minimum stability, a comparison with an earlier theory and with existing practice. *International Shipbuilding Progress,* **27**, No. 309, May, 127–32.

Jons, O.P. (1987). Stability-related guidance for the commercial fisherman. *SNAME Transactions,* **95**, 215–37.

Johnson, B., Glinos, N., Anderson, N. *et al.* (1990). Database systems for hull form design. *SNAME Transactions,* **98**, 537–64.

Jordan, D.W. and Smith, P. (1977). *Nonlinear Ordinary Differential Equations.* Oxford: Clarendon Press.

Jorde, J.H. (1997). Mathematics of a body plan. *The Naval Architect,* Jan., 38–41.

Kantorowitz, E. (1958). Calculation of hydrostatic data for ships by means of digital computers. *Ingeniøren International Edition,* No. 2, 21–5.

Kantorowitz, E. (1966). Fairing and mathematical definition of ship surface. *Shipbuilding and Shipping Record,* No. 108, 348–51.

Kantorowitz, E. (1967a). Experience with mathematical fairing of ship surfaces. *Shipping World and Shipbuilder,* **160**, No. 5, 717–20.

Kantorowitz, E. (1967b). *Mathematical Definition of ship surfaces.* Danish Ship Research Institute, Report No. DSF-14.

Kastner, S. (1969). Das Kentern von Schiffen in unregelmäßiger längslaufender See. *Schiffstechnik,* **16**, No. 84, 121–32.

Kastner, S. (1970). Hebelkurven in unregelmäßigem Seegang. *Schiffstechnik,* **17**, No. 88, 65–76.

Kastner, S. (1973). Stabilität eines Schiffes im Seegang. *Hansa,* **110**, No. 15/16, 1369–80.

Kastner, S. (1989). On the accuracy of ship inclining experiments. *Ship Technology Research – Schiffstechnik,* **36**, No. 2, 57–65.

Kat de, J.O. (1990). The numerical modeling of ship motions and capsizing in severe seas. *Jr. of Ship Research,* **34**, No. 4, Dec., 289–301.

Kat de, J.O. and Paulling, R. (1989). The simulation of ship motions and capsizing in severe seas. *SNAME Transactions,* **97**, 139–68.

Kauderer, H. (1958). *Nichtlineare Mechanik.* Berlin: Springer-Verlag.

Kehoe, J.W., Brower, K.S. and Meier, H.A. (1980). The Maestrale. *Naval Engineers' Journal,* Oct., **92**, 60–2.

Kerwin, J.E. (1955). Notes on rolling in longitudinal waves. *International Shipbuilding Progress,* **2**, No. 16, 597–614.

Kim, C.H. Chou, F.S. and Tien, D. (1980). Motions and hydrodynamic loads of a ship advancing in oblique waves. *SNAME Transactions*, **88**, 225–56.

Kiss, R.K. (1980). Mission analysis and basic design. In *Ship Design and Construction* (R. Taggart, ed.). New York: SNAME.

Kouh, J.S. (1987). Darstellung von Schiffoberflächen mit rationalen kubischen splines. *Schiffstechnik*, **34**, 55–75.

Kouh, J.-S. and Chen, S.-W. (1992). Generation of hull surfaces using rational cubic Bézier curves. *Schiffstechnik – Ship Technology Reasearch*, **39**, 134–44.

Krappinger, O. (1960). Schiffstabilität und Trim. In *Handbuch der Werften*, 13–82. Hamburg: Schiffahrts-Verlag "Hansa" C. Schroedter & Co.

Kupras, L.K. (1976). Optimisation method and parametric study in precontracted ship design. *International Shipbuilding Progress*, May, 138–55.

Kuo, Ch. (1971). *Computer Methods for Ship Surface Design*. London: Longman.

Leparmentier, M. (1899). Nouvelle méthode pour le calcul des carènes inclinées. *Bulletin de l'Association Technique Maritime*, **10**, 45 and following.

Letcher, J.S., Shook, D.M. and Shepherd, S.G. (1995). Relational geometric synthesis: Part 1 – framework. *Computer-Aided Design*, **27**, No. 11, 821–32.

Lewis, E.V. (ed.) (1988). *Principles of Naval Architecture – Second Revision*, Vol. I – Stability and Strength. Jersey City, N.J.: The Society of Naval Architects and Marine Engineers.

Lindemann, K. and Skomedal, N. (1983). Modern hullforms and parametric excitation of the roll motion. *Norwegian Maritime Research*, **11**, No. 2, 2–20.

Little, P.E. and Hutchinson, B.L. (1995). Ro/ro safety after the Estonia – A report on the activities of the ad hoc panel on ro/ro safety. *Marine Technology*, **32**, No. 3, July, 159–63.

McGeorge, H.D. (2002). *Marine Auxiliary Systems*. Oxford: Butterworth-Heinemann.

McLachlan, N.W. (1947). *Theory and Application of Mathieu Functions*. Oxford: Clarendon Press.

Magnus, K. (1965). *Vibrations*. London: Blackie & Son Limited.

Manning, G.C. (1956). *The Theory and Technique of Ship Design*. New York: The Technology Press of M.I.T. and John Wiley & Sons.

Maritime and Coastguard Agency (1998). *The code of practice for safety of small workboats & pilot boats*. London: The Stationery Office.

Maritime and Coastguard Agency (2001). *The code of practice for safety of large commercial sailing & motor vessels*, 4th impression. London: The Stationery Office.

Marsh, D. (1999). *Applied Geometry for Computer Graphics and CAD*. London: Springer.

Merriam-Webster (1990). *Webster's Ninth New Collegiate Dictionary*. Springfield, MA: Merriam Webster.

Merriam-Webster (1991). *The Merriam-Webster New Book of Word Histories*. Springfield, MA: Merriam-Webster.

MoD (1999a). *Naval Engineering Standard NES 109 – Stability standard for surface ships – Part 1, Conventional ships*, Issue 4.

MoD (1999b). *SSP 24 – Stability of surface ships – Part 1 – Conventional ships*. Issue 2. Abbey Wood, Bristol: Defence Procurement Agency. Unauthorized version circulated for comments.

Morrall, A. (1980). The GAUL disaster: an investigation into the loss of a large stern trawler. *Transactions RINA*, 391–440.

Mortenson, M.E. (1997). *Geometric Modeling*. New York: John Wiley and Sons.

Myrhaug, D. and Dahle, E.Aa. (1994). Ship capsize in breaking waves. In *Fluid structure interaction in Ocean Engineering* (S.K. Chakrabarti, ed.), pp. 43–84. Southampton: Computational Mechanics Publications.

Nayfeh, A.H. and Mook, D.T. (1995). *Nonlinear Oscillations*. New York: John Wiley and Sons.

Nicholson, K. (1975). Some parametric model experiments to investigate broaching-to. In *The dynamics of marine vehicles and structures in waves* International Symposium (R.E. Bishop and W.G. Price, eds). London: The Institution of Mechanical Engineers, Paper 17, pp. 160–6.

Nickum, G. (1988). Subdivision and damage stability. In *Principles of Naval Architecture*, 2nd revision (E.V. Lewis, ed.). Vol. 1, pp. 143–204. Jersey: SNAME.

Norby, R. (1962). The stability of coastal vessels. *Trans. RINA*, **104**, 517–44.

Nowacki, H., Bloor, M.I.G., Oleksiewicz, B. *et al.* (1995). *Computational Geometry for Ships*. Singapore: World Scientific.

Paulling, J.R. (1961). *The transverse stability of a ship in a longitudinal seaway. Jr. of Ship Research*, **5**, No. 1, March, 37–49.

Pawlowski, M. (1999). Subdivision of ro/ro ships for enhanced safety in the damaged condition. *Marine Technology*, **36**, No. 4, Winter, 194–202.

Payne, S. (1994). Tightening the grip on passenger ship safety: the evolution of SOLAS. *The Naval Architect*, Oct., E482–7.

Pérez, N. and Sanguinetti, C. (1995). *Experimental results of parametric resonance phenomenon of roll motion in longitudinal waves for small fishing vessels. International Shipbuilding Progress*, **42**, No. 431, 221–34.

Piegl, L. (1991). On NURBS: a survey. *IEEE Computer Graphics & Applications*, Jan., **11**, 55–71.

Piegl, L.A. and Tiller, W. (1997). *The NURBS Book*, 2nd edition. Berlin: Springer.

Pigounakis, K.G., Sapidis, N.S. and Kaklis, P.D. (1996). Fairing spatial B-Splines Curves. *Journal of Ship Research*, **40**, No. 4, Dec., 351–67.

Pnueli, D. and Gutfinger, Ch. (1992). *Fluid Mechanics*. Cambridge: Cambridge University Press.

Poulsen, I. (1980). *User's manual for the program system ARCHIMEDES 76*, ESS Report No. 36. Hannover: Technische Universität Hannover.

Price, R.I. (1980). Design for transport of liquid and hazardous cargos. In *Ship design and construction* (R. Taggart, ed.). New York: SNAME, pp. 475–516.

Rabien, U. (1985). Integrating patch models for hydrostatics. *Computer-Aided Geometric Design*, **2**, 207–12.

Rabien, U. (1996). Ship geometry modelling. *Schiffstechnik – Ship Technology Research*, **43**, 115–23.

Rao, K.A.V. (1968). Einfluß der Lecklänge auf den Sicherheitsgrad von Schiffen. *Schiffbautechnik*, **18**, No. 1, 29–31.

Ravn, E.S., Jensen, J.J. and Baatrup, J. *et al.* (2002). Robustness of the probabilistic damage stability concept to the degree of details in the subdivision. Lecture notes for the Graduate Course *Stability of Ships* given at the Department of Mechanical Engineering, Maritime Engineering, of the Technical University of Denmark, Lyngby, 10–18 June.

Rawson, K.J. and Tupper, E.C. (1994). *Basic Ship Theory*, Vol. 1, 4th edition. Harlow, Essex: Longman Scientific & Technical.

Reich, Y. (1994). Information Management for Marine Engineering Projects. In *Proceedings of the 25th Israel Conference on Mechanical Engineering*. Technion City, Haifa, May 25–26, pp. 408–10.

RINA (1978). *ITTC Dictionary of Ship Hydrodynamics*. London: The Royal Institution of Naval Architects.

Rogers, D.F. (2001). *An Introduction to NURBS with Historical Perspective*. San Francisco: Morgan Kaufmann Publishers.

Rogers, D.F. and Adams, J.A. (1990). *Mathematical Elements for Computer Graphics*, 2nd edition. New York: McGraw-Hill Publishing Company.

Rondeleux, M. (1911). *Stabilité du Navire en Eau Calme et par Mer Agitée*. Paris: Augustin Challamel.

Rose, G. (1952). *Stabilität und Trim von Seeschiffen*. Leipzig: Fachbuchverlag GMBH.

Ross, C.T.F., Roberts, H.V. and Tighe, R. (1997). Tests on conventional and novel model ro-ro ferries. *Marine Technology*, **34**, No. 4, Oct., 233–40.

Rusås, S. (2002). Stability of ships: probability of survival. Lecture notes for the Graduate Course *Stability of Ships* given at the Department of Mechanical Engineering, Maritime Engineering, of the Technical University of Denmark, Lyngby, 10–18 June.

Saunders, H.E. (1972). *Hydrodynamics in Ship Design*, Vol. 2, 2nd printing of the 1957 edition. New York: SNAME.

Schatz, E. (1983). *User's guide for the program DAMAGE*. Haifa: Techion – Department of Computer Sciences amd Faculty of Mechanical Engineering.

Schneekluth, H. (1980). *Entwerfen von Schiffen*, 2nd edition. Herford: Koehler.

Schneekluth, H. (1988). *Hydromechanik zum Schiffsentwurf*. Herford: Kohler.

Schneekluth, H. and Bertram, V. (1998). *Ship Design for Efficiency & Economy*, 2nd edition. Oxford: Butterworth-Heinemann.

Schumaker, L.L. (1981). *Spline Functions: Basic Theory*. New York: John Wiley and Sons.

Semyonov-Tyan-Shanski, V. (no year indicated). *Statics and Dynamics of the Ships*, translated from the Russian by Konyaeva, M. Moscow: Peace Publishers.

Sjöholm, U. and Kjellberg, A. (1985). RoRo ship hull form: stability and seakeeping properties. *The Naval Architect*, Jan., E12–14.

Söding, H. (1978). *Naval Architectural Calculations*. In *WEGEMT 1978* (I.L. Buxton, ed.), pp. E2, 29–50.

Söding, H. and Tonguc, E. (1989). Archimedes II – A program for evaluating hydrostatics and space utilization in ships and offshore structures. *Schiffstechnik*, **36**, 97–104.

Söding, H. (1990). Computer handling of ship hull shapes and other surfaces. *Schiffstechnik*, **37**, 85–91.

Söding, H. (2002). Water ingress, down- and cross-flooding. Lecture notes for the Graduate Course *Stability of Ships* given at the Department of Mechanical Engineering, Maritime Engineering, of the Technical University of Denmark, Lyngby, 10–18 June.

SOLAS (2001). *SOLAS Consolidated Edition 2001 – Consolidated text of the International Convention for the Safety of Life at Sea, 1974, and its Protocol of 1988, Articles, Annexes and Certificates. Incorporating all amendments in effect from 1 January 2001*. London: International Maritime Organization.

Sonnenschein, R.J. and Yang, Ch. (1993). One-compartment damage survivability versus 1992 IMO probabilistic damage criteria for dry cargo ships. *Marine Technology*, **30**, No. 1, Jan., 3–27.

Spyrou, K. (1995). Surf-riding, yaw instability and large heeling of ships in following/quartering waves. *Schiffstechnik/Ship Technology Research*, **42**, 103–12.

Spyrou, K.J. (1996A). Dynamic instability in quartering seas: the behavior of a ship during broaching. *Jr. of Ship Research*, **40**, No. 1, March, 46–59.

Spyrou, K.J. (1996B). Dynamic instability in quartering seas – Part II: Analysis of ship roll capsize for broaching. *Jr. of Ship Research*, **40**, No. 4, Dec., 326–36.

Stoker, J.J. (1950). *Nonlinear Vibrations*. New York: Interscience Publishers.

Stoker, J.J. (1969). *Differential Geometry*. New York: Wiley Interscience.

Stoot, W.F. (1959). Some aspects of naval architecture in the eighteenth century. *Transactions of the Institution of Naval Architects*, **101**, 31–46.

Storch, R.L. (1978). *Alaskan king crab boats*. *Marine Technology*, **15**, No. 1, Jan., 75–83.

Struik, D.J. (1961). *Lectures on Classical Differential Geometry*. Reading MA: Addison-Wesley Publishing Company.

Susbielles, G. and Bratu, Ch. (1981). *Vagues et Ouvrages Pétroliers en Mer*. Paris: Éditions Technip.

Svensen, T.E. and Vassalos, D. (1998). Safety of passenger/ro-ro vessels: lessons learned from the North-West European R&D Project. *Marine Technology*, **35**, No. 4, Oct., 191–9.

Talib, A. and Poddar, P. (1980). *User's manual for the program system ARCHIMEDES 76*, translated from the original of Poulsen. Technical University of Hannover, ESS Report No. 36.

The New Encyclopedia Britannica (1989). Vol. 18. Chicago: Encyclopedia Britannica.

Tuohy, S., Latorre, R. and Munchmeyer, F. (1996). Developments in surface fairing procedures. *International Shipbuilding Progress*, **43**, No. 436, 281–313.

Wagner, P.H., Luo, X. and Stelson, K.A. (1995). Smoothing curvature and torsion with spring splines. *Computer-Aided Design*, **27**, No. 8, Aug., 615–26.

Watson, D.G. (1998). *Practical Ship Design*. Amsterdam: Elsevier.

Wegner, U. (1965). Untersuchungen und Überlegungen zur Hebelarmbilanz. *Hansa*, **102**, No. 22, 2085–96.

Wendel, K. (1958). Sicherheit gegen Kentern. *VDI-Zeitschrift*, **100**, No. 32, 1523–33.

Wendel, K. (1960a). Die Wahrscheinlichkeit des Überstehens von Verletzungen. *Schiffstechnik*, **7**, No. 36, 47–61.

Wendel, K. (1960b). Safety from capsizing. In *Fishing boats of the world: 2* (J.O. Traung, ed.). London: Fishing News (Books), pp. 496–504.

Wendel, K. (1965). Bemessung und Überwachung der Stabilität. *Jahrb. S.T.G.*, **59**, 609–27.

Wendel, K. (1970). Unterteilung von Schiffen. In *Handbuch der Werften*, Vol. X, pp. 17–37.

Wendel, K. (1977). Die Bewertung von Unterteilungen. In *Zeitschrift der Technischen Universität Hannover*, Volume published at 25 years of existence of the Department of Ship Technique, pp. 5–23.

Zigelman, D. and Ganoni, I. (1985). *Frigate seakeeping – A comparison between results obtained with two computer programs*. Haifa: Technion – Department of Computer Sciences and Faculty of Mechanical Engineering.

Ziha, K. (2002). Displacement of a deflected hull. *Marine Technology*, **39**, No. 1, Jan., 54–61.

Zucker, S. (2000), *Theoretical analysis for parametric roll resonance in trimaran*. MSc work, University College of London.

Index

Note: Page numbers in *italics* refer to tables and figures

Lightning Source UK Ltd.
Milton Keynes UK
UKOW020931040212

186643UK00001B/24/P